Temperature Measurement, 1975

536.5

LIBRARY
No. B 6507
18 JUN 1976
R.P.E. WESTCOTT

European Conference on Temperature Measurement, Teddington, April 1975

Temperature Measurement, 1975

Invited and contributed papers from the European
Conference on Temperature Measurement held at the
National Physical Laboratory, Teddington, 9-11 April 1975

Edited by B F Billing and T J Quinn

Conference Series Number 26
The Institute of Physics
London and Bristol

Copyright © 1975 by The Institute of Physics and individual contributors.
All rights reserved. Multiple copying of the contents or parts thereof without
permission is in breach of copyright but permission is hereby given to copy
titles of papers and names of authors. Permission is usually given upon
written applications to the Institute to copy illustrations and short extracts
from the text of individual contributions provided that the source (and,
where appropriate, the copyright) is acknowledged.

ISBN 0 85498 116 0
ISSN 0305 2346

The European Conference on Temperature Measurement was sponsored
by The Institute of Physics, The Institute of Measurement and Control,
The Institution of Electrical Engineers, and the European Physical Society.

International Committee
    R E Bedford (Canada), F Cabannes (France), L Crovini (Italy),
    M Durieux (Netherlands), R P Hudson (USA), T P Jones (Australia),
    H Kunz (Germany)

Local Organizing Committee
    T J Quinn (Chairman), B F Billing (Secretary), R Barber, F R Bareham,
    E Duncombe

Honorary Editors
    B F Billing and T J Quinn

Published by The Institute of Physics, 47 Belgrave Square, London SW1X 8QX,
and Techno House, Redcliffe Way, Bristol BS1 6NX, in association with the
American Institute of Physics, 335 East 45th Street, New York, NY 10017, USA.

Set in 10/12 Press Roman, and printed in Great Britain by Adlard and Son Ltd,
Dorking, Surrey.

Opening remarks

It gives me much pleasure to open your conference on Temperature Measurement. This is the third of a series of meetings devoted to temperature which have been sponsored by the Materials and Testing Group of The Institute of Physics in collaboration with The Institute of Measurement and Control, The Institution of Electrical Engineers and, for the first time, the European Physical Society. Now is a very appropriate time for such a meeting since we are this year celebrating the centenary of the signing of the Convention of the Metre in Paris in 1875. In a few weeks' time the 15th General Conference on Weights and Measures is due to convene in Paris and one of the items which will be presented for ratification is an amended edition of the International Practical Temperature Scale. This has resulted from work, over the past four years, of the Consultative Committee for Thermometry. Many of the members of this committee are here today and I extend to them a special welcome as I do to all who have come from other countries to this conference.

Thermometry has occupied an important place in the work of NPL from the very earliest days of the Laboratory when, on its foundation in 1900, research on gas thermometry and platinum resistance thermometry was transferred to Teddington from the Kew Observatory, a few miles away. This was followed in about 1909 by the transfer to Teddington of the calibration and testing of all types of thermometers. It was at the Kew Observatory in the Old Deer Park at Richmond that thermometer testing on a regular basis had been started in 1851 following a gift by the famous French research worker Regnault of a calibrated thermometer together with the associated equipment.

During the discussions which led up to the foundation of the National Physical Laboratory various sites were considered. The first choice of the committee under Lord Rayleigh was Kew itself alongside the old Observatory. We owe our presence here at Teddington not to any fine balance of technical argument regarding the relative suitability of the two sites but to a deputation of local inhabitants of Kew who went to see the Financial Secretary to the Treasury to say that under no circumstances would they have a new laboratory in the Park. The comments of the residents of Teddington are not recorded, but Her Majesty Queen Victoria was graciously pleased to offer the Royal residence in Bushy Park on loan as a nucleus for the new Laboratory. I have the privilege as Director of living in the House and you are most welcome to visit the garden during the luncheon break.

Looking at the programme of your conference I notice that gas thermometry and platinum resistance thermometry figure prominently. The fact that research on these subjects continues seventy-five years after first appearing at Teddington is a sure indication of the importance which the outside world places upon the ultimate results of such work.

Those who argued about a hundred years ago for the establishment of a national laboratory, among whom we find the name not unfamiliar to you of 'Kelvin', recognized the necessity of maintaining sound standards of physical measurement if engineering and international trade were to flourish.

Nothing has happened since those days to make this need any less pressing. It is for this same reason that we are all here today.

I wish you every success and I look forward to meeting you all informally at your Conference Dinner.

Dr J V Dunworth CB CBE
Director of the National Physical Laboratory

Preface

This volume contains the invited and contributed papers given at the third of a series of conferences on temperature whose main sponsor has been the Materials and Testing Group of The Institute of Physics. The proceedings of the earlier conferences, held in 1964 and 1970, were not published.

The extent of new work and developments in thermometry which have taken place since the last conference is illustrated by the number and range of the papers presented. In collating them for publication, we have commenced with an introductory chapter reviewing standards followed by six chapters covering specific temperature ranges or methods, each containing a review paper and papers on standards and techniques.

Of the 154 registered delegates, 77 came from the United Kingdom and 77 from among thirteen countries in Europe and four countries elsewhere.

The conference organizing committee wish to express their gratitude to the Director of NPL for the provision of conference facilities, to NPL staff who so ably assisted in making the conference a success, to the Director of RAE for the provision of technical secretarial assistance and to the staff of the meetings and publishing offices of The Institute of Physics for their help and guidance.

B F Billing
T J Quinn

Contents

   v   Opening remarks

 vii   Preface

xiii   Index of institutional abbreviations

xiv   Note regarding references

Chapter 1: Standards

1–16   Temperature standards
T J Quinn

Chapter 2: Cryogenic thermometry

17–31   Cryogenic thermometry between 0·1 K and 100 K
M Durieux

32–37   Gas therometry at low temperatures
K H Berry

38–43   A gas thermometer for low temperatures
R L Anderson and W Neubert

44–48   A proposal for dielectric-constant (or refractive-index) gas thermometry in the range 90–373 K
R L Rusby

49–56   Solid state phase transitions as thermometric fixed points
J F Schooley

57–64   Crystalline transformations of oxygen
J Ancsin

65–69   Silicon and gallium arsenide diodes for low-temperature thermometry
L Janšák, P Kordoš and M Blahová

70–79   Realization of the IPTS-68 between 54·361 and 273·15 K and the triple points of oxygen and argon
F Pavese

Contents

Chapter 3: Resistance thermometry

80–90   Resistance thermometry
J S Johnston

91–98   International comparison of low-temperature platinum resistance thermometers
J P Compton and S D Ward

99–106   Control of oxygen-activated cycling effects in platinum resistance thermometers
R J Berry

107–116   Characteristics of platinum resistance thermometers up to the silver freezing point
P Marcarino and L Crovini

117–124   Thermal drift correction and precision evaluation by data processing of resistance thermometer comparisons
F Pavese and G Cagna

125–130   Resistance thermometry using rhodium–iron, 0·1 K to 273 K
R L Rusby

131–134   Resistance thermometers with MOS field effect transistors
I Eisele and G Dorda

135–143   The 220 Ω Allen–Bradley resistor as a temperature sensor between 2 and 100 K
B W A Ricketson

Chapter 4: Thermocouple thermometry

144–161   Thermocouple thermometry
G W Burns and W S Hurst

162–171   Nicrosil and Nisil: their development and standardization
N A Burley, G W Burns and R L Powell

172–180   Practical performance of Nicrosil–Nisil thermocouples
N A Burley and T P Jones

181–187   Improved compensating lead systems for platinum-base thermocouples
W G Bugden, J A Tomlinson and G L Selman

188–194   Thermocouple referencing
G R Sutton

195–202   High-integrity, small-diameter mineral-insulated thermocouples
A Thomson and A W Fenton

203–210   Experimental evidence of erasable EMFs induced by thermal gradients in sheathed chromel–alumel thermocouples during long-term exposure
F Mathieu, R Meier, J Brenez and A Falla

Contents

211–218 The automatic calibration of thermocouples in the range 0–1100 °C
         T P Jones and T M Egan

Chapter 5: Radiation thermometry

219–237 Non-contact determination of temperatures by measuring the infrared radiation emitted from the surface of a target
         W Heimann and U Mester

238–243 The NPL photon-counting pyrometer
         P B Coates

244–255 Photoelectric direct current standard pyrometers and their calibration at PTB
         H Kunz and H J Kaufmann

256–263 Standard lamps and their calibration
         J W Andrews and P B Coates

264–272 Increasing precision in two-colour pyrometry
         G Ruffino

273–277 On the state of ratio pyrometry with laser absorption measurements
         H Kunz

278–286 Determination of the difference between the thermodynamic fixed-point temperatures of gold and silver by radiation thermometry
         H J Jung

287–296 Radiance temperature of molybdenum at its melting point
         A Cezairliyan, L Coslovi, F Righini and A Rosso

297–305 Temperature measurement of filaments above 2500 K applying two-wavelength pyrometry
         W Lechner and O Schob

306–314 Temperature measurement using a Plumbicon camera tube
         J R de Bie and W G Klomp

315–320 Time-resolved spectroscopic measurements on flashbulbs
         L W van der Meer and J de Vries

321–328 Heat flux pyrometer
         I R Ashcroft and P A Norris

Chapter 6: Plasma (arc) thermometry

329–340 Plasma (arc) thermometry
         J Richter

341–347 A fully automatic system for determining temperature distributions in thermal plasmas
         K C Lapworth and L A Allnutt

348–357 A plasma standard of temperature and vacuum ultraviolet radiation
R C Preston

358–367 Temperature scale and thermal radiation standards in the visible and vacuum ultraviolet spectral regions between 10 000 and 20 000 K
B Wende

368–374 Theoretical prediction of temperature profiles in a wall-stabilized argon arc
R E Shawyer

Chapter 7: Miscellaneous thermometry

375–388 Thermal imaging: a technical review
J Agerskans

389–397 The freezing point of aluminium as a temperature standard
G T Furukawa, W R Bigge, J L Riddle and M L Reilly

398–408 Systematic errors in high-temperature noise thermometry
L Crovini and A Actis

409–414 Non-contact temperature measurement using forced air convection
I R Fothergill

415–421 Ionic thermometers
M Strnad

422–427 Measuring the temperature maxima and distribution on the surface of dynamically heated moving systems, using ^{85}Kr techniques
J Fodor, L Léb and G Muk

428–438 The gas-controlled heat pipe: a temperature–pressure transducer
C A Busse, J P Labrande and C Bassani

439–445 Intercalibration of temperature transducer with a heat pipe furnace
P Coville and A Laurencier

446–452 Heat pipes for the realization of isothermal conditions at temperature reference sources
G Neuer and O Brost

453–454 Author Index

Index of institutional abbreviations

The text makes reference to several institutions and laboratories, and, for convenience, abbreviations are used. The corresponding full titles are given below.

ANSI	American National Standards Institute
ANSL	Australian National Standards Laboratory, Sydney, Australia (now NML)
ASTM	American Society for Testing and Materials, Philadelphia, Pa, USA
BIPM	Bureau International des Poids et Mésures, Paris
BSI	British Standards Institution
CCT	Comité Consultatif de Thermometrie
CGPM	Conférence Général des Poids et Mésures
CIPM	Comité International des Poids et Mésures
IEC	International Electrotechnical Commission
IMGC	Instituto di Metrologia 'G Colonnetti', Torino, Italy
ISA	Instrument Society of America
ISU	Iowa State University, USA
KOL	Kamerlingh–Onnes Laboratory, University of Leiden, Netherlands
NBS	National Bureau of Standards, Washington, DC, USA
NML	National Measurement Laboratory, Sidney, Australia (previously ANSL)
NPL	National Physical Laboratory, Teddington, Middlesex
NRC	National Research Council of Canada, Ottawa, Canada
PRMI	Physico-Radio-Technical Measurement Institute, Moscow, USSR
PSU	Pennsylvania State University, USA
PTB	Physikalisch–Technische Bundesanstalt, Braunschweig, Federal Republic of Germany
RAE	Royal Aircraft Establishment

Note regarding references to the Proceedings of the Symposia on Temperature

Five symposia have been held under the general title of 'Temperature, its Measurement and Control in Science and Industry'. The proceedings of the first were not published; those of the second were published in 1941 by Reinhold Publishing Co. and are now known as: 'Temperature, its Measurement and Control in Science and Industry Volume 1 1941' (Reinhold).

The third symposium was held in 1954 and its proceedings were published as: 'Temperature, its Measurement and Control in Science and Industry Volume 2 1955' published by Reinhold (New York) and Chapman and Hall (London), edited by H C Wolfe.

The fourth symposium was held in 1961 and its proceedings were published as: 'Temperature, its Measurement and Control in Science and Industry Volume 3, parts 1, 2 and 3, 1962' published by Reinhold (New York) and Chapman and Hall (London), edited by C M Herzfeld.

The fifth symposium was held in 1971 and its proceedings were published as: 'Temperature, its Measurement and Control in Science and Industry Volume 4, parts 1, 2 and 3, 1972' published by Instrument Society of America, edited by H H Plumb.

Since these symposia have brought together so much important work in thermometry, references to them are very frequent in this volume. We have therefore made each reference in an abbreviated form. For example:

Bedford R E 1972 *TMCSI* 4 part 1 15–25.

Temperature standards

T J Quinn
Division of Quantum Metrology, National Physical Laboratory, Teddington, Middlesex, England

Abstract. In this review the activities of the Consultative Committee for Thermometry leading up to the Amended Edition of 1975 of IPTS-68 are described. The differences between the new edition of IPTS-68 and the old are outlined along with the main areas of future activity in the field of international standards.

1. Introduction

It is now four years since the last major conference on temperature measurement, the Fifth Symposium on Temperature, held in Washington. Since then there has been considerable activity over the whole field of temperature standards both in experimental work and in the formulation of temperature scales. This, the third in a series of conferences held in this country on temperature and whose main sponsor has been The Institute of Physics, therefore comes at an opportune time.

During the course of the conference many of the new results and advances in technique which have taken place since 1971 will be described. In an attempt to place all this work in its context of a coordinated international programme aimed at improved practical thermometry I shall outline the results and recommendations which have stemmed from the 1971 and 1974 meetings of the Consultative Committee for Thermometry (CCT)† of the International Committee of Weights and Measures. In addition I will present some of the problems which need further experimental work for their resolution.

One important outcome of the 1971 CCT meeting was the setting up of four working groups to advise the CCT on various aspects of thermometry. It had become clear as a result of the experience gained during the drafting of IPTS-68 that thermometry had become so complex that any future revision of the scale would require very much more prior discussion. This was brought out very clearly by Preston-Thomas (President of the CCT) in his introductory address to the 1971 Symposium on Temperature. The different areas of thermometry assigned to the four working groups were as follows:

(I) The revision of IPTS-68.
(II) The secondary reference points of IPTS-68 and the evaluation of secondary-level calibration standards.
(III) Thermodynamic temperature above 100 K.
(IV) Thermodynamic temperature below 100 K.

Following the 1974 CCT meeting a fifth working group was set up:
(V) Practical low-temperature thermometers for the range below 30 K.

† See bibliography for the references to the Proceedings of the 5th Symposium and the Proceedings of the CCT.

Each working group was required to report and, if necessary, make recommendations to the CCT at the beginning of each year. The 1974 CCT meeting showed the fruits of this preparation in that it was possible at just one three-day meeting to approve the amended edition of IPTS-68. This new edition of IPTS-68, which I shall discuss in some detail later, was approved by the International Committee of Weights and Measures in October 1974 and ratified by the 15th General Conference on Weights and Measures at its meeting to celebrate the centenary of the Convention of the Metre in May 1975. No numerical changes in temperature are heralded in the new edition of the scale and temperatures, T_{68} and t_{68}, measured on it are unchanged. (In table 1 are listed the defining fixed points of IPTS-68.) The changes lie in the text, supplementary information, secondary reference points and in the suppression of the ^{3}He and ^{4}He vapour pressure scales. The new scale originates in the work of working group I aided by the recommendations of the other working groups. Before considering the text of the new edition of IPTS-68 in more detail it will be useful here to indicate some of the main features of the thermometry work carried out since 1971.

At low temperatures an international comparison of germanium thermometers has been carried out in Australia at the National Measurement Laboratory (NML) and a similar intercomparison of platinum capsule thermometers here at NPL (see Compton and Ward 1975). The results of the germanium thermometer intercomparison (see Durieux 1975) are at present being evaluated with a view to recommending a provisional scale between 0·9 K and 30 K pending the downward extension of IPTS-68. Proposals for novel interpolating thermometers based upon the variation with temperature of the refractive index or dielectric constant of a gas have been made (Colclough 1974) and experimental work is underway to explore the techniques (see Rusby 1975). The use of superconducting transitions as low temperature fixed points has been further developed with the provision by NBS of a series of such fixed points as part of their standard reference materials programme. The rhodium–iron resistance thermometer is now becoming established as a very suitable instrument for the most precise work of temperature scale realization and maintenance below 30 K (see Rusby 1975).

At high temperatures further work on extending the range of the photoelectric pyrometer to lower temperatures has led to very good agreement on the size of the silver–gold interval (see Jung 1975). In addition the presence of deviations of IPTS-68 from thermodynamic temperature in the range 700 °C to the gold point has become established. Results which are likely to have a profound influence on future temperature scales are those of Guildner and his collaborators (Guildner and Edsinger 1973) at the National Bureau of Standards (NBS). Their gas thermometry is leading to a complete reappraisal of our knowledge of thermodynamic temperature above 0 °C. I am hopeful that by the end of 1975 results will be available up to 630 °C thus allowing a start to be made on the link between photoelectric pyrometry and modern gas thermometry.

New work at the National Research Council (NRC) in Ottawa is leading to a much better understanding of the causes of instabilities in platinum–rhodium thermocouples and to the possibility of greatly improved performance in this respect. At the 5th Symposium a joint paper (Bedford et al 1972) was presented by NBS, NRC and NPL announcing the results of a cooperative programme on new reference tables for Pt 10% Rh–Pt (type S) and Pt 13% Rh–Pt (type R) thermocouples. Since then both the ASTM in the USA and the BSI in the UK have adopted these tables as new national standards,

Table 1. Defining fixed points of the IPTS-68[†].

Equilibrium state	Assigned value of International Practical Temperature	
	T_{68} (K)	t_{68} (°C)
Equilibrium between the solid, liquid and vapour phases of equilibrium hydrogen (triple point of equilibrium hydrogen)[‡]	13·81	−259·34
Equilibrium between the liquid and vapour phases of equilibrium hydrogen at a pressure of 33 330·6 Pa (25/76 standard atmosphere)[‡] [§]	17·042	−256·108
Equilibrium between the liquid and vapour phases of equilibrium hydrogen (boiling point of equilibrium hydrogen)[‡] [§]	20·28	−252·87
Equilibrium between the liquid and vapour phases of neon (boiling point of neon) [§]	27·102	−246·048
Equilibrium between the solid, liquid and vapour phases of oxygen (triple point of oxygen)	54·361	−218·789
Equilibrium between the solid, liquid and vapour phases of argon (triple point of argon)[∥]	83·798	−189·352
Equilibrium between the liquid and vapour phases of oxygen (dew point of oxygen) [§][∥]	90·188	−182·962
Equilibrium between the solid, liquid and vapour phases of water (triple point of water)	273·16	0·01
Equilibrium between the liquid and vapour phases of water (boiling point of water)[¶]	373·15	100
Equilibrium between the solid and liquid phases of tin (freezing point of tin)[¶]	505·1181	231·9681
Equilibrium between the solid and liquid phases of zinc (freezing point of zinc)	692·73	419·58
Equilibrium between the solid and liquid phases of silver (freezing point of silver)	1235·08	961·93
Equilibrium between the solid and liquid phases of gold (freezing point of gold)	1337·58	1064·43

[†] Except for the triple points and one equilibrium hydrogen point (17·042 K) the assigned values of temperature are for equilibrium states at a pressure p_0 = 1 standard atmosphere (101 325 Pa). The effects of small deviations from this pressure are shown in table 5. In those cases where differing isotopic abundances could significantly affect the fixed point temperature, the abundances specified in section III must be used.
[‡] The term equilibrium hydrogen is defined in section III, 5.
[§] Fractionation of isotopes or impurities dictates the use of boiling points (vanishingly small vapour fraction) for hydrogen and neon, and dew point (vanishingly small liquid fraction) for oxygen (see section III).
[∥] The triple point of argon may be used as an alternative to the dew point of oxygen.
[¶] The freezing point of tin (t' = 231·9292 °C, see equation (10)) may be used as an alternative to the boiling point of water.

ASTM E230-72 and BS 4937 respectively. In addition new reference tables for thermocouple types B, E, T and K proposed by NBS have been incorporated into these two same national standards. Agreement was reached in 1974 in the relevant working group of the International Electrotechnical Commission (IEC) to propose all these new reference tables as international IEC standards. Formal approval is expected during 1975. Worldwide adoption of the new tables will, inevitably, be slow since many national standards organizations will need to change their existing standards. Until this has happened manufacturers will still be called upon to make material to a number of different specifications — the very situation the whole programme was designed to eliminate. Nevertheless the final result will be an excellent example of the positive economic benefits to industry resulting from collaboration between standards laboratories of different countries.

2. IPTS-68, amended edition of 1975

It must be emphasized that this new edition of IPTS-68 constitutes only an amendment and not a replacement, and any measured temperature T_{68} or t_{68} remains unchanged.

The main differences between the two editions occur in the Introduction, where the relationship between thermodynamic temperature and Celsius temperature and their units is clarified, in the Supplementary Information where much of the original material has been removed, in the list of secondary reference points and in the deletion of the ^{3}He and ^{4}He vapour pressure scales which are now known to be in error. The amended Introduction reads:

I. Introduction

The unit of the fundamental physical quantity known as thermodynamic temperature, symbol T, is the kelvin, symbol K, defined as the fraction 1/273·16 of the thermodynamic temperature of the triple point of water.[†]

For historical reasons, connected with the way temperature scales were originally defined, it is common practice to express a temperature in terms of its difference from that of a thermal state 0·01 kelvins lower than the triple point of water.[†] A thermodynamic temperature, T, expressed in this way is known as a Celsius temperature, symbol t, defined by

$$t = T - 273 \cdot 15 \text{ K}. \tag{1}$$

The unit of Celsius temperature is the degree Celsius, symbol °C, which is, by definition, equal in magnitude to the kelvin. A difference of temperature may be expressed in kelvins or degrees Celsius.

The International Practical Temperature Scale of 1968 (IPTS-68) has been constructed in such a way that any temperature measured on it is a close approximation to the numerically corresponding thermodynamic temperature. Moreover, such measurements are easily made and are highly reproducible; in contrast, direct measurements of thermodynamic temperatures are both difficult to make and imprecise.

The IPTS-68 uses both International Practical Kelvin Temperatures, symbol T_{68}, and International Practical Celsius Temperatures, symbol t_{68}. The relation between T_{68} and t_{68} is the same as that between T and t, in other words

$$t_{68} = T_{68} - 273 \cdot 15 \text{ K}. \tag{2}$$

[†] Comptes Rendus des Séances de la Treizième Conférence Générale des Poids et Mesures (1967–1968), Resolutions 3 and 4, p104.

The units of T_{68} and t_{68} are the kelvin, symbol K, and the degree Celsius, symbol °C; that is, the names of the units are the same as those used for the thermodynamic temperatures T and t.

The IPTS-68 was adopted by the International Committee of Weights and Measures at its meeting in 1968 in accordance with the power given to it by Resolution 8 of the 13th General Conference of Weights and Measures. That scale replaced the International Practical Temperature Scale of 1948 (amended edition of 1960).

It is now made clear that a Celsius temperature is a thermodynamic temperature expressed as a scale having simply a shift of the zero point. The unit of Celsius temperature, the degree Celsius, although equal in magnitude to the kelvin is not intended to be used interchangeably with the kelvin. It is not intended, for example, that a Celsius temperature be expressed in kelvins or vice versa since this would depart so much from common usage that it would lead to great confusion.

The text of the definition of the scale in the amended edition does not differ in any major way from that in the original edition of IPTS-68. New work at NBS and NPL on the realization of the boiling points of gases (Compton 1970a, 1970b) has led to small changes being made in the way the boiling point is defined. Also the triple point of argon at 83·798 K is introduced as an alternative to the boiling point of oxygen. This follows the general opinion that in principle triple points should be used whenever possible. The text now makes it clear that in using the Planck law for the definition of the scale above the gold point, the wavelength referred to is the vacuum wavelength rather than that used previously which was the air wavelength. This was made explicit following a note by Blevin of the NML (Blevin 1972) in which he pointed out that up until then thermometrists and photometrists had taken no account of the fact that the Planck law when expressed in terms of wavelength includes the refractive index of the medium, n. The correct formulation of the Planck law is, in the usual notation:

$$L_\lambda = \frac{2hc^2/n^2\lambda^5}{\exp(hc/kn\lambda T) - 1}.$$

Taking the refractive index of air to be about 1·0028, the errors incurred in omitting n are shown in table 2. Since optical pyrometry is concerned only with values of L_λ relative to that of the black body at the reference temperature of the gold point (1337·58 K), it is only the appearance of n in the exponential which needs to be taken into account.

In that part of the scale defined by the Pt 10% Rh–Pt thermocouple the relations which the EMF must satisfy at 630·74 °C, the silver (961·93 °C) and the gold points have been modified to be compatible with the new international reference tables. The equations in the text now read as follows:

$$E(t_{68}(\text{Au})) = 10\,334 \,\mu\text{V} \pm 30 \,\mu\text{V} \tag{13}$$

$$E(t_{68}(\text{Au})) - E(t_{68}(\text{Ag})) = 1186 \,\mu\text{V} + 0\cdot 17[E(t_{68}(\text{Au})) - 10\,334 \,\mu\text{V}] \pm 3 \,\mu\text{V} \tag{14}$$

$$E(t_{68}(\text{Au})) - E(630\cdot 74\,°\text{C}) = 4782 \,\mu\text{V} + 0\cdot 63[E(t_{68}(\text{Au})) - 10\,334 \,\mu\text{V}] \pm 5 \,\mu\text{V}. \tag{15}$$

Table 2. Comparison of values for T_{68} calculated from equation (3) when the term λ is set equal to (a) λ_a, the wavelength in air, and (b) $n_a\lambda_a$, the wavelength in vacuum. $T_{68}(\text{Au}) = 1337\cdot58$ K, $c_2 = 0\cdot014388$ K m, $\lambda_a = 655$ nm, $n_a = 1\cdot00028$.

$\lambda = \lambda_a$	$\lambda = n_a\lambda_a$
1000·00	999·93
1500·00	1500·05
2000·00	2000·28
2045·00	2045·30
2500·00	2500·61
3000·00	3001·04

There was some discussion during the drafting of the new edition as to the purpose served by these relations since they severely limit the rhodium content in the alloy arm yet it is known that in practice quite large variations in rhodium content make little difference to interpolated temperatures.

For example if Pt 10% Rh–Pt and Pt 13% Rh–Pt thermocouples are calibrated at the three fixed points and a quadratic interpolation equation is used as specified by IPTS-68, then both types of thermocouple lead to interpolated temperatures which hardly differ by 0·1 °C. It was eventually decided that equations of this type serve a useful purpose in eliminating the use of contaminated thermocouple wire which would be unlikely to fit the equations.

That section of the IPTS-68 devoted to supplementary information has undergone major revision. It was the general view that much of the detailed information given in the first edition of IPTS-68 was now out of date and in some instances misleading since new work had brought to light problems not appreciated at the time of drafting. It was agreed therefore that the supplementary information should be much shorter and be confined to essentials. As a result, the section on the platinum resistance thermometer now contains very little information on construction and is largely confined to annealing procedures. In the section on the standard thermocouple it is now pointed out that in general use an accuracy better than ±0·2 °C cannot be expected because of the continually changing chemical and physical inhomogeneities in the wires in the region of temperature gradients.

The description of the boiling points of hydrogen and neon now contains a note about the difference, of about 0·4 mK, between the boiling point at vanishing liquid fraction (dew point) and that at vanishing vapour fraction (boiling point). The normal isotopic composition of neon is also given, namely 2·7 mmol of ^{21}Ne and 92 mmol of ^{22}Ne per 0·905 mol of ^{20}Ne. Most of the experimental detail previously given in the description of the boiling and triple points along with that relating to the freezing points of metals has been removed and the remaining information is mostly concerned with possible sources of error.

In the original version of IPTS-68 the reference function for the region below 0 °C was expressed in terms of a 20th degree polynomial in the logarithm of the resistance ratio. The coefficients of the twenty terms were each specified to sixteen significant figures and because of the form of the polynomial all twenty terms to sixteen significant figures have to be used even if the full accuracy of 0·1 mK is not required. This is necessary because the evaluation of a polynomial of this form involves the cancellation of large terms in the higher order coefficients. The omission of one term or even the truncation of one coefficient can lead to unacceptable errors. In the amended edition of IPTS-68 a more efficient formulation of the same reference table has been adopted. It has been obtained by the normalization of the independent variable ($W_{\text{CCT-68}}(T_{68})$) so that for any particular temperature range the normalized variable lies between 1 and −1. The result is that the unit of each coefficient is the kelvin hence the number of significant digits needed in each coefficient is just the same as the desired precision in T_{68}. It is still necessary to use all twenty terms of the polynomial, but the computation no longer involves double length arithmetic and can be carried out without difficulty on a desk calculator. In view of the fact that, for even the highest precision, the number of significant figures in T_{68} never exceeds seven, a polynomial involving twenty coefficients each to sixteen significant figures appears anomalous. A further step in mathematical elegance and simplicity was discussed but in the end not adopted. This would have required the use of a Chebyshev polynomial form of reference function in which not only can the number of digits in the coefficients be truncated depending upon the accuracy required, but also the number of terms can be truncated. Because of the relative unfamiliarity of Chebyshev polynomials, however, the CCT decided that their advantages were not so great as to justify their use at this stage. For comparison purposes the coefficients of the three types of polynomial are listed in table 3. The

Table 3.†

i	A_i	B_i	C_i
0	0·273 150 000 000 000 0 E +3	+38·59276	+161·82457
1	0·250 846 209 678 803 3 E +3	+43·44837	+101·20451
2	0·135 099 869 964 999 7 E +3	+39·10887	+51·52727
3	0·527 856 759 008 517 2 E +2	+38·69352	+24·77384
4	0·276 768 548 854 105 2 E +2	+32·56883	+10·11559
5	0·391 053 205 376 683 7 E +2	+24·70158	+3·53282
6	0·655 613 230 578 069 3 E +2	+53·03828	+0·91392
7	0·808 035 868 559 866 7 E +2	+77·35767	+0·14666
8	0·705 242 118 234 052 0 E +2	−95·75103	−0·00327
9	0·447 847 589 638 965 7 E +2	−223·52892	+0·00265
10	0·212 525 653 556 057 8 E +2	+239·50285	+0·01172
11	0·767 976 358 170 845 8 E +1	+524·64944	+0·01013
12	0·213 689 459 382 850 0 E +1	−319·79981	+0·00366
13	0·459 843 348 928 069 3 E +0	−787·60686	+0·00007
14	0·763 614 629 231 648 0 E −1	+179·54782	−0·00071
15	0·969 328 620 373 121 3 E −2	+700·42832	−0·00066
16	0·923 069 154 007 007 5 E −3	+29·48666	−0·00085
17	0·638 116 590 952 653 8 E −4	−335·24378	−0·00028
18	0·302 293 237 874 619 2 E −5	−77·25660	+0·00034
19	0·877 551 391 303 760 2 E −7	+66·76292	+0·00026
20	0·117 702 613 125 477 4 E −8	+24·44911	+0·00005

† The numbers in the second column of the above table are written in the form: decimal number E signed integer exponent, where E denotes the base 10.

coefficients A_i are those of the original IPTS-68, B_i are those of the normalized variable polynomial of the 1975 edition of IPTS-68 and C_i are those of the equivalent Chebyshev polynomial. Each set is sufficient to calculate T_{68} to an accuracy of 0·1 mK.

In the list of secondary reference points a number of changes have been made on the recommendation of working group II. The most significant change is that concerning the freezing point of platinum previously given at 1772 °C. The new value, 1768·5 °C, is the mean of the NPL value of 1971 (Quinn and Chandler 1972), 1767·9 °C ± 0·3 °C and a new value, as yet unpublished, from NML of 1769·5 °C ± 0·5 °C. Both results are calculated on the basis of vacuum wavelengths, the omission of which would lead to an error of 0·3 °C. Despite extensive correspondence and discussion the reason for the difference of 1·6 °C between the two results remains obscure. Until the results are available of measurements being made of the platinum point at the Physikalisch-Technische Bundesanstalt (PTB) Braunschweig and the Istituto di Metrologia 'G Colonnetti' (IMGC) in Turin, the value of 1768·5 °C must be provisional.

In the first edition of IPTS-68 a table was given (table 7) listing the assigned values of the defining fixed points together with their estimated uncertainties with respect to thermodynamic temperature. In the amended edition this table has been deleted and in its place the following paragraph has been inserted in the text:

> Table 7 has been deleted from the text. The uncertainties listed in table 7 of the original version of IPTS-68 were assigned on the basis of the best available evidence at the time IPTS-68 was devised. Since then new work has indicated that IPTS-68 probably departs from thermodynamic temperature above 10 m °C, both at the defining fixed points and between them, by rather more than the uncertainties shown. Revised estimates of these uncertainties will, therefore, be published from time to time by the Comité Consultatif de Thermométrie.

The decision to delete table 7 was only reached after a great deal of discussion both at the 1974 CCT meeting and at earlier meetings of working group III. It had become apparent that the position of table 7 was going to be difficult to resolve when working group III first came to consider the uncertainties listed therein in the light of the new NBS gas thermometry. The original uncertainty of 5 mK at the steam point was clearly inconsistent with the NBS work which had already given a difference of some 30 mK between IPTS-68 and their realization of thermodynamic temperature. Strong views were expressed that despite this the original uncertainty should stand pending confirmation of the NBS work. In any case a simple change in the uncertainty of the steam point alone to accommodate the NBS result would leave some of the remaining higher temperature fixed points in an impossible position. The question of table 7 led to a discussion at the 1974 CCT meeting of the whole problem of the status of new metrological work. The view can be taken that any new result must simply be added to the set of those already existing and allowed to influence the overall mean and that no significant changes should be made on the basis of just one result. The general view was that this is not a tenable position. If it is evident that a new measurement is more accurate than the older ones and that systematic errors which were not appreciated at the time of the older measurements have been avoided in the new measurement then one should not average the old with the new but use the new result. One of the roles of the CCT in thermometry is to evaluate new experimental work and decide whether or not it needs confirmation or can be accepted alone. However, as every experimental metrologist is aware, the evaluation of another's work is very difficult indeed except in the rare cases when it reveals sources of systematic error which have been neglected by the experimenter.

Another deletion from the original text of IPTS-68 is the reference to the T_{58} and T_{62} helium vapour pressure scales as the recommended low-temperature scales. This deletion was recommended by working group IV in the light of recent work which has shown substantial errors in the scales relative to thermodynamic temperature. The background to this decision is discussed in more detail in the next paper (see Durieux 1975); here I need only remark that the intention is to revise the helium vapour pressure scales at the earliest opportunity, if possible in time for ratification by the 16th General Conference on Weights and Measures which will probably be held in 1979. Reference to them was deleted from the amended edition of IPTS-68 because the opportunity for revision would not otherwise have arisen until IPTS-68 itself was replaced, and that is not likely before 1983.

We hope to be able to publish the amended 1975 edition of IPTS-68 in this country before the end of 1975. The official French text will be available in the Proceedings of the 15th General Conference on Weights and Measures.

3. Future work

3.1. The low-temperature range

Having discussed the revisions which it was possible to make to the text of IPTS-68 without changing numerical values of T_{68} it is worth speculating on the likely form of its successor. There is little doubt that any new scale will extend to temperatures below the present limit of 13·81 K but the form of this extension is by no means clear at present. There has, however, been intensive activity in national standards laboratories aimed at laying the groundwork for this future extension downwards. By means of gas and acoustic thermometry primary measurements of thermodynamic temperature have been made in the range from 2 K to 30 K, and magnetic thermometer scales have been set up extending down below 2 K. These have been based upon the helium vapour pressure scales, the NBS-55 scale and, at higher temperatures, IPTS-68. There are a number of possible ways of providing a low-temperature scale none of which appears, at this stage, to have over-riding advantages over all the others. It has yet to be decided, for example, whether or not the new scale is to be one based upon a primary thermometer or a secondary thermometer. A method in which a primary thermometer is used was proposed by Barber (Barber 1972) in which a constant-volume gas thermometer is used as an interpolating instrument. The thermometer would be calibrated at two fixed points having defined temperatures, such as the helium and equilibrium hydrogen boiling points, and interpolation would then be made using specified virial coefficients. Secondary thermometers could then be calibrated at will over any part or all of the range. An example of a scale based upon a secondary thermometer is, of course, IPTS-68 above 13·81 K. The extension of this method below 13·81 K poses problems because the platinum thermometer is too insensitive. The germanium and rhodium–iron thermometers are likewise unsuitable for different reasons. The germanium thermometer is not sufficiently stable in the long term and its resistance–temperature relation varies so much from one instrument to another that a single reference function is impracticable. The poor long-term reproducibility is thought to be a result of sudden changes in the contact resistance between the leads and the germanium. The rhodium–iron thermometer on the other hand appears to be very reproducible in the long term but the spread of the resistance–temperature characteristic, though less than that of the germanium thermometer, is still inconveniently large, a point discussed in more detail in the paper by Rusby (1975). The magnetic thermometer requires calibration against other thermometers to enable the constants in the relation between susceptibility and temperature to be evaluated but it also suffers from serious experimental difficulties if it is to be used up to 30 K. The balance of opinion at the moment is tending towards a scale based upon a primary thermometer though of what kind has yet to be decided. If a new scale is to be introduced even in 1983 then urgent work is required in this range to enable national laboratories to evaluate the various proposals before a final choice is made.

3.2. Primary thermometry above 100 K

At the higher temperatures the successor to IPTS-68 will be built upon the results of primary thermometry at present in progress or planned at national laboratories. Since the activity in this area is quite extensive it is worth summarizing the present position so that an overall view can be obtained.

3.2.1. Gas thermometry.
Three national laboratories are actively engaged upon gas thermometry above 0 °C, these are NBS, NRC and PTB. The work at NBS has been conducted with strong emphasis on both the improvement of instrumentation and the investigation of possible systematic effects of sorption. The most recent published results are those of Guildner and Edsinger (1973) in which they give thermodynamic values obtained up to 140 °C. At the steam point, the difference $t_{\text{therm}} - t_{68}$ is -0.030 ± 0.0035 K and this difference is nearly linear with t_{68} over the range of measurements. Provisional results as yet unpublished show the deviations of t_{therm} from t_{68} continuing to increase up to 420 °C. It is hoped to extend measurements up to 630 °C by the end of 1975.

The gas thermometry underway at PTB is directed at the steam point. The constant-volume gas thermometer used for this work has a silica bulb (800 ml volume) and a piston gauge is being used for the pressure measurement. It is hoped that the first results will be available during 1975. At the 1971 meeting of the CCT, PTB announced previously unpublished measurements of the steam point made in 1956. These showed differences at 100 °C between t_{48} (=t_{68} at 100 °C) and the gas thermometer of 18 mK or 28 mK ± 10 mK, depending upon how the reference temperature was calculated.†

A constant-volume gas thermometer is also under construction at NRC. This thermometer uses a stainless steel bulb of 1400 cm³ capacity, which can be baked at 650 °C,

---

† A recalculation by Haar (1972) of some old calorimetric data indicates a steam point lower than 100 °C by some 0.035 K, though in view of the stated accuracy of the original experimental results the agreement with the new NBS work might be fortuitous.

and is designed for operation at temperatures up to the zinc point. Pressure measurement is also to be by means of a piston gauge. The whole experiment is designed to give an uncertainty in the measurement of thermodynamic temperature of about 5 mK up to the steam point.

3.2.2. Noise thermometry. The results of noise thermometry carried out at the IMGC up to 1971 are described by Actis *et al* (1972). Since then further work, using the ice and tin points as reference temperatures, has shown that the IPTS-68 value for the antimony point is consistent with the assigned values of these reference points to within the estimated uncertainty of the measurement, namely an estimated standard deviation of the mean of 0·2 K. Work is under way to reduce the uncertainty of measurement and then to extend the measurements to the gold point (see Crovini and Actis 1975). New work on noise thermometry is also being undertaken at NML in the temperature range 90 K to 373 K. Results are expected during 1975.

3.2.3. Radiation thermometry. There are two approaches to radiation thermometry: the total radiation method and the spectral radiation method. In both of these the radiation (either total or spectral) emitted by two black bodies, one of which is at a reference temperature, is compared and the ratio of the two allows the unknown to be calculated using either the Stefan–Boltzmann or the Planck law. If total radiation is used then the reference temperature can be made 0 °C (or the triple point of water) and a primary determination of an unknown temperature is then possible. If spectral radiation is used, however, the reference temperature has in practice to be much higher than 0 °C in order to obtain sufficient radiant power, and the reference temperature has to be determined by other methods. An additional uncertainty in spectral radiation measurements arises from uncertainty in the value of the second radiation constant, hc/k. The value of the Boltzmann constant k depends upon the gas constant R. The estimated uncertainty in the gas constant has been 30 ppm, but a new determination of R carried out at NPL has given a value different by 191 ppm (Quinn *et al* 1975). If substantiated then the consequent change in k will lead to changes in temperature amounting to 0·3 K at the freezing point of platinum when measured by reference to the gold point.

Two laboratories are currently engaged upon total radiation measurements, the NBS and NPL. The NBS work was described at the 5th Symposium in 1971 (Ginnings and Reilly 1972), where it was shown that using a reference temperature of 0 °C the sensitivity at 100 °C was approaching 1 mK. A careful examination of possible sources of error in the apparatus is now being conducted. It is intended to operate the apparatus at temperatures up to the gold point. An experiment similar in principle has been started at NPL in which, in addition, a value for the Stefan–Boltzmann constant is to be obtained from an absolute measurement of the radiant power emitted by the reference black body at 0 °C. It is then proposed that thermodynamic temperatures, up to at least the zinc point, will be measured by reference to the ice point, with a precision of a few millikelvins.

Spectral radiation methods are being used by a number of laboratories for the determination of fixed points above the zinc point. At BIPM work by Bonhoure is continuing on the measurement of the antimony–silver–gold intervals by means of a spectral photoelectric pyrometer operating at a series of wavelengths near 950 nm. Latest results were reported at the 1974 CCT meeting. In addition measurements of the difference between 'radiation' temperatures and IPTS-68 between the fixed points indicate a difference of nearly 0·5 °C near 800 °C. This difference is in the same direction and of nearly the same magnitude as that recently reported from NPL, where preliminary results have been obtained down to 725 °C (Quinn et al 1973) using a photoelectric pyrometer working at a wavelength of 662 nm. We are now using a photomultiplier at wavelengths up to 950 nm working in the photon counting mode to provide adequate sensitivity at temperatures down to the zinc point, 419 °C (see Coates 1975). The results so far obtained from gold point to 725 °C include comparison of the radiation scale (based on the IPTS-68 gold point) with IPTS-68 realized by means of a thermocouple and also temperatures obtained by a platinum resistance thermometer using a quadratic interpolation function, based on the ice, zinc and gold points, plus a correction term for the effects on the resistivity of lattice defects. The results show that IPTS-68 deviates from both the radiation and resistance scales by nearly half a degree at 750 °C. Also the silver–gold interval indicated by IPTS-68 appears too large by about 0·15 °C, a result in close agreement with those of Bonhoure and of Jung (of PTB) who presents his first results in a paper at this conference. It is of interest to note here that the IPTS-68 gold point could be interpreted as being too high by 0·3 °C according to the work of Blevin and Brown (1971) following their work on the Stefan–Boltzmann constant.

3.3. The role of the thermocouple in the range 630·74 °C to 1064·43 °C

Between 630·74 °C and 1064·43 °C (gold point) the IPTS-68, in common with its predecessors, is defined by means of a Pt 10% Rh–Pt thermocouple calibrated at 630·74 °C, silver and gold points. Interpolation is by means of a quadratic equation. Since the reproducibility of a thermocouple is, in general, no better than ±0·2 °C there is thus a large step in the reproducibility of IPTS-68 at 630·74 °C, the junction of the resistance thermometer and thermocouple ranges. In addition there is a discontinuity in the first derivative against t_{68} of both resistance and thermal EMF at 630·74 °C amounting to about 0·2%. Because of the poor reproducibility of the thermocouple it has long been the intention to dispense with it by extending upwards the range of the platinum resistance thermometer. To this end, during the decade leading up to the 5th Symposium, many national laboratories devoted considerable effort to developing a thermometer suitable for use up to the gold point. Unfortunately by the time of the 5th Symposium it had become evident that problems related to annealing of quenched lattice defects, mechanical stability and electrical leakage resistance were much more difficult to resolve than anticipated. Consequently, since 1971 there has been very little new work on high-temperature platinum thermometry. There is no thermometer on the market which is considered suitable as a defining instrument for the temperature scale up to the gold point. A thermometer has recently become available for use up to the silver point but as yet insufficient published information is available regarding its use as an interpolation instrument for it to be considered.

The advances made in photoelectric pyrometry during the same decade gave rise to much discussion at the 5th Symposium and afterwards on the possibility of extending to lower temperatures the range of the IPTS-68 defined by the radiation pyrometer. Work has been going on in a number of national laboratories to explore this proposal. There seems little doubt that it would now be possible to set up a radiation scale, based upon a defined value for the gold point and the second radiation constant, which could extend down to at least the aluminium point, 660·46 °C. Preliminary work at NPL has already demonstrated such a scale down to 725 °C having a reproducibility of 0·1 °C (Quinn *et al* 1973). Improvements in technique already made point to a reproducibility of 0·02 °C being within reach.

A new scale could be devised, therefore, in which the thermocouple is replaced by an extension downwards of the radiation scale rather than by the extension upwards of the platinum resistance thermometer scale originally envisaged. An argument against this proposal is that the equipment needed to realize the scale to its full potential would only be available to a few national laboratories. This argument has some force though it applies already to the temperature range above the gold point where this situation has prevailed for some time. Above the gold point day-to-day measurements may be made by thermocouples or other less precise instruments which are themselves calibrated, directly or indirectly, against the defining instrument. There seem few reasons why the same methods should not apply below the gold point as they do above. It should be appreciated that the IPTS, though still called a 'practical' scale, has long since ceased to be a scale which could be easily and simply set up in a laboratory interested in making temperature measurements. The requirements of industry and science are now such that few outside national standards laboratories have either the time or the expertise to realize the scale to its full potential. It was the recognition of this situation that prompted the CCT to ask its working group II to look into ways of providing temperature standards to a secondary level of accuracy. Where requirements are not for the highest precision it should be possible to use a scale which despite its lower accuracy is reproducible and well specified.

A form of temperature scale below the gold point defined by means of a photoelectric pyrometer need not lead to significant differences in practice. For example, a thermocouple could be calibrated at 630·74 °C, and at the silver and gold points as before, but interpolation would be by means of a reference table obtained using a photoelectric pyrometer. We know that if such a scheme were implemented the value obtained for a temperature of 800 °C, say, would differ from the IPTS-68 value by about 0·5 °C (Quinn *et al* 1973). The relation of such a temperature scale to thermodynamic temperature would be well defined and would depend solely upon the temperature assigned to the gold point and to the second radiation constant. The reference table would be based upon the results of work (already under way) in national standards laboratories. A similar reference table could at any stage be produced for a platinum resistance or any other type of thermometer and would in no way involve an alteration to the definition of the scale.

Were a fixed point available in the region of 800 °C then there might be advantages in making it the junction between the radiation and platinum resistance thermometer ranges. At the moment, however, it must be remembered that even up to 800 °C there remain problems in implementing a resistance thermometer as interpolation instrument for IPTS. Recent work on the melting and freezing of binary eutectic alloys (McAllan 1972), some of which having melting points between 660 °C and the silver point, has not given rise to much optimism regarding their suitability as thermometric fixed points. The freezing behaviour of a eutectic is a very rate-dependent process and the melting is strongly influenced by the rate at which the previous freeze took place. Any procedure which took account of these problems would inevitably be involved and time consuming. Even without a fixed point it might be argued that the junction between the resistance thermometer and radiation pyrometer should be near 800 °C to avoid what might appear to be difficulties in radiation pyrometry at the lower temperatures. At present, however, insufficient information is available as to the advantages and disadvantages of the various alternatives and a decision must await the results of more work on resistance and radiation thermometry.

The replacement of the thermocouple as the defining instrument of IPTS-68 in the range between 630·74 °C and the gold point is thus still very much a matter for discussion. The recent unpublished work by McLaren at NRC on the causes of instability in platinum thermocouples has raised the possibility that a proper choice of alloy and annealing may lead to a great improvement in this respect. If this turns out to be the case then it is conceivable that there will be no need to replace the thermocouple as the defining instrument of IPTS-68.

3.4. The range above the gold point

An international comparison of temperature scales above the gold point was carried out in 1971 by circulating a set of tungsten strip lamps among the four participating laboratories, NBS, NML, NPL and PTB (Lee et al 1972). The results were very satisfactory up to a temperature of 1550 °C and the overall spread of results was less than 0·4 °C among all four laboratories and at 1700 °C three laboratories agreed to within ±0·1 °C. The fourth, NML, had instrumental difficulties which precluded them from using the special vacuum lamps provided for use up to 1700 °C. A recent intercomparison between NPL and NML carried out in November 1974 reproduced the relative position of the two laboratories found in 1971 to within ±0·1 °C. In this latest comparison the NML instrumental problem was resolved allowing measurements to be made up to 1770 °C.

In the range between the gold point and 1770 °C, the upper limit of vacuum standard lamps, the reproducibility of the IPTS-68 in many national laboratories is adequate for all practical purposes. At higher temperatures, however, problems remain in the reproducibility of the gas-filled standard lamps used for maintaining the scale. The agreement at high temperatures in the 1971 intercomparison where gas-filled lamps were used was not nearly so good as that below 1700 °C. The only high-temperature point measured was at 2200 °C, but here the total spread among all four laboratories was nearly 2 °C. There remain considerable problems in improving the behaviour of gas-filled standard lamps, most of these being the result of the variable heat loss from the strip by convection. The long-term drift is not, by itself, a limiting factor in the use of the gas-filled lamp. Drift rates as low as 1 °C per hundred hours operation at 2200 °C can be achieved. The problem lies in the sensitivity of the lamp to changes in orientation and surroundings. There is a clear need for further investigation into the behaviour of gas-filled standard lamps aimed at a better understanding of the convection effects. Despite these difficulties gas-filled lamps continue to serve a useful purpose. Table 4

Table 4. Luminance temperature calibrations: before and after each experiment.

Set current (A)	1968 IPTS luminance temperature (K)		ΔT (K)
	initial	final	
100/10/gas	September 1968	March 1970	—
56·518	2843·5	2845·1	+1·6
54·061	2765·6	2765·8	+0·2
44·670	2454·4	2452·9	−1·5
40·950	2323·9	2321·8	−2·1
32·906	2021·9	2021·4	−0·5
31·288	1957·1	1957·1	0·0
100/11/gas	October 1970	December 1972	
53·869	2760·8	2761·9	+1·1
44·688	2451·0	2451·6	+0·6
40·977	2320·7	2321·1	+0·4
32·978	2020·4	2020·5	+0·1

shows the results of calibrations of two gas-filled black-body lamps at temperatures up to 2845 K made at intervals of two years (Jones and Gordon-Smith 1973). Among the total of ten pairs of calibrations six show differences less than 1 K.

4. Conclusion

My intention in this review has been to outline in general terms the progress that has been made in some aspects of thermometry directly related to the IPTS-68 and to highlight those areas where further work is needed. Many of the points I have raised are discussed in greater detail in subsequent papers of this conference (see also Quinn and Compton 1975). I hope that the overall effect of this paper when read alongside all the others will be to give a balanced view on how fundamental thermometry stands today.

References

Actis A *et al* 1972 *TMCSI* **4** part 1 335
Barber C R 1972 *TMCSI* **4** part 1 99–103
Bedford R E *et al* 1972 *TMCSI* **4** part 3 1585–602
Blevin W R 1972 *Metrologia* **8** 146–7
Blevin W R and Brown W J 1971 *Metrologia* **7** 15–29
Coates P B 1975 this volume
Colclough A R 1974 *Metrologia* **10** 73–4
Compton J P 1970a *Metrologia* **6** 69–74
—— 1970b *Metrologia* **6** 103–9
Compton J P and Ward S D 1975 this volume
Crovini L and Actis A 1975 this volume
Durieux M 1975 this volume
Ginnings D C and Reilly M L 1972 *TMCSI* **4** part 1 339
Guildner L A and Edsinger R E 1973 *J. Res. NBS* **77A** 383–9

Haar L 1972 *Science* **176** 1293
Jones O C and Gordon-Smith G W 1973 *Proc. R. Soc. Lond.* A **335** 369–86
Jung H J 1975 this volume
Lee R D *et al* 1972 *TMCSI* **4** part 1 377–94
McAllan J V 1972 *TMCSI* **4** part 1 265–74
Quinn T J and Chandler T R D 1972 *TMCSI* **4** part 1 295–309
Quinn T J *et al* 1973 *Metrologia* **9** 44–6
Quinn T J, Colclough A R and Chandler T R D 1975 to be published
Quinn T J and Compton J P 1975 *Rep. Prog. Phys.* **38** 151–239
Rusby R L 1975 this volume

Inst. Phys. Conf. Ser. No. 26 © 1975: Chapter 2

Cryogenic thermometry between 0·1 K and 100 K

M Durieux
Kamerlingh Onnes Laboratory of the University of Leiden, The Netherlands

Abstract. A review of thermometry between 0·1 K and 100 K is given. The emphasis is on precision thermometry and temperature scales. Results of a recent experiment by Cetas in which the consistency of the IPTS-68 is investigated with a magnetic thermometer are given. Further, the emphasis is on the range between 1 K and 30 K. The principles of gas thermometry and magnetic thermometry are mentioned. Then the international comparison of temperature scales (1−30 K) which is presently being carried out at the National Measurement Laboratory of Australia and the plan to publish a 'best estimate for a smooth low-temperature scale between 1 K and 30 K' under the auspices of the Comité Consultatif de Thermométrie is discussed. Such a scale would solve the present difficulty that no internationally agreed scale between 4 K and 14 K exists.

The need for a stable sensitive resistance thermometer which does not require calibration at too many points is emphasized.

1. Introduction

In this paper some recent developments in cryogenic thermometry between 0·1 K and 100 K will be mentioned. The emphasis will be on temperature scales and on those methods of temperature measurement which are used to deduce or to realize the scales. No claim of completeness is made.

More detailed information on cryogenic thermometry can be found in the review papers of Rubin (1970) ($T < 100$ K), of Swenson (1970) (1−30 K), and of Quinn and Compton (1975). A wealth of information is, of course, available from the Proceedings of the Fifth Symposium on Temperature in Washington (see Rubin et al 1971). Recently a review paper on diode and resistance thermometry was published by Swartz and Swartz (1974). Fundamental work on thermometry in the national standards laboratories is published in *Metrologia*.

As an introduction to the subject treated in this paper several methods of temperature measurement between 0·1 K and 100 K are listed in table 1.

For the determination of thermodynamic temperatures the classical gas thermometer is still the most common instrument. Recently, however, attention has been given to the use of measurements of the dielectric constant (Sherman 1971, Rusby 1975) or the refractive index (Colclough 1974) of the gas in the gas thermometer bulb for determining the density of the gas, thus avoiding the cumbersome determination of the volume of the bulb (and the dead volume) of the gas thermometer and of the number of moles in it. The acoustic thermometer has been used in at least two thermometric studies (Plumb and Cataland 1966, Colclough 1971). Below about 1·5 K the condensation pressure of available gases is too low to make the use of the properties of these gases feasible for thermometry. An interesting possibility for thermometry below 0·1 K

Table 1. Methods of thermometry between 0·1 K and 100 K.

Thermometers for thermodynamic temperature measurements

    Pressure–volume gas thermometer $pV_m = RT$ V_m = molar volume of the gas

    Pressure–dielectric constant gas thermometer
$$pV_m = RT \qquad \frac{\epsilon_r - 1}{\epsilon_r + 2} = \frac{1}{3} N\alpha, \; V_m = N_A/N$$

        α = polarizability of an atom, N = number of atoms in a certain volume divided by that volume, N_A = Avogadro's number

    Pressure–refractive index gas thermometer
$$pV_m = RT \qquad \frac{n^2 - 1}{n^2 + 2} = \frac{1}{3} N\alpha$$

    Acoustic thermometer $w^2 \sim T$ w = velocity of sound

(Ideal gas laws are given; in practice, experimental data for real gases are extrapolated to zero pressure. Properties of gases other than those listed are not (yet) used for thermometry.)

    Osmotic pressure of a liquid ^{3}He/^{4}He solution p = osmotic pressure, V_m = partial molar
$$pV_m = RT$$
    volume of ^{3}He in the solution

The radiation thermometer (ideal photon gas) is not used below 100 K.

    Noise thermometer $V^2 = 4kTR\Delta\nu$ V = noise voltage, R = resistance, $\Delta\nu$ = frequency bandwidth

Thermometers for semi-thermodynamic temperature measurements

    Magnetic thermometers

based on the paramagnetic susceptibility	$\chi = C/T$ (Curie's law)
electronic paramagnetism	0·01–100 K
nuclear paramagnetism	10^{-6}–0·1 K

(The ratio of the sensitivities of magnetic thermometers based on the electronic and on the nuclear paramagnetism is given by the relation (Bohr magneton/nuclear magneton)^2 = $1800^2 = 3 \times 10^6$.)

    Vapour pressure thermometers

vapour pressure of ^{4}He	1–5 K (1958 ^{4}He scale)
vapour pressure of ^{3}He	0·7–3 K (1962 ^{3}He scale)

    Thermometer based on the melting pressure of ^{3}He   0·003–0·9 K

Practical thermometers

    Resistance thermometers

platinum	Above 13·81 K (IPTS-68) and with less sensitivity down to about 4 K
germanium (doped)	0·015–100 K
carbon	0·015–100 K
rhodium–iron	0·1–30 K
carbon-in-glass	1–30 K
metal-oxide–semiconductor transistors	1–30 K

    Diodes

silicon and gallium arsenide	Above 1 K

    Thermocouples

conventional and gold – 0·07 atom %/chromel	9 μV K^{-1} at 1 K

(These thermometers need calibration at several reference temperatures. A set of such reference temperatures and an interpolation procedure for a particular thermometer form a practical temperature scale.)

is to put ^{3}He atoms in a background of liquid ^{4}He, thus increasing the available pressures of the ^{3}He 'gas' by several orders of magnitude, and to measure the pressure of this 'gas' as the osmotic pressure of the ^{3}He–^{4}He solution (see Lounasmaa 1974). Noise thermometry is now used below 0·1 K (Lounasmaa 1974, Soulen 1974) and is expected to be useful for the determination of thermodynamic temperatures above 0·1 K also.

The magnetic thermometer and the vapour pressure thermometer, and also a thermometer for very low temperatures based on the melting pressure of ^{3}He, are listed in table 1 as instruments for semi-thermodynamic temperature measurements. For these thermometers certain, but not complete, *a priori* information is available on the relations between the thermometric parameters and the thermodynamic temperature. When measurements of the dielectric constant or the refractive index are feasible for routine precision determinations of gas densities at low temperatures, vapour density thermometers could also be used in which the density of the saturated vapour is the thermometric parameter instead of the pressure. Several types of magnetic thermometers based on nuclear paramagnetism are used below 0·1 K. They will not be discussed in this paper.

Finally, the most well known practical laboratory thermometers are given in table 1.

The platinum thermometer, which is the standard instrument to realize the IPTS-68 down to 13·81 K, can give a precision of about 5 mK down to about 4 K (Tiggelman 1973). The lower limit of usefulness of germanium and carbon thermometers given in the table is approximate and is not so much determined by the intrinsic properties of these thermometers as by the increasing difficulty at the lower temperatures of obtaining good thermal contact between the thermometers and the objects whose temperature must be measured.

Several of the thermometers listed in table 1 are discussed in the 13 papers on cryogenic thermometry presented at this conference. These include the gas thermometer and the magnetic thermometer, platinum, carbon and rhodium–iron thermometers, metal–oxide–semiconductor transistors and diodes. Resistance thermometry with emphasis on technical thermometers is discussed at this conference in a review paper by Johnston (1975).

Present work on the establishment of new and the improvement of existing international practical low-temperature scales can be illustrated by the following recommendations and plans which were made at the latest session of the Comité Consultatif de Thermométrie (CCT) (see Quinn 1975) in 1974.

(i) Recommendation of an 'Amended Edition of 1975' of the IPTS-68.
(ii) Recommendation to continue the international comparison of standard platinum resistance thermometers (13·81 K to 273·15 K).
(iii) Recommendation to continue the international comparison of temperature scales in the range from 1 K to 30 K.
(iv) A plan to publish as soon as possible, under the auspices of the CCT, a best estimate for a smooth low-temperature scale for the range 1–30 K.
(v) Plans to revise the 1958 ^{4}He scale and the 1962 ^{3}He scale.
(vi) It was envisaged by the CCT that the IPTS-68 should be revised around 1983.

These recommendations and plans will be used as a guide for the further discussions on cryogenic thermometry in this paper.

2. Reproducibility and accuracy of the IPTS-68

The IPTS-68 and its 'Amended Edition of 1975' have been mentioned in the preceding paper (Quinn 1975). Only a few remarks on the reproducibility of the realization of the scale and on its accuracy in terms of thermodynamic temperatures will be made here.

Much work has been done in recent years in the national standards laboratories to realize the defining fixed points of IPTS-68 and to calibrate standard platinum thermometers at these points with mK or even sub-mK precision (see, eg, Pavese 1975). In order to compare the realizations of the fixed points in different institutes an international comparison of calibrated standard capsule-type platinum thermometers was proposed by W R G Kemp of the National Measurement Laboratory of Australia and is now being carried out under the auspices of the CCT. The CCT has gratefully accepted the offer of the National Physical Laboratory to make the actual comparison measurements. First results of these measurements will be presented in the paper by Compton and Ward (1975). At this conference the NPL has also expressed its willingness to continue the comparison measurements; the national standards laboratories which have not yet sent their thermometers to NPL are invited to do so. In conformity with the principles for these comparisons set out by the CCT, the national standards laboratories which have not yet realized the IPTS-68 fixed points may also send platinum thermometers to NPL. Through the comparisons they will obtain access to the IPTS-68 as realized in other standards laboratories.

Apart from the precision of the calibration of platinum thermometers at the defining fixed points of the IPTS-68 there has always been the question of how reproducible the scale is between these fixed points, ie, what the spread of temperatures measured with different platinum thermometers is. In table 2 the data of Tiggelman (1973), based on seven thermometers, are given. (One other thermometer used in this research differed

Table 2. Spread of temperatures measured with platinum thermometers, calibrated on the IPTS-68, between the defining fixed points. Data from Tiggelman (1973) for seven thermometers.

T_{68} (K)	Total spread (mK)
90·188	1·6
54·361	1·4
27·102	0·6
20·28	0·7
17·042	1·2
13·81	

below 54·361 K from the mean of the seven other thermometers by amounts up to 2·0 mK; this thermometer had an α coefficient of barely 0·003925 K^{-1}, while the other seven all had values of α higher than 0·003926 K^{-1}.) The data of Tiggelman are in agreement with those of Compton and Ward (1975). However, substantially larger differences between thermometers were found by Bedford and Ma (1970) for 48 thermometers.

As to the accuracy of the IPTS-68 in terms of the thermodynamic temperature, it was estimated at the time that the scale was introduced that the differences between the international temperature T_{68} and the thermodynamic temperature T were not larger than 0·01 K.

The smoothness of the IPTS-68 has recently been checked with magnetic thermometers. Figure 1, taken from Cetas (1975)†, shows the differences between his smooth magnetic temperature scale and the IPTS-68. It presents the first independent

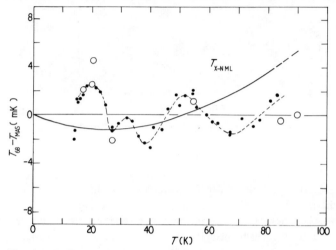

Figure 1. Differences between the IPTS-68 and a magnetic temperature scale, T_{MAS}, based on a fit of susceptibility data for manganese ammonium sulphate (MAS) to the IPTS-68 from 13·8 K to 83 K. For the points marked ● T_{68} was obtained from temperatures measured on the NBS-55 scale and values for $T_{\text{NBS-55}} - T_{68}$ given by Bedford et al (1969); this procedure introduces uncertainties up to 5 mK at 13·81 K. Above 17 K these uncertainties can be estimated from the differences with the points marked ○, for which T_{68} was obtained in a more direct way. The figure has been taken from T C Cetas (1975) (the broken curve through the points marked ● was hand drawn for the present paper). The solid line gives the smooth differences $T_{\text{X-MAS}} - T_{\text{MAS}}$ ($T_{\text{X-NML}}$ is another smooth magnetic temperature scale derived by Cetas.)

check of the IPTS-68 between 30 K and 83 K. (For the temperature range below 30 K the smoothness of the IPTS-68 was investigated earlier with other magnetic thermometers; this will be discussed later in this paper.) For the range between 30 K and 83 K the data of Cetas show that 'wiggles' in the IPTS-68 are not larger than ±2 mK and ±0·05% in the derivative $d(T - T_{68})/dT_{68}$.

The smoothness of the IPTS-68 can, to a certain extent, also be judged from the smoothness of the derivatives $d^n W_{\text{CCT-68}}(T_{68})/(dT_{68})^n$ in which $W_{\text{CCT-68}}(T_{68})$ is the reference function for $W(T_{68}) = R(T_{68})/R(0\,°\text{C})$ used in the definition of the IPTS-68. This was done in the past when the scale was developed. Figure 2 shows the fourth derivative as calculated at the National Research Council in Ottawa at that time. It was then considered that this derivative was sufficiently smooth, but now it would be

† I would like to thank Dr Thomas C Cetas and Professor Clayton A Swenson for sending a preprint of this paper and the publisher of *Metrologia* and the author for their permission to use the figure here.

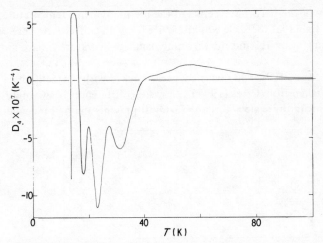

Figure 2. The fourth derivative $d^4 W_{CCT-68}(T_{68})/(dT_{68})^4$ of the $W_{CCT-68}(T_{68})$ reference function of the IPTS-68.

interesting to see if further mathematical smoothing of $d^4 W_{CCT-68}(T_{68})/(dT_{68})^4$ would produce the same 'wiggles' with respect to T_{68} as the experimental smoothing by the magnetic thermometry of Cetas. (For example, it can easily be calculated that between 50 K and 90 K wiggles of the order of those found by Cetas are just possible; much larger wiggles would introduce visible irregularities in the fourth derivative.)

3. International comparison of low-temperature scales (1–30 K)

An international comparison of temperature scales between 1 K and 30 K by means of a comparison of germanium thermometers calibrated on these scales is presently being carried out under the auspices of the CCT. In this case, the actual comparison measurements are made by the National Measurement Laboratory of Australia. The scales which are being compared include the NBS Provisional Temperature Scale 2–20 K, based on the acoustic thermometer of Plumb and Cataland (1966), the magnetic temperature scales of Iowa State University (ISU), the National Measurement Laboratory (NML) of Australia, the Physicotechnical and Radiotechnical Measurements Institute (PRMI) in Moscow and the Kamerlingh Onnes Laboratory (KOL).

All of the comparisons depend heavily on the stability of the germanium thermometer calibrations and on the ability of these thermometers to carry a temperature scale. Therefore, and because germanium thermometers are not discussed in any other paper at this conference, a brief discussion about these thermometers will be given.

4. Stability of germanium thermometers

The stability of good germanium thermometers is illustrated in figures 3(a–c). The data are for the four KOL germanium thermometers used in the international comparison of low-temperature scales at NML. The figures show that differences between two thermometers at each temperature do not spread over more than 2 mK

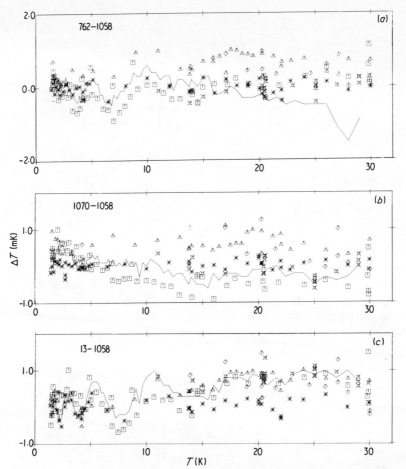

Figure 3. Differences between germanium thermometers. Each figure gives a comparison of two thermometers. All thermometers were calibrated at the same time in a copper comparison block (data marked ∗) at KOL in 1970 on the magnetic scale of KOL. Points marked differently indicate different comparisons series carried out at KOL; the final and most complete series before sending the thermometers to the NML is marked ▫. All points were calculated from the calibration polynomials. The systematic wiggles in the data marked ▫ and the NML data, with respect to the original data, indicate that the polynomials did not give a sufficiently accurate representation of the resistance against temperature data.

including all KOL measurements during about 3 years and the NML measurements. The differences between two thermometers according to the latest and most complete series of measurements at KOL before they were sent to NML do not differ by more than 0·5 mK from the data measured at NML (except above 25 K where the differences in some cases are 1 mK). Certainly, the average of the four thermometers is the same for this last series of measurements at KOL and the measurements at NML within 0·3 mK between 1 K and 25 K.

However, not all germanium thermometers show such a good stability. In table 3 the changes in the calibrations of 15 germanium thermometers are given; it can be seen

Table 3. Stability of 15 germanium thermometers.

Thermometer	Total spread at 4·2 K (mK)	Total change at 20 K (mK)	δT (mK)
Cryocal 13†	0·7	< 1·0	0·5
762	0·6	< 1·0	0·7
1058	< 0·5	< 1·0	0·5
1070	< 0·5	< 1·0	0·5
Cryocal 914†	1·0	2	1·5
Honeywell 705†	1·6	1·5	1·5
CGE 311†	1·0	< 1	2·5
Cryocal 761	1·4	3	1·5
915†	1·0	6	2·0
Honeywell 1086†	–	7	–
1126	1·0	7	3·5
61	1·0	3	2·0
Cryocal 775	1·0	< 1	1·0
777	1·0	8	2·0
1046	1·0	5	2·0

Most thermometers were used during three years (1970–1973) in different apparatus.
δT = reproducibility with thermal cycling but during short periods (3 months) without demounting.
† These thermometers have been used since 1967; three of them changed by some 20 mK (at 30 K) before 1970.

that five out of these 15 thermometers changed by more than 5 mK at 20 K. Such changes in the calibration present the most serious problem in germanium thermometry. It is therefore necessary that the thermometers be repeatedly checked against other thermometers or at certain reference points (see also Van Rijn *et al* 1972).

5. Gas thermometer measurements between 2 K and 30 K

The equation of state of a gas can be written as

$$pV = NRT[1 + B(T)N/V + C(T)(N/V)^2 + \ldots] \tag{1}$$

where p is the pressure of the gas, V the volume of the gas thermometer and N the number of moles of gas in the gas thermometer. B and C are the (molar) virial coefficients of the gas. T is the thermodynamic temperature.

The most straightforward method in gas thermometry is to measure isotherms, ie, to determine pV/NR for various values of N/V. From such measurements one can obtain T by extrapolation of the value of N/V to zero and, in addition, the virial coefficients from the shape of the isotherm. In many cases, however, one relies on values for B and C as functions of T from the literature and one determines the temperature ratios from the pressure ratios for a constant value of N. In this case the amount of gas and the volume of the gas thermometer do not have to be known precisely because they enter only in the correction terms $B(T)N/V$ and $C(T)(N/V)^2$.

In the past when conventional mercury manometers read with cathetometers were used, the accuracy of the pressure measurements presented a serious problem. The use

of sophisticated mercury manometers with which it was possible to measure pressures with an accuracy of a few mT solved this problem but these instruments were expensive and not very convenient to use. Several investigators now prefer to use piston gauges which combine good accuracy, and a linearity of the order of 1 part per million, with reasonable cost and convenience in use. A thorough discussion of modern low-
with an accuracy of a few mTorr solved this problem but these instruments were expensive and not very convenient to use. Several investigators now prefer to use piston gauges

Several laboratories have recently set up, or are now preparing, gas thermometer experiments between 2 K and 30 K. Of these the NPL has done and the KOL has just started isotherm measurements, the others will use a gas thermometer with a constant amount of gas to realize a low-temperature scale. Results of the NPL measurements and the design of the gas thermometer at the PTB are presented at this conference by Berry (1975) and Anderson and Neubert (1975) respectively.

6. Acoustic thermometer measurements

The velocity of sound w in helium gas is given by the relation

$$w^2 = \frac{5}{3} \frac{RT}{M} (1 + \alpha p + \beta p^2 \ldots) \qquad (2)$$

where M is the molar mass of helium and α and β are functions of the virial coefficients and of their derivatives with respect to temperature.

Plumb and Cataland (1966) at the NBS were the first to build an accurate acoustic thermometer. With it they showed that the 1958 ^{4}He scale is in error by about 0·2% (the temperature T_{58} being too low) and, later, they used the instrument to establish the NBS Provisional Temperature Scale 2–20 K. An advantage of the acoustic thermometer over the gas thermometer was that the density of the gas does not occur in the leading term of the equation for w^2.

Measurements with an acoustic thermometer were also made by Colclough (1973) at NPL. He used a sound frequency of 3 kHz which is much lower than the frequency used by Plumb and Cataland (1 MHz) (Colclough 1973).

7. Magnetic thermometers

In its usual form the magnetic thermometer consists of a set of mutual inductance coils in which a paramagnetic salt is situated (figure 4). The coefficient of mutual inductance M of the coil system is related to the susceptibility of the salt χ by the equation

$$M = a + b\chi. \qquad (3)$$

The susceptibility of the salt for ellipsoidal samples is written as

$$\chi = C/(T_m + \Delta + \gamma/T_m) \qquad (4)$$

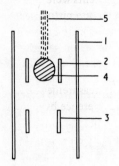

Figure 4. Principle of a magnetic thermometer. 1, primary coil; 2, main secondary coil; and 3, compensating secondary coil of the mutual inductance. 4, Paramagnetic salt sample; 5, copper wires to make thermal contact between the salt sample and germanium or vapour pressure thermometers. (These, and metal parts in general, have to be placed sufficiently far from the salt and coil assembly in order not to influence the inductance.)

so that

$$M = A + [B/(T_m + \Delta + \gamma/T_m)]. \tag{5}$$

T_m is the temperature derived from the magnetic data. A and B are constants which depend on the geometry of the magnetic thermometer and on the salt; they must be determined by calibration of the magnetic thermometer at at least two known temperatures. The constants Δ and γ are usually small for the salts which are used; in general they must be determined empirically by calibration of the magnetic thermometer at known temperatures.

Magnetic thermometers were used at ISU, KOL and PRMI to establish so-called magnetic temperature scales between about 1 K and 30 K (see Cetas and Swenson (ISU) 1972, Swenson 1973, Van Rijn and Durieux (KOL) 1972, Astrov *et al* (PRMI) 1975).

Recently Cetas (1975) used a magnetic thermometer up to temperatures of 83 K; results of this research have already been given in figure 1. The magnetic temperature in this figure was obtained by Cetas from an expression such as equation (5), the constants Δ and γ being determined from data between 1 K and 4 K, and A and B being calculated from a least-squares fit of the magnetic temperature to T_{68} between 13·81 K and 83 K.

In the other magnetic thermometer experiments mentioned above the magnetic temperature was usually fitted to the IPTS-68 at some temperatures between 13·81 K and 30 K. Simple extrapolation from these temperatures down to lower temperatures would lead to large uncertainties in the magnetic temperature but it was shown by Cetas and Swenson (1972) that such an extrapolation can be made sufficiently accurate if one adds the additional assumption that the magnetic temperature is proportional to T_{58} (the 1958 ^{4}He scale) between 1 K and 3 K.

To give an impression of the deviations from the ideal Curie law behaviour, values of Δ and γ for the paramagnetic salts used in the various measurements are shown in table 4.

8. Comparison of various temperature measurements between 1 K and 30 K

Figure 5 shows the differences between several low-temperature laboratory scales based on acoustic and magnetic thermometers, IPTS-68 and the 1958 ^{4}He scale. Data of thermodynamic temperature measurements from K H Berry, A Colclough and R H Sherman (private communications) are also given. Apart from small changes the

Cryogenic thermometry between 0·1 K and 100 K   27

Table 4. Values of Δ and γ in the equation $M = A + B/[T + \Delta + (\gamma/T)]$ used in magnetic thermometry.

Salt	Δ (K)	γ (K^2)	Temperature range (K)	Authors
CeMg$_{1.5}$(NO$_3$)$_6$12H$_2$O	0	0	0·9–2·6	Cetas and Swenson (1972)
CrCH$_3$NH$_3$(SO$_4$)$_2$12H$_2$O	0·0019	0·0028	0·9–30	Cetas and Swenson (1972)
Mn(NH$_4$)$_2$(SO$_4$)$_2$6H$_2$O	−0·0359†	0·0083	0·9–30	Cetas and Swenson (1972)
	−0·001	0	2–30	Van Rijn and Durieux (1972)
	−0·017	0·0065	1–83	Cetas (1975)
Gd$_2$(SO$_4$)$_3$8H$_2$O	−0·0087	0·0762	2–30	Van Rijn and Durieux (1972)
	−0·022	0·064	1–83	Cetas (1975)

All data, except those for cerium magnesium nitrate, are for powders.
† This value is for a non-spherical sample; recalculated to a sphere, using the single crystal density, the value of Δ becomes −0·009 K.

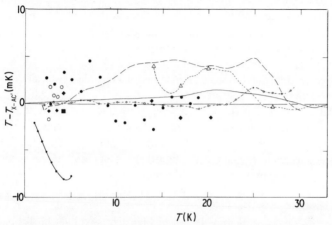

Figure 5. Differences between various low-temperature scales and determinations of thermodynamic temperatures. The base line represents a magnetic ISU temperature scale $T_{X-AC'}$, which is very nearly equal to the scale T_{X-AC} given by Swenson (1973) (it is defined by the relation $T_{X-AC'} = T_X(1 + 0·3 \times 10^{-3})$. The differences between the KOL magnetic scale $\overline{T_m}(III)$ and $T_{X-AC'}$ are provisional results of the scale comparisons at NML. $\overline{T_m}(III)_{reduced}$ denotes the KOL magnetic scale reduced to $T_{X-AC'}$ at 2, 4, 14 and 20 K. The IPTS-68 points and curve are deduced from the $T_{68} - \overline{T_m}(III)$ data of Van Rijn and Durieux (1972) (slightly recalculated). The NBS 2–20 K points (●) are deduced from the differences between the ISU magnetic scale and the NBS 2–20 K scale given by Swenson (1973). The data of Sherman's ^{3}He isotherm measurements (○), Berry's ^{4}He isotherm and gas thermometer measurements (◆) and Colclough's acoustic isotherm measurements (■) are deduced from the authors' data with respect to T_{58} and T_{68}. ----, magnetic scale $\overline{T_m}(III)$ (KOL); --○--○--○, $\overline{T_m}(III)_{reduced}$; ——, magnetic scale (Cetas 1975) (NML); △--△--△, IPTS-68; ●●●●, T_{58}.

graph is identical to the figure given by Kemp (1974), and shows the provisional results of the scale comparisons at NML. It can be seen that the magnetic scales of ISU and KOL when reduced to the same reference points at 2, 4, 14 and 20 K agree with each other to within 0·5 mK, and that these magnetic scales agree within ±2 mK with most of the thermodynamic temperature determinations. Further it may be noticed that the irregularity in the IPTS-68, with respect to the magnetic scale found by Cetas (1975) between 17 and 27 K (see figure 1), closely resembles the irregularity in the IPTS-68 with respect to the magnetic scales in figure 5. No particular significance should be given to the exact values of the points and curves since they are for a large part provisional and the graph should not be used as a proposal for an average low-temperature scale. More precise data will be published under the auspices of the CCT when the scale comparisons at NML are completed.

Then, data of magnetic thermometry at PRMI also, and hopefully more data from the NPL gas thermometer and isotherm measurements will be included. It is expected that from all these data a low-temperature scale can be deduced which is equal to the thermodynamic temperature within an amount varying from 2 to 3 mK at 27 K to 1 mK at 1 K, and which is smooth within ±0·02% in the slope.

The scale will then be published under the auspices of the CCT. It will be defined by: (i) the calibrated germanium thermometers (and in the future rhodium–iron thermometers also) which were compared at NML and which will be distributed over various standards laboratories; (ii) the differences of the scale from the various existing laboratory scales; (iii) the differences of the scale from the international scales (the IPTS-68 and the helium scales); (iv) the temperatures assigned to the superconductive transition points of lead, indium and aluminium (see below).

NML is prepared to continue the scale comparisons to include new scales or results of thermodynamic temperature measurements when these become available. Also, in accordance with the principles of the comparison programme laid out by the CCT, NML will accept uncalibrated thermometers for the comparison programme (provided that these are checked for stability) from those national standards laboratories which would like to obtain access to the new scale and other current scales. This service from NML to the thermometric community is highly appreciated.

There will be the inconvenience that the new scale differs above 13·81 K from the IPTS-68 and below 4·2 K from the helium scales. For the helium scales this problem will be resolved when new helium scales are established (as discussed later in this paper). For the range above 13·81 K the differences will remain until the IPTS-68 is revised, which will possibly be around 1983.

9. The 1958 ^{4}He scale and the 1962 ^{3}He scale

The 1958 ^{4}He scale is a vapour pressure against temperature table recommended by the International Committee of Weights and Measures in 1958. This table was based on gas thermometry, smoothed with magnetic data and, below about 2·2 K, by thermodynamical calculations. As mentioned before, measurements have now shown that the scale is in error by about 0·2% in the temperature, the thermodynamic temperature being higher than T_{58}. (It may be remarked that about one-third of the error is due to the choice of the temperature of the hydrogen boiling point (20·26 K was used in 1958 instead of the present best value of about 20·275 K) which was taken as a reference point for most of the gas thermometry on which T_{58} was based.) Therefore, the CCT has planned that the 1958 ^{4}He scale should be revised. A simple method of doing this would be to publish an equation giving the differences between T_{58} and the best known present approximation of the thermodynamic temperature. In fact, the differences between T_{58} and the smooth scale for the 1–30 K range mentioned before could be used for this purpose. However, it is intended to define the revised ^{4}He scale by equations instead of by a table and certain additional work will be done to ensure that these equations will be consistent with the thermodynamically calculated vapour pressure equation.

The 1962 ^{3}He scale is a vapour pressure against temperature relation for liquid ^{3}He, valid for the range 0·2 to 3·3 K. The scale was established by S G Sydoriak, R H Sherman and T R Roberts at the Los Alamos Scientific Laboratory. It is above 0·9 K based on the 1958 ^{4}He scale through comparisons of vapour pressures of ^{3}He and ^{4}He. These comparisons were estimated to be accurate within ±0·2 mK, which was confirmed later by the measurements of Gonano and Adams (1970). Below 0·9 K no accurate direct measurements of the vapour pressure–temperature relation were (and are) available and the ^{3}He scale at these temperatures was therefore calculated by extrapolation of the data above 0·9 K, using the thermodynamic vapour pressure equation. It was estimated that no errors greater than 2 mK were introduced by this extrapolation.

The CCT has planned that, together with the new ^{4}He temperature scale, a new ^{3}He temperature scale will be derived†.

10. Superconductive transition points as reference points for thermometry

An important development in low-temperature thermometry in recent years has been the work on superconductive transition points for use as reference points for thermometry by Schooley at the National Bureau of Standards in Washington (see Schooley 1975). At present a device, referred to as SRM 767, is available from NBS which incorporates samples of lead, indium, aluminium, zinc and cadmium (with transition points at about 7·2, 3·4, 1·2, 0·8 and 0·5 K) enclosed in a small set of mutual inductance coils; the transition points can be determined from discrete changes in the mutual inductance at these temperatures. The device is used in several laboratories and has great potentialities for the calibration of resistance or magnetic thermometers. Present results indicate that the transition points measured with each device are reproducible to better than 1 mK and that the transition temperatures determined with different devices are certainly equal within ±1 mK. For further developments in this field see Schooley (1975).

11. Temperature measurements between 1 K and 30 K: concluding remarks

It was mentioned before that three laboratories have used the magnetic thermometer to set up low-temperature scales and that at least two of these scales agree within 1 mK when reduced to the same reference temperatures. There is little doubt that the same accuracy can be achieved with modern gas thermometers and indeed several standards laboratories are setting up such experiments. In these standards laboratories the realization of temperature scales which are smooth and in good agreement with the thermodynamic temperature is thus progressing in a very satisfactory way.

† At the same time that the CCT made these preparations to revise the helium scales R H Sherman and S G Sydoriak of the Los Alamos Scientific Laboratory had independently started work in this direction and cooperation between the two groups has been established. My thanks are due to Dr R S Sherman and Dr S G Sydoriak for discussions on the helium scales and to the former for providing his unpublished isotherm data given in figure 5.

This cannot be said for the development of practical thermometers for precision temperature measurements in this temperature range. Only resistance thermometers are, at present, considered for this purpose; germanium and rhodium–iron thermometers are available. Germanium thermometers are not always stable and, moreover, their calibration is rather complicated (see, eg, Van Rijn *et al* 1972) and rhodium–iron thermometers are still rarely used. There is little doubt that the development of a satisfactory laboratory thermometer would at present be the most important step towards better low-temperature thermometry. Because germanium thermometers are so widely used (to say nothing in disregard of the rhodium–iron thermometer) it seems that the production of a stable germanium thermometer, if this is at all feasible, is the most direct way to improve the present situation.

The question whether a formal international practical temperature scale should be defined for the range between 4 K and 14 K will probably not arise before 1983. By that time the fixed points at the lower end of IPTS-68, the superconductive transition points of lead, indium and aluminium and the liquid helium scales will all be defined in close agreement with the thermodynamic temperature. Gas and magnetic thermometers could then both be allowed as interpolation instruments†.

Little would be gained, however, by such a definition of a scale when it could not be transferred to the users in the form of reliable calibrated resistance thermometers.

12. Thermometry between 0·1 K and 0·7 K

The ^{3}He vapour pressure thermometer can be used with good precision down to temperatures of about 0·7 K ($p = 1·38$ Torr, $dp/dT = 0·011$ Torr mK^{-1}). (Gonano and Adams (1970) suggested that it could be used at low temperatures if one measures the vapour pressure *in situ* with a diaphragm gauge.)

The magnetic thermometer, with cerium magnesium nitrate as the paramagnetic salt, is the most prominent thermometer between 0·1 K and 0·7 K. It is calibrated at higher temperatures against a ^{3}He vapour pressure thermometer or at the superconductive transition points of cadmium (0·5 K), zinc (0·8 K) and aluminium (1·2 K) (see Soulen 1974). The accuracy of this magnetic thermometer, in terms of thermodynamic temperatures, is certainly 0·5 mK or better down to 0·1 K (errors due to uncertainties in the calibration temperatures are not included).

The melting pressure of solid ^{3}He has been suggested as a thermometric parameter but at present the uncertainties in the thermal and pVT data for liquid and solid ^{3}He are too large to make an accurate calculation of the melting pressure against temperature relation possible and accurate direct measurements of this relation have not yet been made.

Resistance thermometers useful for this range are the germanium, the carbon and the rhodium–iron thermometer.

† Which of these thermometers is the most convenient in use is not easy to say. In an impromptu discussion at this conference (present were K H Berry, D Gugan, W Neubert, R L Rusby, C A Swenson and S D Ward) it appeared that the gas thermometer was considered to be the most solid instrument ('nothing can go wrong with it'). The magnetic thermometer has the advantage of smaller departures from ideal law behaviour than the gas thermometer (for a pressure of say 400 Torr at 20 K) and of purely electrical measurements instead of pressure measurements. The use of it as a precision thermometer is, however, still somewhat of a (difficult) art.

For a further discussion of thermometry below 0·7 K we refer to the recent book of Lounasmaa (1974) and to a review of cryogenic thermometry at NBS (Cataland et al 1974), and, for magnetic thermometry, also to Hudson (1972).

Acknowledgments

The low-temperature scale comparisons which are now being carried out at the National Measurement Laboratory of Australia and, in general, work towards the improvement of low-temperature scales was commissioned by the CCT to its working group 4. This group consists of W R G Kemp (NML), C A Swenson (ISU), Mrs M P Orlova (or D N Astrov) (PRMI) and this author. The sections on thermometry between 1 K and 30 K in this paper depend heavily on the exchange of opinions within this working group, which is gratefully acknowledged; the results of the scale comparisons presented in figure 5 are those of W R G Kemp and L M Besley of NML.

References

Anderson R L and Neubert W 1975 this volume
Astrov D N et al 1975 Metrologia to be published
Bedford R E, Durieux M, Muijlwijk R and Barber C R 1969 Metrologia **5** 47
Bedford R E and Ma C K 1970 Metrologia **6** 89
Berry K J 1972 PhD Thesis University of Bristol
—— 1975 this volume
Cataland G, Hudson R P, Mangum B W, Marshak H, Plumb H H, Schooley J F, Soulen R J and Utton D B 1974 NBS Technical Note 830
Cetas T C 1975 Metrologia to be published
Cetas T C and Swenson C A 1972 Metrologia **8** 46
Colclough A R 1971 Proc. CCT 9^e session p117
—— 1973 Metrologia **9** 75
—— 1974 Metrologia **10** 74
Compton J and Ward S D 1975 this volume
Gonano R and Adams E D 1970 Rev. Sci. Instrum. **41** 716
Hudson R P 1972 Principles and Application of Magnetic Cooling (Amsterdam: North-Holland)
Johnston J S 1975 this volume
Kemp W R G 1974 Proc. CCT 10^e session document 33
Lounasmaa O V 1974 Experimental Principles and Methods Below 1 K (London: Academic Press)
Pavese F 1975 this volume
Plumb H H and Cataland G 1966 Metrologia **4** 127
Quinn T J 1975 this volume
Quinn T J and Compton J P 1975 Rep. Prog. Phys. **38** 151
Rubin L G 1970 Cryogenics **10** 14
Rubin L G, Powell R L and Anderson A C 1971 Cryogenics **11** 489
Rusby R L 1975 this volume
Schooley J F 1975 this volume
Sherman R H 1971 5th Symposium on Temperature, its Measurement and Control in Science and Industry, unpublished
Soulen R J 1974 NBS Technical Note 830 Appendix 2
Swartz D L and Swartz J M 1974 Cryogenics **14** 67
Swenson C A 1970 Crit. Rev. Solid St. Sci. **1** 99
—— 1973 Metrologia **9** 99
Tiggelman J L 1973 PhD Thesis University of Leiden
Van Rijn C and Durieux M 1972 TMCSI **4** part 1 73
Van Rijn C, Nieuwenhuys-Smit M C, Van Dijk J E, Tiggleman J L and Durieux M 1972 TMCSI **4** part 2 815

Gas thermometry at low temperatures

K H Berry

Division of Quantum Metrology, National Physical Laboratory, Teddington, Middlesex, England

Abstract. The need for new measurements of temperature below 30 K is outlined and the methods of $P-V$ isotherms and constant-volume gas thermometry used at the NPL are described. Some of the more important results of these experiments are reported.

1. Introduction

In recent years several experimental techniques have been exploited to measure temperature in the range extending from around 30 K down to the helium vapour pressure region, the most common approach being that of constant-volume gas thermometry (Astrov et al 1969, Rogers et al 1968). The uncertainty, however, in the published values of the second virial coefficient of ^{4}He, the thermometric gas, demands that a gas thermometer should be operated at a low pressure to minimize the uncertainty in the non-ideality correction to the gas scale. The difficulty of operating at these low pressures is reflected in the scatter of low-temperature gas thermometry results where the precision is often of the order of several millikelvins. In sharp contrast to these results are those derived from magnetic thermometry measurements where a tenfold improvement in precision has been reported (Cetas and Swenson 1972, van Rijn and Durieux 1972). Neither of these methods is, however, strictly absolute as they do not rely directly on the definition of the kelvin in terms of the temperature assigned to the triple point of water. A magnetic scale is a function of the reference scale used in the calibration of the magnetic thermometer in the same way that a gas thermometer scale is a function of some reference temperature. In order to overcome this problem and produce an absolute temperature scale, measurements have been made using acoustic thermometry where the derived temperature scale is only a function of the value chosen for the universal gas constant. Unfortunately, the precision of the NPL low frequency acoustic thermometer is only of the order of 3 mK whereas that at NBS suffers from certain systematic errors which are difficult to quantify (A R Colclough 1973 and private communication, Plumb and Cataland 1966).

2. Experimental method

Clearly a new approach or improvements to an existing technique were needed to be brought to this field of measurement. At NPL the problem has been approached using a combination of constant-volume gas thermometry and a method which has been used most recently by Keller in 1955 and before that by Keesom in 1947, namely $P-V$ isotherm thermometry (Keesom and Walstra 1940, 1947, Keller 1955). This latter technique has been used to determine the normal boiling point of equilibrium hydrogen relative to the triple point of water, whereas gas thermometry has been used to relate temperatures below 20 K to this isotherm temperature. Isotherm thermometry has also been employed at 13·8 K (the triple point of hydrogen), 7·2 K (the superconducting transition temperature of lead), 4·2 K (the normal boiling point of ^{4}He), 3·3 K, 2·8 K and 2·6 K in order to obtain sufficiently accurate values for the second and third virial coefficients of ^{4}He required in the gas thermometry calculations.

In $P-V$ isotherm thermometry ^{4}He gas is admitted from a reference volume immersed in a bath of melting ice into another bulb at the unknown low temperature, the number of moles (N) being calculated from the pressure decrement of gas in the reference volume caused by the transfer of gas from it to the gas bulb. The resulting gas pressure (P) in the gas bulb is measured and a graph of (P/N) against (N) is drawn. When this isotherm is extrapolated to zero density its intercept yields a value for the product of the isotherm temperature and the ratio of the volumes of the reference volume and gas bulb. Values for the second and third virial coefficients of ^{4}He are calculated from the first and second derivatives of the isotherm at zero density.

Once an isotherm has been plotted at the normal boiling point of equilibrium hydrogen it is more expeditious to measure other temperatures using the technique of constant-volume gas thermometry than to repeat an isotherm at each temperature as the problems of volume measurement, gas purity and long time delays due to gas flow along fine bore tubes are then not severe. The gas bulb is simply filled with a fixed mass of gas and any unknown temperature is calculated from a measurement of the ratio of the gas pressure at the unknown temperature to that at the isotherm temperature. Allowances must be made for the departures from ideal behaviour of the thermometric gas and corrections are applied using the measured values of the second and third virial coefficients.

3. Apparatus

A detailed description of the cryostat has already been given (Berry 1972b) and so it will suffice to report here that the cryostat is of a conventional design as may be seen from figure 1. The copper gas bulb has a large volume, of about 1 litre, thus facilitating accurate measurement of the deadspace correction and reducing the surface area to volume ratio of the bulb to minimize any possible gas adsorption effects. Inserted in wells around the wall of the gas bulb are one platinum and three rhodium—iron resistance thermometers. The platinum thermometer has been calibrated against the hydrogen fixed points and the rhodium—iron thermometers have been calibrated against T_{58}. The results of the current gas-thermometry measurements are also recorded on the rhodium—iron thermometers. Gas in the gas bulb communicates with the pressure measuring equipment at room temperature via a 1 mm bore stainless steel tube along which there are six gold—iron/chromel thermocouples used to measure the temperature distribution along the tube and hence the deadspace correction. From figure 2 it can be seen that the volume of the gas thermometer is defined by a capacitance diaphragm pressure transducer which isolates the thermometric gas from that in the pressure balance. The gas pressure in the bulb is therefore equal to the sum of that set up by

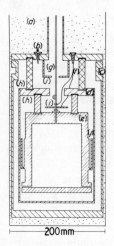

Figure 1. The gas-thermometer cryostat showing (a) liquid helium reservoir, (b) stycast seal for leads, (c) stainless steel vacuum jacket, (d) copper isothermal shield, (e) copper gas bulb, (f) heat links to liquid helium reservoir, (g) stainless steel tube, (h) vacuum space, (i) radiation shields, (j) resistance thermometer wells.

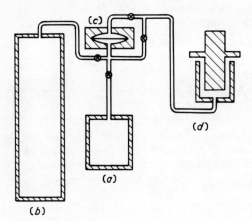

Figure 2. The arrangement of (a) the gas bulb, (b) the reference volume, (c) the diaphragm gauge, and (d) the pressure balance.

the pressure balance assembly, the pressure difference across the diaphragm gauge and the small pressure gradient corrections along the pressure-sensing tube due to aerostatic head and thermomolecular pressure effects. The reference bulb used in the isotherm measurements has a volume of about 6 litres and consists of a cylindrical stainless steel bulb whose volume was determined by weighing the amount of water needed to fill it. The bulb is immersed in a bath of crushed ice which is melting and whose temperature is measured at the top and the bottom of the reference volume using two platinum resistance thermometers.

4. Results

In the preliminary isotherm work already reported the volume of the gas bulb was also measured using the method of filling with water (Berry 1972a). More recently, however, an alternative approach has been used where the ratio of the two volumes was determined directly by immersing the gas bulb in melting ice and expanding gas from it into the reference bulb. The ratio of the volume was found to be 5·67158 ± 0·00003, where this uncertainty is the standard error in the mean of three independent measurements. This ratio was then corrected for the change in volume of the gas bulb due to thermal contraction (C A Swenson private communication concerning new measurements of the thermal expansivity of copper).

An isotherm at the boiling point of hydrogen has recently been completed which consists of twelve points covering a density range from 0.8×10^{-4} to 6.2×10^{-4} mol cm^{-3}. The problem of calculating values for the virial coefficients and the isotherm temperature from this data is not a trivial exercise in curve fitting (Berry 1972c). First-, second- and third-order power series have been fitted to the data and the residuals of each of these fits are shown in figure 3 and the coefficients of the power series in table 1.

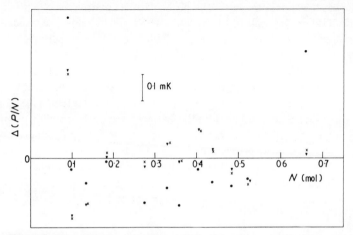

Figure 3. The residuals of the fits to the isotherm data at the boiling point of hydrogen. The standard deviations of the residuals of the first-order (●), second-order (▼) and third-order (×) fits are equivalent to 0.2_4 mK, 0.1_6 mK and 0.1_6 mK respectively.

Table 1. Coefficients of the power series fits to the isotherm data at the boiling point of hydrogen. The uncertainties are expressed as standard errors.

Order of fit	T (K)	B (cm^3 mol^{-1})	C (cm^3 mol^{-1})^2
1	20·2747 ± 0·0001$_5$	−2·33 ± 0·02	—
2	20·2752 ± 0·0001$_6$	−2·53 ± 0·06	312 ± 80

Clearly the first-order fit is inadequate and so the second-order fit has been adopted. The normal boiling point of equilibrium hydrogen is, therefore, 20·2752 K[†] with a random error (1 standard error) of ±0·1$_6$ mK. A detailed discussion of the possible systematic errors in these measurements will be reported later but it is unlikely that such errors when compounded will be larger than 1 mK. This new value of the hydrogen

[†] The ratio of the volumes of the gas and reference bulbs has still to be remeasured after the completion of the isotherm measurements. It is possible, therefore, that this quoted value of the boiling point of hydrogen will be modified.

boiling point compares well with the preliminary value of 20·2746 K which was quoted as having a random error of ±0·2 mK and an estimated systematic error of ±1·0 mK (Berry 1972a). It now seems clear, therefore, from these results that the value assigned to the normal boiling point of equilibrium hydrogen on IPTS-68 is too high by 5 mK.

Although isotherms have now been completed at six other temperatures below 20 K the results are still to be calculated. The gas-thermometer measurements have, however, been analysed. Three separate runs were completed at a sensitivity of 0·82 kPa K^{-1}, each run consisting of 21 measurements covering the range 2·6 K to 20·3 K. It is found that eighth-order fits to the rhodium−iron resistance/temperature data are required and that the standard deviation of the residuals of these fits is 0·1$_8$ mK. The standard error in the fits for all 63 points is 0·07 mK which may be considered to be a measure of the precision of the gas-thermometer scale. The problem of estimating the effect of physical adsorption of ^{4}He on the walls of the gas bulb when cooled below the critical temperature has in previous work proved to be a difficult one. In this experiment the effect was studied by repeating some of the gas-thermometer measurements after the surface area to volume ratio had been increased by a factor of five by inserting twenty-two discs into the bulb. No significant effect was observed and so it is concluded that any systematic error in the gas scale resulting from adsorption effects is less than 0·2 mK.

At present the NPL gas scale is based on the values of the normal boiling point of equilibrium hydrogen and the second virial coefficient data of the preliminary isotherm work. The scale will, of course, be based ultimately on the most recent definitive isotherm measurements when it is estimated that the gas scale will have a precision of ±0·2 mK (3 standard errors) and will not deviate by more than 1 mK from thermodynamic temperature.

References

Astrov D N, Orlova M P and Kytin G A 1969 *Metrologia* 5 111
Berry K H 1972a *Metrologia* 8 125
—— 1972b *TMCSI* 4 part 1 323−31
—— 1972c *PhD Thesis* University of Bristol
Cetas T C and Swenson C A 1972 *Metrologia* 8 46
Colclough A R 1973 *Metrologia* 9 75
Keesom W H and Walstra W K 1940 *Physica* 7 985
—— 1947 *Physica* 13 225
Keller W E 1955 *Phys. Rev.* 97 1
Plumb H H and Cataland G 1966 *Metrologia* 2 127
van Rijn C and Durieux M 1972 *TMCSI* 4 part 1 73−84
Rogers J S, Tainsh R J, Anderson M S and Swenson C A 1968 *Metrologia* 4 47

Note added in proof, August 1975

Since this paper was submitted at the conference the ratio of the volumes of the gas and the reference bulbs has been redetermined. It is now possible, therefore, to publish the final values for the three fixed points in the hydrogen vapour pressure region (see table on following page). A complete summary of the results and their analyses will appear later.

Table 2. Recent results for three fixed points in the hydrogen vapour pressure region.

Fixed point	NPL-75 (K)	IPTS-68 (K)	ΔT (mK)
Normal boiling point of equilibrium hydrogen	20·2748	20·2800	5·2
'17 K' point	17·0386	17·0420	3·4
Triple point of equilibrium hydrogen	13·8062	13·8100	3·8

A gas thermometer for low temperatures

R L Anderson[†] and W Neubert
Physikalisch-Technische Bundesanstalt, Braunschweig, W Germany

Abstract. In order to establish a practical temperature scale between 3 K and 20 K a constant-volume gas thermometer has been constructed. Details of the design and some results of performance tests are reported. From these it is inferred that an overall precision of ±1 mK should be achieved.

1. Introduction

A constant-volume gas thermometer has been constructed at PTB to serve primarily as an interpolating instrument (Barber 1972) for a practical temperature scale between the helium-4 vapour pressure scale T_{58} and IPTS-68. Originally a magnetic thermometer was also considered for this purpose. The final decision to embark on gas thermometry was made after a visit to NPL several years ago. Two features of the PTB gas thermometer were adopted from the NPL design: a large (1 litre) bulb volume and employment of a pressure balance instead of a mercury manometer (Berry 1972) because both would contribute to an increased reliability of the results.

The PTB gas thermometer is now in the stage of performance tests. In this paper constructional details of the instrument and some preliminary results are given.

2. The experimental apparatus

The PTB low-temperature gas-thermometer cryostat is shown in figure 1. The bulb (A) has a volume of approximately 1 litre and was machined from a solid bar of high-purity OFHC copper (supplied by Norddeutsche Affinerie, Hamburg). It is connected by a small diameter tube and a bellows valve to a capacitive differential pressure transducer (made by Datametrics, Wilmington, Massachusetts). For minimum surface to volume ratio, the bulb height and diameter are about the same. Both ends of the cylinder are equipped with removable lids so that the surface area may be increased for adsorption experiments. The wall and lids are 14 mm thick. In the wall of the bulb there are wells for eight germanium and four platinum resistance thermometers. They are equally spaced around the top and bottom, with six wells on either end of the cylinder. This allows checks to be made for thermal gradients in the bulb. A heater of 0·1 mm manganin wire is bifilarly wound on the side of the bulb and lacquered in place. Surrounding the bulb there is a thermal shield (B) made of copper. The bulb is suspended from the shield by three thin-walled stainless-steel tubes. There is a moderate thermal link from the bulb to the shield provided by the 32 electrical leads consisting of

[†] Present address: Instruments and Controls Division, Oak Ridge National Laboratory, Oak Ridge, Tenn 37830, USA.

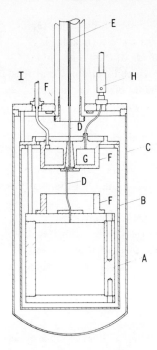

Figure 1. The gas thermometer cryostat. A, thermometer bulb (copper); B, thermal shield (copper); C, vacuum can (brass); D, stainless-steel capillaries; E, copper capillary; F, thermal anchors for electrical leads; G, liquid helium chamber; H, throttle valve; I, pumping line.

0·15 mm diameter copper wires. At least 20 cm of these electrical leads are wound around and lacquered onto two copper thermal anchors (F), one located on the bulb, the other on the shield.

The 1·5 mm inner diameter stainless-steel pressure-transmitting tube (D) is soldered to a heavy copper cone which is clamped snugly into a mating cone in the shield. Thermal contact is increased by a film of vacuum grease applied to the cone before it is clamped into place. The pressure-transmitting tube is offset in the cone so that there is no direct path for radiation from the room-temperature parts of the apparatus to reach the bulb. Included in the shield, and constructed as part of the thermal anchor for the electrical leads, is a chamber (G) into which liquid helium from the bath can be admitted through a throttle valve (H). A 6 mm diameter pumping line (I) is connected to this chamber so that by suitable adjustments of pumping and liquid helium admission different amounts of heat can be extracted to cool the shield. A germanium resistance thermometer in the wall of the chamber serves as the sensing element in the temperature regulating system. Two heaters are installed at the shield. One is connected to the electronic regulation system and the other can be manually controlled for constant supplementary heating.

The brass vacuum can (C) is in the form of a cylinder with the bottom made as a spherical section. Six 20 mm diameter ports are provided in the lid with removable covers. The various pumping lines extending through the lid of the vacuum can are thus soldered to the port covers which in turn are sealed to the lid of the vacuum can. This avoids unnecessary heating of the vacuum can lid in case of repairs or changes such as those contemplated when adding another tube to the system to directly determine the thermomolecular pressure. All of the removable seals on the vacuum can as well as on the bulb are made with confined gold 'O-rings'.

The lid is attached to a thin-walled 50 mm diameter stainless-steel tube which serves as a support for the vacuum can, a pumping line, and a path for the electrical leads and the pressure-transmitting tube. The 50 electrical leads, made of 0·09 mm copper wire, are assembled on a 32 mm diameter stainless-steel tube which is terminated at the lower end by a copper heat anchor (F) thermally linked to the lid of the vacuum can. The portion of the pressure-transmitting tube (E) which extends from the room-temperature region outside the cryostat to the level of the top of the lid of the vacuum can is made of 1 mm inner and 2 mm outer diameter copper capillary. The tube is provided with a heater at the lower end and is kept at approximately room temperature. Five copper—constantan thermocouples are attached at equally spaced intervals along the tube to measure the thermal gradient. To minimize heat radiation, the tube is wrapped with several layers of aluminized mylar foil.

Between the copper capillary and the shield there is a 1·5 mm ID thin-walled stainless-steel tube. The major part of the thermal gradient in the gas-thermometer pressure-transmitting tube is thus confined to a relatively small piece of tubing about 50 mm long. In this way, the uncertainty usually associated with accurate determination of the thermal gradient in the dead space is considerably reduced.

Figure 2 shows schematically the pressure measuring system. A gas-operated pressure balance (made by CEC, Monronia, California) provides a precise pressure reference.

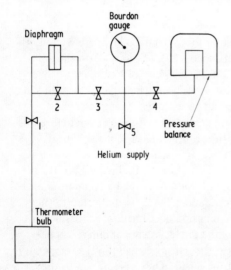

Figure 2. Schematic of the valve assembly.

The deflection of the metal diaphragm is converted into a DC voltage and indicates any pressure differences between the gas thermometer and the piston balance. The resistance measurements are made using a self-balancing AC potentiometer (made by ASL, Leighton Buzzard, Bedfordshire) employing inductive voltage dividers and working at half the line frequency. The two six-digit numbers displayed are proportional to the voltage drops across the unknown and the reference resistor connected in series.

The data from the AC potentiometer, from a digital voltmeter connected alternatively to the capacitance manometer or to the thermocouples on the pressure transmitting tube, and information about various switch positions are recorded on punched paper tape for computer processing.

3. Experimental procedure and performance tests

The operating procedure for the apparatus is similar to that described by Guildner (1962) for the NBS gas thermometer. The volume of the gas thermometer extends only to the first valve in the valve assembly (see figure 2), but not to the diaphragm. This way, dead space is minimized. To describe the procedure take as starting point the situation just after recording resistance values at a certain temperature. Valve 2 is closed, valves 1 and 3 are open. The pressure is the same on both sides of the diaphragm. Before changing to another temperature — and pressure — valve 1 is closed. Valve 2 is then opened and the null of the diaphragm checked. Then the pressure generated by the pressure balance is changed to a new pre-selected value and the temperature is adjusted to generate this pressure as closely as can be estimated. Subsequently, valve 2 is closed, then valve 1 is opened. This will usually produce a deflection of the diaphragm. Therefore, the bulb temperature has to be readjusted until the pressure is again the same on both sides of the diaphragm. At this point the mass of gas in the bulb and the capillary is the same as it was at the starting point.

It should be mentioned that the precision of the final results would be about the same if the dead space had included the volume up to the diaphragm because the correction for the additional volume is small and can be calculated with sufficient accuracy (Berry 1972). But it was considered to be advisable that the equilibrium position of the diaphragm be checked at each measuring point and excessive differential pressures across the diaphragm will not be supplied.

Temperature control is provided by two separate electronic regulators for the shield and the bulb. The shield is cooled either by pumping liquid helium into the chamber G or, at temperatures significantly above the normal boiling point of helium, by admitting continuously small amounts of cold gas through this chamber. One regulator employing a germanium temperature sensor on the shield supplies current to one of the heaters at the shield to compensate for the cooling. To measure the resistance of the sensor, a special AC bridge circuit was designed (Neubert 1975). It contains operational amplifiers arranged in such a way as to keep the voltage across the sensor constant and to eliminate errors due to lead resistance. Originally the bulb temperature was controlled only by changing the shield temperature. As this was a rather time-consuming procedure, a second regulator was installed for the bulb. The error signal is taken from the output of the capacitance manometer. The shield temperature is set slightly below the required bulb temperature.

Figures 3(a) and (b) show the response of the regulator to a small temperature deviation at 4·4 K (a) and 20 K (b). These figures are reproduced from actual plots of the pressure difference at the diaphragm against time. The total pressures were (a) 13·3 kN m^{-2} and (b) 64 kN m^{-2}. The sensitivity was 4 N m^{-2} full scale in both cases. The bursts, more pronounced in figure 3(b), reflect oscillations of the piston balance caused by accidental mechanical shocks in the laboratory as, for example, closing a door. Since it is somewhat inconvenient to interrupt the regulation process every time the piston balance stops rotating, especially in the low-pressure region where the inertia is low, use of an error signal derived from one of the resistance thermometers was attempted. But with the available equipment the results were not as good as those obtained with the capacitance manometer.

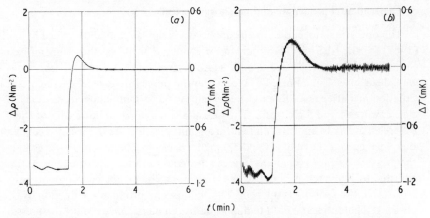

Figure 3. Pressure difference Δp across the diaphragm, plotted against time at 4·4 K (a) and 20 K (b). The temperatures were allowed to drift away before the regulator was switched on.

At the present time, four major test runs have been completed, each consisting of 20 to 36 points between 3 K and 20 K. The results, however, are not directly comparable because the apparatus has been modified; two resistance thermometers have been replaced because of excessive noise and the experimental procedure has been changed between runs. The two-stage regulating system was used only for the last run.

In general, the temperature between successive points was increased because reasonable cooling rates were possible only by admitting exchange gas to the vacuum space. At the beginning and at the end of each measuring day the fiducial point which was chosen at a temperature slightly above the normal boiling point of helium was repeated. The reproducibility was 0·1 mK except for one case when it was 0·3 mK. In the last run, the pressure was 64 kN m^{-2} at 20 K. This is the lowest possible pressure to cover the whole temperature range when using the piston–cylinder combination of 0·8 cm^2 cross sectional area which works better than one with 8 cm^2 area. The temperature drift of the equilibrium position of the diaphragm when left exposed to room temperature was equivalent to less than 0·16 N m^{-2} which would cause an error of 0·05 mK.

The end of the copper capillary settles at a temperature of 180 K when not heated. Heating to 300 K decreases the bulb temperature, measured at the top and the bottom, by 0·3 mK when the bulb is at 20 K. This is the change of the dead-space correction expressed in terms of temperature since the pressure is constant. At 4·4 K this effect should not be noticeable because the portion of the gas in the capillary is too small. In fact, the average temperature changes less than 0·1 mK, but the temperature rises by 0·3 mK at the top and falls by 0·3 mK at the bottom. Presumably, the thermal contact between the capillary and the shield is not as good as it should be. The problem of thermal gradients in the bulb will be investigated in more detail by interchanging thermometers between top and bottom.

4. Conclusion

There are some problems to be solved and perhaps some modifications to be made before the gas thermometer performs satisfactorily. The valves 1 and 2 (figure 2) are small bellows sealed valves — commercially available — but they should be of the constant-volume type as for instance described in Anderson *et al* (1970). The resistance thermometers have to be checked individually for self-heating effects. Furthermore, the boiling point of equilibrium-hydrogen or neon has to be realized. But from the results already obtained, it appears that a practical temperature scale can be established between 3 K and 20 K with an overall precision of ±1 mK.

References

Anderson R L, Guildner L A and Edsinger R E 1970 *Rev. Sci. Instrum.* **41** 1076
Barber C R 1972 *TMCSI* **4** part 1 99
Berry K H 1972 *TMCSI* **4** part 1 313
Guildner L A 1962 *TMCSI* **3** part 1 151
Neubert W 1975 to be published

A proposal for dielectric-constant (or refractive-index) gas thermometry in the range 90–373 K

R L Rusby
Division of Quantum Metrology, National Physical Laboratory, Teddington, Middlesex, UK

Abstract. Conventional constant-volume gas thermometry is bedevilled by the requirement that the number of moles of gas present should be constant, and particularly by the possible occurrence of reversible sorption processes. The difficulty is overcome if the density of the gas is measured *in situ*, and it has recently been suggested that modern techniques of capacitance measurement should allow the density to be obtained with useful accuracy by measurement of the dielectric constant of the gas. Alternatively in an optical analogue the refractive index of the gas could be measured by laser interferometry. In either case a precision of better than 10^{-8} is required if the method is to be of value in the range 90–373 K. In this paper the problems of the method and some design considerations are discussed.

1. Introduction

The gas thermometry measurements of Guildner and Edsinger (1973) gave a value for the steam point 0·030 °C below the IPTS-68 value of 100 °C. They showed that reversible sorption effects are only eliminated by thorough baking of the evacuated system such as only they have undertaken, and they suggest that previous work was subject to systematic errors due to sorption. It has therefore become important that new experiments to measure thermodynamic temperature should be conducted, based if possible upon a variety of methods. One such method is total radiation calorimetry (Ginnings and Reilly 1972), and since the quantity measured is proportional to T^4, this method is also well suited for application at temperatures above 100 °C.

The possibility discussed here is that of dielectric-constant (or refractive-index) gas thermometry. To obtain temperature by gas thermometry one must either measure both the gas pressure and its density, or measure just one and hold the other constant. In conventional gas thermometry it is usual to maintain a constant density, but in dielectric-constant gas thermometry both the pressure and the density are variable, the value of the latter being obtained at each point by capacitance measurement from the dielectric constant of the gas. The method has only recently been developed and although R H Sherman (1973 unpublished communication to CCT working group 4) has reported some measurements in the range below 5 K, no results are yet published. D Gugan (also unpublished) has analysed the potentiality of a thermometer intended for operation below 30 K, and Colclough (1974) has discussed the optical equivalent, a refractive-index gas thermometer, for use in the same range. That these activities all relate to cryogenic thermometry is no accident since the uncertainty $\delta T/T$ is inversely proportional to the density and hence is proportional to temperature. It is suggested here, however, that useful measurements may be made in the range 90–373 K.

2. The method

The relationship between the dielectric constant ϵ and the density, or rather the molar volume v, is expressed by the Clausius–Mossotti equation which, allowing for interactions between molecules, may be written

$$\frac{\epsilon - 1}{\epsilon + 2} = \frac{4\pi N_0 \alpha}{3v}\left(1 + \frac{b}{v} + \ldots\right)$$

where $N_0 \alpha$ is the molar polarizability and b is the second dielectric virial coefficient. Using the virial equation of state

$$P = \frac{RT}{v}\left(1 + \frac{B}{v} + \ldots\right)$$

the molar volume may be eliminated, giving

$$\left(\frac{\epsilon - 1}{\epsilon + 2}\right)\left(\frac{1}{P}\right) = \frac{A}{RT} + P\left(\frac{A(b-B)}{(RT)^2}\right) + P^2(\ldots) + \ldots \tag{1}$$

where $A = (4\pi/3)N_0\alpha$.

To measure ϵ a capacitor is included in the gas cell such that the gas fills the space between the electrodes. In an ideal capacitor, ϵ is given by the ratio of the value of the capacitance in the presence of the gas, C, to that under vacuum, C_0. In practice it is necessary to make a small correction to this to allow for the unavoidable dimensional changes which occur whenever the gas pressure is changed. Assuming that the capacitor experiences a hydrostatic compression and that it responds isotropically we obtain

$$\frac{C}{C_0} = \epsilon\left(1 + \frac{P\kappa(T)}{3}\right)$$

or

$$\frac{\Delta C}{C_0} = \frac{C - C_0}{C_0} = (\epsilon - 1) + \epsilon\frac{P\kappa(T)}{3}$$

where $\kappa(T)$ is the isothermal compressibility. This is not the only possible condition, but in any experiment the capacitor must be carefully designed so that the dimensional changes are well understood and are of known magnitude. Including the compressibility effect to first order only we now write equation (1) in the form obtained by Gugan:

$$X \equiv \left(\frac{\Delta C/C_0}{\Delta C/C_0 + 3}\right)\left(\frac{1}{P}\right) = \left(\frac{A}{RT} + \frac{\kappa(T)}{9}\right) + P\left(\frac{A(b-B)}{(RT)^2}\right) + P^2(\ldots) + \ldots \tag{2}$$

where X is calculated from measured quantities. Thus an isotherm of X against P extrapolated to $P = 0$ gives the temperature provided that A is known and that the compressibility correction can be made.

Since the measurement of gas density is part of the method, it follows that we are able to admit gas and evacuate the cell as desired without measuring the quantity added or removed. This freedom greatly simplifies the isotherm procedure and also allows frequent checks of C_0 and gas purity to be made. Furthermore as there are no dead space effects there is no restriction in principle on the diameters of the pressure sensing or pumping tubes, and the difficulty of obtaining a clean system is as a result less severe than in a constant-volume gas thermometer. On the other hand it is the polarizability of an impurity which is important and it is therefore more necessary to exclude highly polarizable molecules from the cell.

The problem of eliminating A from equation (2) is in practice best solved by taking an isotherm at a known temperature, so that in effect ratios of temperatures are measured. Ideally the calibration would be at the triple point of water and this would also be convenient in an operating range of 90–373 K. A temperature ratio T_1/T_2 is given by

$$\frac{T_1}{T_2} = \left(\frac{X_2}{X_1}\right)_{P \to 0} \left(1 + \frac{RT_1}{A}\frac{\kappa(T_1)}{9} - \frac{RT_2}{A}\frac{\kappa(T_2)}{9}\right)$$

$$= \left(\frac{X_2}{X_1}\right)_{P \to 0} \left[1 + \left(\frac{P}{\epsilon - 1}\right)_{T_1}\frac{\kappa(T_1)}{3} - \left(\frac{P}{\epsilon - 1}\right)_{T_2}\frac{\kappa(T_2)}{3}\right]$$

with sufficient accuracy. The compressibility correction, in the square brackets, is now seen to be the difference between two terms, although where T_1 and T_2 are very different the cancellation is not good. In the experiments envisaged, the magnitude of each term would be $\lesssim 5 \times 10^{-4}$ and a knowledge of $\kappa(T)$ to 1% or better is desirable. A further advantage of using a calibration temperature is that we are now interested in measurements of pressure ratios rather than absolute pressures, and it should be possible to use a pressure balance for this purpose with a precision of 3 ppm or better.

3. Assessment

We now consider the magnitudes of the quantities involved in order to assess the likely accuracy of an experiment. By far the most problematical measurement is that of ΔC since this is the difference between two similar capacitances. We may therefore write

$$\frac{\delta T}{T} \simeq \frac{\delta(\Delta C)}{\Delta C} = \frac{\delta(\Delta C)}{C_0}\frac{1}{(\epsilon - 1)} = \frac{\delta(\Delta C)}{C_0}\frac{RT}{3AP} \quad (3)$$

where δ is used to indicate the random uncertainty in a quantity. From the penultimate stage we see that $\delta(\Delta C)/C_0$ must be very small indeed; in other words the precision of the capacitance measurement must be very high. Using helium at NTP for example, $(\epsilon - 1)$ is about 7×10^{-5} so that a precision of 1 mK in temperature requires $\delta(\Delta C)/C_0 \simeq 2 \cdot 5 \times 10^{-10}$. Clearly it is desirable to use a more polarizable thermometric gas, and an eightfold improvement would be achieved by using argon. In table 1 we give uncertainties in the ratio $T/273 \cdot 16$ for the rare gases calculated on the basis of single measurements at 100 kPa at the unknown temperature T and 273·16 K, and

Dielectric constant gas thermometry, 90–373 K

Table 1. Uncertainties in temperature T obtained from ratios of $(T/273 \cdot 16)$, based upon single measurements at T and 273·16 K with P = 100 kPa, and assuming $\delta(\Delta C)/C_0 = 10^{-8}$.

T (K)	Xe		Kr		Ar		Ne		He	
	ppm	(mK)	ppm	(mK)	ppm	(mK)	ppm	(mK)	ppm	(mK)
373	12	(4·5)	20	(7·5)	30	(11)	130	(48)	242	(90)
200	8·8	(1·8)	15	(3·0)	22	(4·4)	95	(19)	177	(35)
150	–		14	(2·0)	20	(3·1)	87	(12)	163	(24)
90	–		–		19	(1·7)	81	(7)	150	(14)
$(\epsilon - 1)_{\mathrm{NTP}}$ $\times 10^5$	140		85		56		13		7	

combining the random uncertainties in quadrature. The precision of the capacitance measurement is assumed to be 1 part in 10^8.

The figures in the table serve as a guide to the measurement requirements rather than as a forecast of the outcome of an experiment. In particular they ignore the need to take isotherms, and here we gain precision by taking many points, but lose on extrapolation to zero pressure. It is also evident from the last step in equation (3) that the uncertainty is reduced by operating at higher pressures, and a convenient limit for our pressure balance would be 300 kPa, provided that this is not close to the saturated vapour pressure of the gas. Computer simulations suggest that for a 40 point isotherm extending to 300 kPa with $\delta(\Delta C)/C_0 = 10^{-8}$ the uncertainties in temperature may be reduced by a factor of 2 below those given in the table.

It is clear, however, that a precision in capacitance measurement of better than 10^{-8} is desirable and it is necessary to consider how this may be achieved. Experience gained at standards laboratories suggests that ratio transformer bridges may be used to compare capacitances with a resolution approaching 10^{-9} pF. Values of capacitance of several pF may be obtained from a parallel plate or coaxial cylindrical geometry, but these capacitors are probably not capable of the stability required and are somewhat sensitive to the adsorption of gas on the surfaces of the electrodes. This problem may be circumvented by using a cross-capacitor of the type used in calculable standards (Clothier 1965, Rayner 1972), in which a thin uniform layer of dielectric makes no contribution to the capacitance measured. Such a capacitance should also possess excellent dimensional stability, as this is almost entirely dependent on the stability of the length. On the other hand the value of capacitance is only 2 pF per metre of length and in order to achieve a suitable value of C_0 it would be necessary to build up an array of electrodes; for instance a 4 × 4 array would give nine cross-capacitors and at 0·2 m length parallel connection would provide 3·6 pF. The precision of the measurements should then be only a few parts in 10^9, and in practice the experiment may be limited by the stability of the defining length.

We now compare this method with the alternative, proposed by Colclough (1974), of measuring the refractive index n of the gas by laser interferometry. Since $\epsilon = n^2$ and $(\epsilon - 1) \simeq 2(n - 1)$ the two experiments are directly analogous. In the optical experiment the gas cell would form part of one arm of a Michelson or another suitable interferometer and the change in optical path Δl on admission of the gas gives $(n-1)$ and T from

$$\frac{\Delta l}{l_0} = \frac{N\lambda}{l_0} = 2(n-1) = \text{const.} \frac{T}{P}$$

λ being the laser wavelength, N the fringe count and l_0 the geometrical length of the cell; the factor of two results from passing the beam through the cell twice. Then for $\lambda = 0.6\,\mu\text{m}$ and $l_0 = 0.18$ m we achieve $\delta(\Delta l)/l_0 = 10^{-8}$ and uncertainties in temperature ratios as given in the table provided that the fringe counter can be used to discriminate 3×10^{-3} of a fringe. In practice it may be possible to improve on this by a factor of 3, and the precision then would be comparable with that of the capacitance method.

The practical advantages and disadvantages of the two instruments are very different and finely balanced. Thus for instance in the dielectric constant experiment, small changes in temperature which may occur during an isotherm would cause changes in dimension and therefore in capacitance, while in the optical experiment the second arm of the Michelson interferometer could be used to compensate for this effect and reduce it by a factor $(n-1)$, and the effect of any instability in laser wavelength would be similarly reduced. On the other hand there is no simple way of measuring l_0 as there is of measuring C_0. As we are required to measure ratios of temperature, however, we need only measure ratios of $l_0(T)$ to $l_0(273 \cdot 16\text{ K})$. Since these will be within 0·3% of unity it is only necessary to measure changes in length as a function of temperature (using the interferometer, with the cell evacuated) and the absolute magnitude of l_0 at one temperature with a precision of 0·1%. These data may be obtained from preliminary experiments. An added complication of the optical experiment is the need to provide windows in the cell and the effect of pressure and temperature on the optical path in these is likely to be at the level of significance. Changes in l_0 due to the pressure of gas enter the analysis as they do in C_0 and must be corrected for in the same way.

In conclusion it may be said that although neither of these experiments has been performed, either is probably capable of giving useful measurements of temperature in the range 90–373 K. The precision is unlikely, initially at least, to equal that of constant-volume gas thermometry; using Ar one might hope to achieve standard deviations in the range 2 to 4 mK at 373 K and 0·5 to 1 mK at 90 K. Since the systematic errors would be very different, however, such results would be of great value.

References

Clothier W K 1965 *Metrologia* **1** 36–55
Colclough A R 1974 *Metrologia* **10** 73–4
Ginnings D C and Reilly M L 1972 *TMCSI* **4** part 1 339–48
Guildner L A and Edsinger R E 1973 *J. Res. NBS* **77A** 383–9
Rayner G H 1972 *Trans IEEE* **IM-21** 361–5

Solid state phase transitions as thermometric fixed points

J F Schooley

Institute for Basic Standards, National Bureau of Standards Temperature Section, Heat Division, Washington, DC, 20234 USA

Abstract. The general question of the usefulness of thermometric fixed-point devices based upon phase transitions in solids is examined. It appears that both the method of detection and the intrinsic character of the transition relate to thermometric precision and practical utility of such devices. Promising systems which are discussed are the AC mutual inductance detection of superconductive transitions and the AC heat capacity detection of antiferromagnetic and ferroelectric transitions.

1. Introduction

It is interesting to the neophyte in thermometry to notice that, of the eleven defining fixed points and twenty-seven secondary reference points of the IPTS-68, every one involves the liquid state (CIPM 1969). Indeed, one finds only occasional references to solid state fixed points in the literature. Kemp and Pickup (1972), and Orlova (1962), in the two most recent *TMCSI* symposia suggested that the α–β and β–γ transitions in solid oxygen might suffice for use as fixed points. However, each discussion relegates these transitions to second-rate status, owing principally to their relative thermometric imprecision.

The present author, in collaboration with R J Soulen Jr, suggested, at the Fifth Temperature Symposium, that devices employing superconductive transitions might be used as thermometric fixed points below 10 K (Schooley and Soulen 1972). Since then, such devices have been fabricated and distributed (Schooley et al 1972), and some work has been done to relate the device transitions to thermodynamic properties (Cataland and Plumb 1975). Whether or not these particular devices are successful, it seems useful at this time to examine the whole class of phase transitions in solids from a thermometric point of view.

In discussing the question of the general utility of solid state transitions as thermometric fixed points, we have found that professional thermometrists nearly always stress the fact that solid state transitions are notorious for thermometric unreliability. The origins of such unreliability, we are told, are many: impurities, which are usually present in substantial quantity, broaden and change the temperature of the transition; strains, occurring naturally or induced by sample preparation, produce similar effects; lattice or magnetic domains serve as sources of hysteresis and broadening of the corresponding transitions; and finally, many transitions such as those in solid oxygen are difficult to use because the sample is characterized by very poor thermal conductivity. In addition to these problems, we also are instructed occasionally on the convenience of fixed-point transitions which involve large latent heats and thus can serve as temperature-equilibrating environments with only minimal assistance from temperature controllers. (We cannot refrain from pointing out that, when we ask whether supercooling is ever a problem in freezing-point cell thermometry, thermometrists note the necessity to briefly remove tin and antimony cells from their ovens until a freeze commences.)

The battle line is drawn, then — in order to succeed in advancing the thermometric art by the use of solid state transitions as fixed points, we must find some unique combination of transition-plus-detection technique which is miraculously precise. Furthermore, we should demonstrate that the use of such a device is both inexpensive and as enjoyable as a day in the park.

In the succeeding sections of this paper, we shall discuss some experiments which appear to us to provide potential solid state fixed-point schemes. We note, in turn, the AC mutual inductance technique applied to superconductivity, the AC heat capacity detection of antiferromagnetic transitions, and the heat capacity of ferroelectrics. We shall then point out changes in thermometric technique which must accompany the use of solid-state-transition fixed points, and finally, we shall ask whether there are occasionally special circumstances in which solid state thermometric fixed points might be quite attractive despite a reduced level of attainable precision.

2. Superconductivity and AC mutual inductance

Probably the most comprehensive examination of superconductive fixed points up to the present time is that mentioned in § 1. This work originated at NBS, although the resulting devices are now under study at several national and university laboratories.

In considering the combination of superconductors and detection method for thermometric fixed points, we deliberately selected 'soft' or type 1, low-melting elemental superconductors which were available in high purity, and we chose the AC mutual inductance detection method with which we had some familiarity (Soulen *et al* 1973). We were aware both that the soft superconductors were known for narrow transitions and that the transition as measured by calorimetric, electrical resistivity, or magnetic detection techniques should in principle yield the same temperature. We also knew that well annealed, high-purity single crystals were most likely to produce narrow, reproducible superconductive transitions.

Table 1 lists the superconductive samples presently offered with the NBS fixed-point device, along with some of their properties. We should like to note here only a few features regarding these samples: first, they are prepared for the device by simple vacuum casting rather than with the care required to produce single crystals and, excepting the hexagonal metals, seem to need no special annealing to guarantee the listed transition width; second, all samples of a given element can be prepared so as to yield the same transition midpoint temperature (T_c) within ±1 millikelvin.

It may be worthwhile to note here that the intrinsic width of a superconductive phase transition should be quite small; theoretical estimates of the width in some cases are as small as one microkelvin (Hohenberg 1968). In an effort to define more closely the ultimate thermometric precision available with cadmium samples, we have examined transitions occurring in single crystals and compared them with those in the polycrystal-

Table 1. Properties of superconductive elements in NBS fixed-point device.

Element	Superconductive transition temperature (K)	Typical transition widths (mK)	Melting point (K)	Crystal system
Lead	7·201†	1–2	600·6	cubic
Indium	3·416†	1–2	429·8	tetragonal
Aluminium	1·175‡	2–4	933·5	cubic
Zinc	0·844§	2–5	692·7	hexagonal
Cadmium	0·515§	1–4	594·3	hexagonal

† Temperature derived by magnetic thermometry interpolation of NBS 2–20 K acoustic temperature scale.
‡ Temperature derived from direct measurements on the 1962 ^{3}He vapour pressure scale (T_{62}).
§ Temperature derived by magnetic thermometry extrapolation of T_{62}.

line samples used in the NBS device (Schooley 1973). The results of this study indicate that widths of 10–20 μK can be obtained in cadmium. However, the sample-to-sample reproducibility at that level has not been established; at the present time the level of precision in cadmium stands at ±0·5 mK for transitions of 2 mK width or less.

As noted above, there is no theoretical basis for choosing one particular detection technique for superconductivity. Soulen and Colwell (1971) were able to show experimentally that, within ±0·5 millikelvin, the heat capacity, electrical resistivity, and magnetic induction methods yield identical results for indium. The results of these experiments are shown in figure 1.

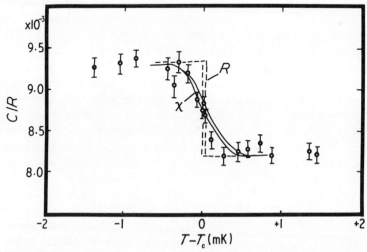

Figure 1. Temperature dependence of heat capacity (open circles), magnetic susceptibility (solid lines), and electrical resistivity (broken lines) near the superconductive transition of an indium sample, determined in two separate experiments. The equality of the experimental temperature axes was carefully established using a second indium sample. (After Soulen and Colwell 1971.)

Our own choice of the magnetic-induction method was prompted by the following considerations: no leads need be attached to the sample, so that contamination and stray heating effects are minimized; the equipment necessary for measurement was readily available and easily used; and the technique lends itself readily to automatic temperature control. Figure 2 is a block diagram illustrating the use of a diode circuit to regulate the temperature of the tempering block at T_c. The transition is detected by use of

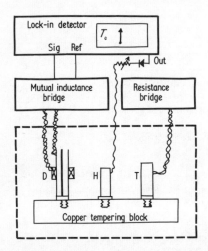

Figure 2. Superconductive fixed-point calibration scheme. $D \equiv$ fixed-point device; $H \equiv$ heater; $T \equiv$ resistance thermometer to be calibrated. The diode circuit on the detector output applies power to the heater only when the sample temperature is below T_c.

mutual inductance coils in a bridge circuit. In practice, the bridge imbalance is read on a meter and the transition temperature is maintained by varying the heat to the tempering block until a constant, midpoint meter reading is reached. Then, thermometers can be calibrated while the fixed-point temperature is controlled automatically. The device transition temperatures depend upon magnetic field; since supercooling, roughly proportional to the applied field, often occurs, it is generally the easier course to reduce the ambient field to a negligible value (about 1 μT).

We await the advice of others on the question of the utility of the device as presently constituted. Whether the ±1 mK precision found in our own laboratory will survive the rigours of age and hard travel, only time and use can tell. In the meantime, we are following a dual course in these studies; we are continuing to examine the superconductive transitions in detail to determine the extent to which the transitions occurring in the NBS devices reflect bulk thermodynamic properties of the elements, and we are attempting to expand the application of superconductive fixed points to cover a broader temperature range. This latter point is illustrated by table 2, in which the 'comments' are included to show that not all superconductors are well behaved!

3. Antiferromagnetism and AC heat capacity

We wish now to call attention to an experiment of Garnier (1972) in which the heat capacity of chromium at the Néel temperature was determined by an AC technique. This experiment, which itself is an extension of earlier work (Handler *et al* 1967), illustrates a method which we feel deserves consideration from the point of view of thermometry.

Solid state phase transitions as thermometric fixed points

Table 2. Superconductors under study as possible thermometric fixed points.

Material	Approximate transition temperature (K)	Typical transition widths (mK)	Comments
Nb_3Sn	~18	20–50	Stability, reproducibility uncertain
V_3Ga	~14·5	100–300	Stability, reproducibility uncertain
$AuIn_2$	0·20	1–2	*Uniform T_c* may require elaborate and expensive sample preparation (eg, zone refining); individual samples show sharp, reproducible transitions.
$AuAl_2$	0·15	1–2	
Ir	0·10	1–2	
Be	0·023	0·5–1	
W	0·015	0·5–1	

The sample was a well annealed single crystal in the form of a thin plate (2 × 3 × 0·15 mm). It was weakly coupled to a tempering block by exchange gas, and square-wave pulses of radiant energy were applied to it at a frequency of 11 Hz. This frequency was sufficiently low so that the sample maintained internal thermal equilibrium during the heating pulses, yet sufficiently high so that the sample could equilibrate only slowly with its surroundings. The output of a thermocouple cemented to the sample was fed into a lock-in detector; the detector output then was proportional to the fluctuation of the sample temperature during a heating cycle, ΔT. Since the heat pulse, ΔQ, was constant, the peak in the heat capacity at the Néel point appeared as a minimum in ΔT through the relation $C = \Delta Q/\Delta T$. The inverse of the quantity ΔT is shown as a function of temperature in figure 3.

One must note that, in this experiment, the average sample temperature was higher than that of the tempering block by a δT of about 10 millidegrees. Clearly, in order to build a fixed-point device based on this principle one must obtain reproducible values of δT.

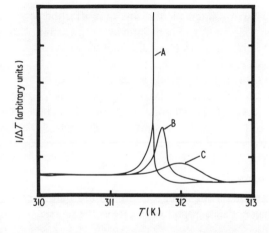

Figure 3. Temperature dependence of the reciprocal of the temperature fluctuations induced in single crystal chromium by 11 Hz heating pulses, showing the striking effect of sample annealing. Sample A, annealed 50 h at 1200 °C; sample B, annealed 2 h at 1000 °C; and sample C, unannealed. (After Garnier 1972.)

The ultimate thermometric precision which one could obtain in this experiment is not clear from the data of figure 3. Besides questions concerning the overall sensitivity of the method, one must determine the effects of small differences among samples resulting from the usual metallurgical problems of preparation. However, we submit that, using the sensitive AC calorimetric technique, one might examine a number of solid state phase transitions with thermometric optimism. The curve of figure 3 can be differentiated by standard electronic techniques; the resulting output, instead of a minimum, becomes its derivative. This output can then be used in a temperature control circuit in the manner of figure 2.

4. Ferroelectricity

As further examples of solid state phase transitions which may be of thermometric interest, we present heat capacity data on the ferroelectric transitions in $NH_4H_2PO_4$ (ADP) (Strukov *et al* 1969), and triglycine sulfate (TGS) (Strukov 1965) which occur at $-123 \cdot 5$ °C and 49 °C, respectively. Figure 4 shows the ADP data and figure 5 that for TGS. The data were obtained by standard DC calorimetry for which the authors claimed submillidegree sensitivity.

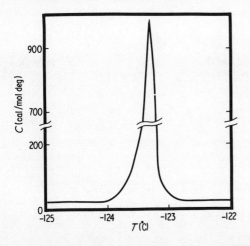

Figure 4. Temperature dependence of the heat capacity of $NH_4H_2PO_4$ (ADP) near the ferroelectric transition. Data obtained by adiabatic calorimetry. (After Strukov *et al* 1969.)

Once again, we see transitions that are narrow in temperature and thus offer the possibility of serving as temperature reference points. It would be interesting to examine the sample-to-sample reproducibility of these transitions using the more sensitive AC technique.

Yet other solid state transitions could be examined. Metal–insulator transitions occur in substances such as VO_2 (Kawakubo and Nakagawa 1964) and the organic conductor commonly identified as (TTF)–(TCNQ) (Cohen *et al* 1974); however, the transitions already mentioned offer some idea of the range of phenomena which are available for study.

5. Thermometry with solid-state fixed points

By now, the changes in experimental technique implied by a shift to solid state fixed points should have become clear. The major change is in physical size of the sample.

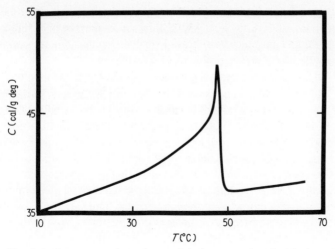

Figure 5. Temperature dependence of the heat capacity of triglycine sulphate single crystal near the ferroelectric transition. (After Strukov 1965.)

Little purpose would be served, for example, in using the same mass for an aluminium superconductive fixed-point device as for a freezing point cell. The zero-field superconductive transition has no latent heat and therefore cannot provide a constant-temperature environment. Thus, instead of relying on a 'freeze' induced in a large sample, the 'solid state thermometrist' must employ a tempering block with its temperature control system after the fashion of figure 2. We have attempted to show, however, that both the magnetic and the thermal detection methods lend themselves to this technique so that the temperature control can be made automatic.

The care which, for freezing point or triple point devices, is devoted to preparing the 'freeze' is invested in solid state devices during their design and manufacture; assuring adequate purity, proper annealing, proper thermal contact, and reproducible detection of the transition. The actual application of the devices to a thermometric calibration involves no handling of the fixed-point sample itself.

6. Who needs solid state fixed points?

One can readily ask why we should bother investigating phase transitions in solids as possible thermometric fixed points when the temperature community already has some thirty-eight primary and secondary reference points. There are several situations which, we submit, call for this consideration.

First, for example, in the region below 13 K, where scientists and superconductivity engineers alike work beyond the bounds of the IPTS-68, few alternatives to solid state transitions are available. Because of the poor thermal properties of boiling helium, only the ^{4}He lambda-point transition can be considered a realistic choice for a low-temperature fixed point.

Second, there is the situation wherein one desires to calibrate a thermometer during the course of an experiment. Many times this is a desirable feature, since it permits the use of a highly sensitive but inherently unstable thermometer to provide not only sensitive, but also accurate thermometric data. Indeed, in some experiments, such as those involving magnetic thermometry, *in situ* calibration is essential. Because of the relatively large size of fixed-point devices which involve the liquid state, they are very poorly suited to such a use.

Third, there is the circumstance in which the economics of a situation militate against the use of highly trained personnel and elaborate calibration equipment in the application of thermometry. Such circumstances occur, for instance, in the control of manufacturing processes, in medical care, (note that the Neél temperature of chromium lies within one degree of body temperature), and in geophysical temperature mapping. Often, in such circumstances, those in charge settle for sensitivity or precision, neglecting accuracy. Since the effort needed to obtain the useful thermometric reproducibility has already been supplied in the preparation of the solid state devices, their use could be extended beyond the group of trained thermometrists – possibly with a consequent improvement in the level of accuracy in those applications.

7. Summary

In this brief communication, we have attempted to point out methods by which solid state thermometric fixed points can be realized and to note changes in technique which their use would entail. We have further suggested circumstances in which such devices, even at reduced levels of thermometric accuracy, might serve quite useful purposes. We close with the hope that these comments might help to generate an evaluation of specific solid state transitions from this point of view.

Acknowledgments

The author gratefully acknowledges enlightening discussions with his colleagues G T Furukawa, D B Utton, R J Soulen, Jr and R S Kaeser during the preparation of this manuscript.

References

Cataland G and Plumb H H 1975 to be published
Cohen M J, Coleman L B, Garito A F, and Heeger A J 1974 *Phys. Rev.* **B10** 2205
CIPM 1969, *IPTS-68, Metrologia* **5** 35
Garnier P R 1972 *Proc. Conf. on Magnetism and Magnetic Materials* ed C D Graham, Jr and J J Rhyne (New York: AIP) p297
Handler P, Mapother D E and Rayl M 1967 *Phys. Rev. Lett* **19** 356
Hohenberg P C 1968 *Proc. Conf. on Fluctuations in Superconductors, Pacific Grove, Cal.* ed W S Goree and F Chilton (Stanford Research Institute, Menlo Park, Ca)
Kawakubo T and Nakagawa T 1964 *J. Phys. Soc. Jap.* **19** 517
Kemp W R G and Pickup C P 1972 *TMCSI* **4** part 1 217
Orlova M P 1962 *TMCSI* **3** part 1 179
Schooley J F 1973 *J. Low Temp. Phys.* **12** 421
Schooley J F and Soulen R J Jr 1972 *TMCSI* **4** part 1 169
Schooley J F, Soulen R J Jr and Evans G A 1972 *NBS Special Publication* 260-44
Soulen R J Jr and Colwell J H 1971 *J. Low Temp. Phys.* **5** 325
Soulen R J Jr, Schooley J F and Evans G A Jr 1973 *Rev. Sci. Instrum.* **44** 1537
Strukov B A 1965 *Sov. Phys.–Solid St.* **6** 2278
Strukov B A, Amin Soliman M and Kopsik V A 1969 *J. Phys. Soc. Jap.* **28** suppl 70 p207

Crystalline transformations of oxygen

J Ancsin
National Research Council of Canada, Division of Physics, Ottawa, Ontario, Canada, K1A 0S1

Abstract. Experimental study of the crystalline transformations of oxygen show that the shape of the $\beta-\gamma$ transformation curve (amount of γ phase formed against temperature) depends upon the rate of freezing and subsequent cooling of the sample. Increasing cooling rates resulted in decreasing slopes for the $\beta-\gamma$ transformation curve, as well as increasing temperatures of transformation. It follows then that accurate comparison of the numerical values obtained by different experiments is meaningless unless experimental procedures are strictly specified. The behaviour of the temperature dependence of Debye θ obtained from the measured specific heats indicates that both the $\alpha-\beta$ and $\beta-\gamma$ crystalline transformations seem to take place gradually over a temperature interval of several kelvin. The latent heats were therefore determined from the heat necessary to warm the sample from the region of constant Debye θ in the initial phase to the region of constant Debye θ in the final phase. The values thus obtained were 103·13 J mol^{-1}, 810·65 J mol^{-1}, and 439·10 J mol^{-1} for the $\alpha \to \beta$ phase change, the $\beta \to \gamma$ phase change and fusion respectively. Values for the $\alpha \to \beta$ and $\beta \to \gamma$ transformation temperatures, for samples specified in the text, were found to be 28·867 ± 0·003 K and 43·8007 ± 0·0003 K respectively.

1. Introduction

Oxygen has three known crystal modifications. Its monoclinic (α) form transforms into rhombohedral (β) at 23·86 K which, in turn, transforms into cubic (γ) at 43·9 K. The γ phase then persists up to the melting point at 54·4 K. Because many physical transformations are potential fixed points for a practical temperature scale, these crystalline transformations are of interest to temperature standards laboratories. One of the most important requirements concerning any transformation that might be used as a thermometric fixed point is that its temperature be highly reproducible. Reproducibilities to within 5 mK and 2·4 mK have been observed for the $\alpha-\beta$ and $\beta-\gamma$ transformations respectively (see Kemp and Pickup 1972). As compared with freezing points, for example, this is a relatively large irreproducibility, yet little is known of the reasons for it. Is it caused by impurities? Is it an inherent property of these types of crystalline transformations? Or is it simply an experimental error which could be considerably reduced? A further question which is of a more general interest concerns the latent heat of the $\alpha-\beta$ transformation. According to some experimenters this transformation takes place with no detectable latent heat (eg Fagerstroem and Hallett 1966, Muijlwijk 1968), yet others claim (eg Giauque and Johnston 1929) to have measured it. While the existence of a latent heat implies a first-order thermodynamic transition, the absence of it implies a phase transition of higher order. In this connection, Barrett *et al* (1967) have pointed out from purely theoretical considerations of the symmetries of the crystal lattices involved that the $\alpha-\beta$ transition cannot be of second order. They therefore conclude, taking into account also the results of Hoge (1950), Giaque and Johnston (1929) and Hörl (1962), that the transformation has to be of first order.

The primary purpose of this work is to study the reproducibility of the temperatures at which the α–β and β–γ crystalline transformations occur and to observe the latent heats of these transformations.

2. β–γ transformation

The β–γ transformation curve (temperature against amount of new phase formed) was obtained from measurements similar to adiabatic heat capacity determinations ie, after supplying some heat to the sample, which is otherwise insulated from its environment, we observed the sample temperature until such time that it no longer changed significantly. Typical observations are shown in figure 1. The onset of heating is always

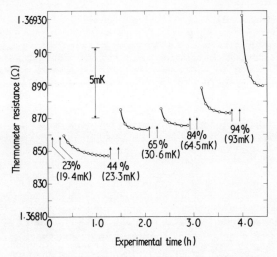

Figure 1. Readings during a typical β → γ transformation experiment (4 April 1974). Sample size 0·0677 mol. The amount of β transformed into γ phase during heating periods between the arrows is indicated, in percentages, and the numbers in parentheses indicate the driving temperatures of the transformation for a β → γ conversion rate of 2%/min.

detected by the thermometer within less than five seconds. (This is true not only during this phase transition but also within the α, β and γ phases.) Figure 1 shows that after heating the sample, its temperature drops toward some equilibrium temperature for a considerable length of time. This relatively long transition period may be contrasted with the transition period of the order of one minute in either the α, β, or γ phases.

The slow, exponential[†], approach toward a stable temperature during the β→γ transformation suggests that the transformation proceeds by diffusion. Diffusion, on the other hand, is the mechanism by which a nucleation and growth crystalline transformation takes place[‡].

[†] Plotting the temperature against time semi-logarithmically yields a straight line.
[‡] This conclusion contradicts the one drawn by Barrett and Meyer (1967) of martensitic transformations.

At first the combined results obtained for several $\beta \rightarrow \gamma$ transformation curves appeared quite irreproducible. Careful re-examination of the details of the experimental procedures, however, revealed that what appeared as a straightforward irreproducibility was, in fact, a dependence upon the rate of cooling of the sample through the freezing, and also through the $\beta-\gamma$ transformation temperature. After this dependency was realized a set of experiments was performed in which the cooling rates of the samples were changed in a systematic manner.† Figure 2 shows the results of some $\beta-\gamma$ transformation experiments, the approximate cooling rate being a parameter. The temperature is displayed as a function of the amount of γ phase formed. This quantity is assumed to be equal to the heating time (measured from the estimated onset of the $\beta-\gamma$ transformation) divided by the total transformation time at constant heating power. It is seen that different cooling rates yield different transformation curves. In

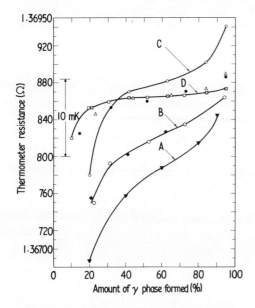

Figure 2. Dependence of the $\beta \rightarrow \gamma$ transformation curves upon the cooling rates ($\dot{T}$) from liquid phase to the indicated final temperature (T_f). Curve A, 1 May 1974, $\dot{T}$ (through freezing) $< 1\,°\text{C min}^{-1}$, $\dot{T}$ (after freezing it) $> 5\,°\text{C min}^{-1}$, $T_f = 4\cdot 2$ K; B, ♦ 2 April 1974, $\dot{T} \approx 1\cdot 8\,°\text{C min}^{-1}$, $T_f = 4\cdot 2$ K, ○ 8 April 1974, $\dot{T} \approx 1\cdot 8\,°\text{C min}^{-1}$, $T_f = 33$ K; C, 9 April 1974, $\dot{T} \lesssim 5\,°\text{C min}^{-1}$ until $T_f = 46\cdot 5$ K and then $\dot{T} \lesssim 2\,°\text{C min}^{-1}$, $T_f = 31$ K; D, × 27 March 1974, $\dot{T} \approx 5\,°\text{C min}^{-1}$, $T_f = 4\cdot 2$ K, □ 28 March 1974, $\dot{T} \approx 5\,°\text{C min}^{-1}$, $T_f = 4\cdot 2$ K, △ 4 April 1974, $\dot{T} \lesssim 5\,°\text{C min}^{-1}$, $T_f = 4\cdot 2$ K, ● 11 April 1974, $\dot{T} \approx 5\,°\text{C min}^{-1}$, $T_f = 28\cdot 5$ K.

† It became evident that our cryostat was not well suited for these particular experiments (ie, dependence of the transformation curve upon the cooling rate) because the cooling rate is a function of the amount of the exchange gas used, and the latter is rather difficult to manipulate quickly and precisely. Cooling by a mechanical heat switch for example, would have been much more suitable because then the heat flow could easily and accurately be controlled by the pressure applied to the heat contacts.

particular the rate of cooling and, to some extent, the final temperature reached, influence the shape of the transformation curve a great deal. This is perhaps not unreasonable because the rate of cooling influences the way the crystal forms, the impurity distribution, and the forming of dislocations, etc in the solid sample. These properties, in turn, will all influence the transformation curve to various degrees. It would be quite interesting to investigate the influence of these properties individually on the phase transformation. The results of such investigations would help a great deal in elucidating the physical processes taking place in these crystalline transitions.

We see from figure 2 that cooling the sample quickly (≈ 5 K min^{-1}) from the liquid state to below the $\alpha-\beta$ transition temperature yields rather 'flat' transformation curves, reminiscent of melting curves of slightly impure O_2 (99·99%) (see curve D of figure 2). In our case, four experiments where identical freezing rates were attempted yielded curve D of figure 2, which exhibits surprisingly good reproducibility of the transformation behaviour.

There are two important numerical values which are of interest concerning crystalline transformations. The latent heat involved, and the temperature at which the transformation takes place. It is seen from figure 2 that the complete transformation seems to take place over a rather indefinite temperature interval; it is difficult, at best, to determine its precise beginning. If one defines the transformation temperature to be that at which the heat capacity of the sample is a maximum, ie the point of inflection of the curves in figure 2, then the transformation temperature is a function of the way the sample was brought to this temperature. From a thermometrist's point of view, however, the transformation temperature defined above can be determined quite precisely for samples which were cooled rapidly from the liquid state to below the $\alpha-\beta$ transformation and yielded transformation curves such that about 50% of the crystal transformation took place within 1 mK (see curve D of figure 2). For such curves the point of inflection can be determined to better than $\frac{1}{4}$ mK. The $\beta \rightarrow \gamma$ transformation temperature corresponding to the point of inflection of curve D of figure 2 was found to be 43·800$_7$ K.

3. $\alpha-\beta$ transformation

A detailed heat capacity determination through the $\alpha-\beta$ transformation is shown in figure 3. Again we noticed that the shape of the heat capacity curve through the transformation region could not be reproduced precisely, presumably because the rate of cooling of the sample was not reproduced precisely. All experiments, nevertheless, produced peak of the anomaly within a temperature interval of 5 mK. The introduction of 190 ppm Ar into the O_2 shifted the peak -15 mK; 385 ppm N_2 shifted the peak -17 mK. The temperature where the peak of the specific heat anomaly due to $\alpha-\beta$ transformation occurs is 23·867 K.

4. Heat capacity determinations

Very careful heat capacity determinations were performed by Giauque and Johnston in 1929 on samples prepared with extreme care and evidently handled with great expertise. Our measurements (open circles) are compared with theirs (full circles) in

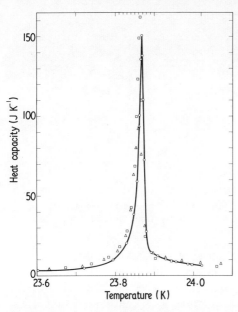

Figure 3. Detailed measurements along the specific heat anomaly associated with the $\alpha-\beta$ crystalline transformation. Curve, 13 February 1974; □, 5 June 1974; △, 6 June 1974.

figure 4. There is overall agreement up to about 40 K; thereafter our results are below theirs by a maximum of 6·5%. In the γ phase our results gradually approach theirs and finally meet at the melting point. A similar trend is seen in the results of Fagerstroem and Hallett (1966) although to a lesser extent. In the liquid state our result at 55 K is higher than that of Giauque and Johnston by about 3·5% whereas Fagerstroem and Hallett obtained values about 1% lower than ours.

A rather interesting thermal behaviour revealed by the Debye θ values deduced from these specific heats is shown in figure 5. The values in figure 5 were obtained

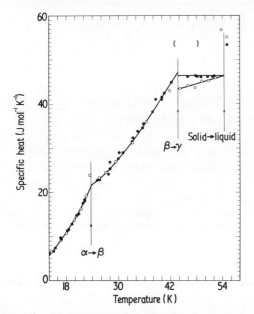

Figure 4. Molal heat capacities obtained for oxygen. ●, values of Giauque and Johnston (1929); ○, our values. The various phase transformations are also indicated.

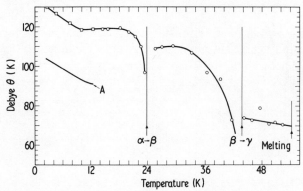

Figure 5. Variation of Debye θ. Curve A, values obtained by Fagerstroem and Hallett (1966) using molal heats; $\bigcirc$, values obtained from our experimental results for atomic heats; $\square$, curve A recalculated for atomic heats.

using atomic heats (one half the values of our molal heats in figure 4) and the Beattie (1926) Debye-function tables[†]. We see that in the α phase the Debye θ is constant from 10 K up to about 18 K and then begins to drop at an increasing rate, indicating that the specific heat is increasing in an abnormal manner[‡]. A similar effect occurs in the β phase, except that the region of non-constant θ, which begins at about 30 K, is spread over a much larger temperature interval. In the γ phase the θ values are nearly constant, changing linearly about 5·5% over the entire γ phase. The gradual increase in the specific heats in the α and β phases relative to the Debye specific heat, beginning much before the actual α–β and β–γ transformation temperatures respectively, seems to indicate that the actual transformation from one phase into the other takes place as if by gradual distortion of some of those lattice properties on which the θ values depend. The β phase can then be considered as a stable final product of the thermal distortion of the α phase. Similarly the β phase seems, by gradual distortion, to yield the γ phase. The γ lattice on the other hand remains reasonably well behaved until it is suddenly destroyed by melting.

A given phase is characterized by a set of properties such as lattice constant, elastic constant etc. On the other hand, figure 5 suggests that these phase transformations take place gradually over a relatively wide temperature range. Therefore the transformation temperature could perhaps be defined as that temperature at which unit energy will cause maximum amount of distortion of at least some of the phase properties. This temperature coincides with the peak of the specific heat anomaly. The latent heat, ie the energy necessary for transforming a given phase into another, can then be obtained by determining the total energy required to raise the temperature of the sample from an initial value, located within the region of θ = constant of the initial phase, to a final value located within the region of θ = constant in the other phase and subtracting the ideal lattice energies calculated with the appropriate θ = constant values of the two phases.

[†] Fagerstroem and Hallett used the molal heats directly in obtaining their values of θ below 14 K, hence their much lower θ values shown by curve A of figure 5.
[‡] Barrett and Meyer (1967) also commented that from crystallographic considerations the elastic constants are expected to change with increasing rapidity as the equal free energy of the α–β phases is approached.

5. Latent heats

The latent heats of the $\alpha \to \beta$ and $\beta \to \gamma$ crystalline transitions and of fusion were determined as discussed above. The total amount of heat q_t required to heat 0·0677 mol of oxygen from an initial temperature T_i in the θ = constant region of one phase to a final temperature T_f in the θ = constant region of the next phase is given in table 1. Of the

Table 1. Some latent heats of oxygen. q_t (J), the total heat supplied in joules, raised the temperature of the sample from an initial temperature T_i to a final temperature T_f. Of the total heat supplied, the sample container absorbed q_c and the well behaved sample would have absorbed q_s. Q_L is the latent heat obtained from $Q_L = (q_t - q_c - q_s)/0.0677$, where 0·0677 is the number of moles of the sample. $Q_{L,G,J}$ is the latent heat obtained by Giauque and Johnston.

	T_i (K)	T_f (K)	q_t (J)	q_c (J)	q_s (J)	Q_L (J mol^{-1})	$Q_{L,G,J}$ (J mol^{-1})
$\alpha \to \beta$	17·800	25·846	25·291	9·441	8·868	103·13	93·85
$\beta \to \gamma$	29·257	45·544	179·751	87·557	37·313	810·65	743·43
Fusion	52·365	54·769	65·836		36·109	439·10	444·97

total heat q_t, the sample chamber itself absorbed an amount q_c determined by numerical integration of the area under the heat capacity curve determined for the empty sample chamber within the appropriate temperature interval. The heat absorbed by the sample, q_s, was similarly obtained from heat capacity curves using the Debye θ values 119·5 K (T < 23·8 K), 110 K (23·8 K < T < 43·8 K), and those of figure 5 (43·8 K < T < 54·351 K).

In the liquid state the heat absorbed by the liquid sample plus the sample chamber was obtained by determining the area under the heat capacity curve determined for the sample chamber containing 0·0677 mol of liquid oxygen. Column 7 of table 1 gives the latent heat per mol, Q_L, deduced from $Q_L = (q_t - q_c - q_s)/0.0677$. These values differ by −1·3%, +8·7%, and +9·4% from those obtained by Giauque and Johnston for the heat of fusion, the β–γ and the α–β transformations respectively. The agreement for the heat of fusion is good. The large differences found for the crystal transitions are partly due to Giaque and Johnston including a portion of the latent heat in the lattice energy; they measured the latent heats within a temperature interval where the Debye θ has already decreased below the θ = constant value. Taking this into account, we should increase their values at the β–γ and α–β transitions by about 4%, leaving a disagreement of approximately 5%. It is quite possible that these residual differences are partly caused by differences in the experimental procedures employed in cooling the samples. Figure 2, for instance, strongly suggests probable differences in thermal behaviour of at least the β phase if different cooling speeds are employed in producing the low-temperature phases. We could have repeated some latent heat determinations for various cooling speeds had we not had to rely on He exchange gas for cooling. Unfortunately, fast cooling speeds unavoidably resulted in some sample condensation inside the sample transfer tube (perhaps as much as a few per cent) and thus made it pointless to attempt cooling-speed-dependent latent heat measurements. We note that our latent heats and heat capacities were measured for samples which were cooled at the slow rate of approximately $-1\cdot2\,°\text{C min}^{-1}$ below the β–γ transformation to avoid sample condensation along the sample transfer tube. At these slow cooling rates it was relatively easy to float the heat exchanger temperature (figure 2 of Ancsin 1973a, b) several degrees above the sample temperature. This, however, became increasingly difficult at higher cooling speeds, resulting in some unavoidable condensation outside the sample chamber. This sample loss, however, would not invalidate in any way the reliability of figure 2.

References

Ancsin J 1973a *Metrologia* 9 26
—— 1973b *Metrologia* 9 147
Barrett C S and Meyer L 1967 *Phys. Rev.* 160 694
Barrett C S, Mayer L and Wasserman J 1967 *Phys. Rev.* 163 851
Beattie J A 1926 *J. Math. Phys.* 6 1
Fagerstroem C H and Hallett A C H 1966 *Ann. Acad. Sci. Fenn.* A 6 210
Giauque W F and Johnston H L 1929 *J. Am. Chem. Soc.* 51 2300
Hoge H J 1950 *J. Res. NBS* 44 321
Hörl E M 1962 *Acta Crystallogr.* 15 845
Kemp W R G and Pickup C P 1972 *TMCSI* 4 part 1 217
Muijlwijk R 1968 *Thesis* University of Leiden

Inst. Phys. Conf. Ser. No. 26 © *1975: Chapter 2*

Silicon and gallium arsenide diodes for low-temperature thermometry

L Janšák, P Kordoš and M Blahová

Electrotechnical Institute of the Slovak Academy of Sciences, 809 32 Bratislava, Czechoslovakia

Abstract. The new types of semiconductor diodes prepared by epitaxial technology from silicon and gallium arsenide were measured in the temperature range 2–350 K. Effects of magnetic field, both transverse and longitudinal, up to 5 T in the temperature range 4·2–35 K were investigated for different operating currents. The repeatability was also tested at liquid helium temperature. The experimental results are compared with commercially available Si and GaAs thermometric diodes.

1. Introduction

The work reported in this note arose from the need for suitable sensors for wide-range thermometry in cryogenic systems, where apart from low cost, interchangeability and reproducibility of the sensors, a linear $V-T$ dependence in the low-temperature region was desired for electronic measurement and control systems. In this contribution we present some experimental results on the new types of semiconductor diodes which seem to be suitable for low-temperature thermometry.

2. Experimental samples and procedure

Two types of diodes were measured in our experiments: the new high-conductance silicon planar diode KB105 manufactured in Czechoslovakia for UHF applications and a GaAs epitaxial diode prepared in our Institute. Both diodes, silicon and gallium arsenide, are characterized by the small depletion width of the p–n junction and the relatively high conductivity base material.

The commercially available silicon diodes KB105 are made by a combination of gas epitaxial technology and diffusion technology. The planar structure of this diode consists of a high conductance silicon substrate with a thin epitaxially grown base layer. Conventional planar double-diffusion technology is used for preparing the p–n junction structure with a very sharp transition profile of the impurity concentration. After the contacting procedure the diode systems are hermetically encapsulated into a plastic mass; before encapsulating, the diode systems are protected by the boron silicate glass to prevent any environmental influence.

Experimental samples of the gallium arsenide diodes were prepared using an epitaxial-growth technique from the liquid phase. The technology of these diodes has followed from earlier experiments (Janšák and Kordoš 1974). A p-type layer Zn-doped with a hole concentration of about 10^{19} cm^{-3} and thickness of about 12 μm was grown on a ⟨100⟩-oriented n-type substrate which had an electron concentration of about 6×10^{16} cm^{-3}. The ohmic contacts were of metal alloy 50% Ag + 50% Sn evaporated on to the n-side and 96% Au + 4% Zn on to the p-side of the GaAs surface at 300 °C. After mesa etching and dicing the diodes were mounted on to a gold-plated sample holder with the gold-plated hex head cap screw case.

The diodes were measured in the temperature range 2–350 K at different forward currents. The temperature dependence of the forward voltage drop was measured using an Oxford Instruments variable-temperature cryostat. The temperature in the range 15–350 K was measured by the standard Pt resistance thermometer, below 20 K by the calibrated germanium resistance thermometer. Special care was devoted to the measuring circuitry to avoid any error due to the induced EMF. At certain temperatures in the range 4·2–35 K the diodes were measured in the presence both of transverse and longitudinal magnetic fields up to 5 T. The effect of magnetic field was measured in a superconducting solenoid with a variable-temperature insert unit.

The stability of the diodes was tested during a few months of cycling between room and liquid helium temperatures.

3. Experimental results

The temperature-dependence of the forward voltage drop for both silicon and gallium arsenide diodes in the temperature range 2–350 K is shown in figure 1. For simplicity only the $V-T$ characteristics at 100 μA exciting current are presented.

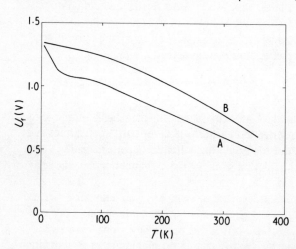

Figure 1. $V-T$ characteristic of Si (KB105, curve A) and GaAs (DT118, curve B) diodes.

The silicon diode KB105 is characterized by two practically linear parts in $V-T$ dependence. In the temperature range from 160–350 K, the sensitivity $S = dU_f/dT = 2\cdot06$ mV K^{-1} and in the temperature range 4·2–20 K, the sensitivity is about 8·6 mV K^{-1}.

The diode made from epitaxial GaAs exhibits a steadily increasing voltage drop with decreasing sensitivity down to low temperatures, where a linear $V-T$ characteristic appears below 30 K. The sensitivity at room temperature of about 2·72 mV K^{-1}

decreases to a sensitivity of about 0·4 mV K^{-1} at low temperatures and remains constant in the temperature range 4·2–28 K.

The temperature dependence of the sensitivity for both silicon and gallium arsenide diodes at 100 μA is shown in figure 2. The detailed parts of the $V-T$ characteristics below 30 K for different exciting currents are shown in figures 3 and 4. Both diodes have the sensitivity practically independent of the operating current in the measured range 10–200 μA.

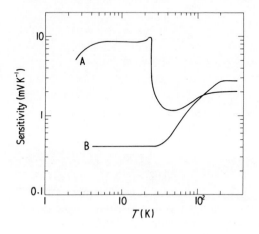

Figure 2. Sensitivity versus temperature of Si (KB105, curve A) and GaAs (DT118, curve B) diodes.

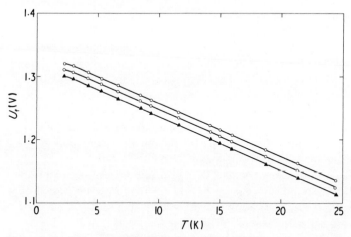

Figure 3. $V-T$ characteristic of the silicon diode KB105 at low temperatures for different operating currents. I_f = 200 μA (full circle), 100 μA (open circle) and 50 μA (full triangle).

Owing to the low-resistivity n-type material used in silicon diodes fabrication and high-density planar technology (a few thousand diode systems are created on one silicon wafer), the low-temperature forward voltage drop dispersion is relatively small. The set of 25 silicon diodes has the mean value of voltage drop at 100 μA exciting current and at liquid helium temperature, equal to 1·3100 V ± 20 mV, while at liquid nitrogen this

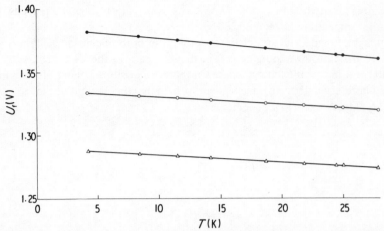

Figure 4. $V-T$ characteristic of the gallium arsenide diode DT118 at low temperatures for different operating currents. I_f = 200 μA (full circle), 100 μA (open circle) and 50 μA (open triangle).

value is almost identical and equal to $1 \cdot 0415$ V $\pm$ 0·5 mV. The sensitivity in the temperature range 4·2–20 K is in the range 8·8 mV K^{-1} $\pm$ 1·2 mV K^{-1}. The mean values vary slightly from one silicon wafer to another. At low temperatures the dependence of the voltage drop on the exciting current dU_f/dI_f is of the order 10^{-2} and so the requirements on current setting and measurement are of about two orders smaller than on voltage-drop measurement.

Preliminary measurements of the reproducibility of silicon diodes KB105 after cycling between the room and liquid helium temperatures for several months have shown a stability of about ±0·02 K at liquid helium temperature.

The epitaxial GaAs diodes exhibit, at low temperatures, a relatively small sensitivity of 0·4 mV K^{-1} due to the highly doped p-side. The sensitivity can be increased up to 0·6 mV K^{-1} or more using a slightly lower impurity concentration in the p-type layer. The reproducibility of GaAs epitaxial diodes after a few months at liquid helium temperature is worse than that of silicon diodes (about 0·1 K) but the present work is towards improved time and temperature cycling stability.

The diodes, both silicon and gallium arsenide, were tested in the presence of a magnetic field of up to 5 T in the temperature range 4·2–35 K. The forward voltage drop of the silicon diode KB105 is very sensitive to the magnetic field in the liquid helium range. The effect of a magnetic field is smaller with increasing temperature and is correlated with the temperature dependence of the forward voltage drop. The effect of the magnetic field is practically independent of the exciting current while strong dependence on the magnetic field direction was observed.

On the other hand, the influence of magnetic field on GaAs epitaxial diodes is practically negligible up to 2·5 T (the temperature error at 2·5 T and 4·2 K is less than 0·1 K). The effect of a magnetic field is slightly temperature dependent in the temperature range 4·2–35 K.

At 4·2 K the ratio of the induced voltage change between transverse and longitudinal magnetic fields is about 4 for the silicon diode KB105 while for the gallium arsenide epitaxial diode this ratio is 1·5. As an illustration the relative voltage change at 5 T versus temperature in the range 4·2–35 K for both diodes is shown in figure 5.

The temperature error due to the magnetic field on the silicon diode KB105 is very large so these diodes can be used only in the absence of a magnetic field. The very small effect of a magnetic field up to 2·5 T on GaAs epitaxial diodes, and isotropy and their temperature independence, could find applications in some special cases.

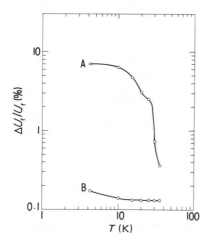

Figure 5. Relative forward voltage change versus temperature at magnetic field 5 T for the silicon diode KB105 (curve A) and the gallium arsenide diode DT118 (curve B).

4. Conclusion

Measurements on the commercially available silicon diode KB105 and the laboratory-prepared epitaxial gallium arsenide diode DT118, both characterized by the sharp transition profile of their impurity concentrations and their high conductivity bases, have shown a linear $V-T$ dependence below 30 K. This result is in agreement with the experimental data on the silicon high-conductance diodes 5FDJ1000 described by Sclar and Pollock (1972).

Although the sensitivity of the diodes described in the previous text is smaller when compared with the commercial diode thermometers type DT500 or TG100 (LSC 1973), their practically constant sensitivity and $V-T$ linearity in the LHe and LH$_2$ range are promising for applications in electronic systems reading temperature directly. Moreover, the interchangeability of the silicon diodes KB105 and their low cost are very important features for their use in a wide range of applications in cryogenic thermometry.

However, the relatively strong effect on them of magnetic fields makes this type of diode unsuitable for temperature measurements in the presence of such fields.

References

Janšák L and Kordoš P 1974 *Czech. J. Phys.* **24** 364
LSC 1973 *Lake Shore Cryotronics short form catalog*
Sclar N and Pollock D B 1972 *Solid St. Electron.* **15** 473

Realization of the IPTS-68 between 54·361 and 273·15 K and the triple points of oxygen and argon

F Pavese
Istituto di Metrologia 'G Colonnetti', Torino, Italy

Abstract. The IPTS-68 between 54·361 and 273·15 K has been realized at the Istituto di Metrologia 'G Colonnetti' (IMGC) using the more practical triple point of argon. The equipment is described. The triple point cell is fully independent of the cryostat, and one cell has been used for each gas. The gas can be sealed in the cell; the materials and the special assembling techniques used make it possible to maintain gas purity for a long time. An outline of the studies on systematic errors due to thermal equipment and gas impurities is given for both gases.

1. Introduction

High-precision pressure measurements are necessary to realize boiling points, two of which are required for calibration on the IPTS-68 between 54·361 and 273·15 K. This complicates measurements in a temperature range where the instrumentation can otherwise be extremely simple and economical, especially when only liquid nitrogen is used as refrigerant.

This range of the scale has therefore been realized at IMGC with the following fixed points, which do not require pressure measurements:

(i) *Freezing point of tin* (505·1181 K). To obtain $W(100\,°C)$ a mean value of $-5·8868 \times 10^{-7}\,°C^{-2}$ (Furukawa *et al* 1973) was used for parameter B (CIPM 1969) as measurements at the freezing point of zinc are inconvenient. The uncertainty with which this value is known produces a ±1 mK uncertainty at 100°C. Since the accuracy of measurements made at 505·1181 K, with capsule thermometers mounted in a long stem, is ±0·5 mK[†] (and this is reflected in a ±0·2 mK uncertainty at 100°C), it follows that this method does not yield on the whole a lesser accuracy (±1 mK) than that obtainable by direct realization of the boiling point of water.
(ii) *Triple point of water* (273·16 K). If precautions like those described by Sparks and Powell (1971) are taken, a ±0·2 mK accuracy is obtained, which is the ultimate value now established by the accuracy of resistance measurements (1 ppm).
(iii) *Triple point of argon* (83·798 K). The temperature of this point, which is proposed as an alternative to the boiling point of oxygen, is sufficiently near the latter, so that errors higher than 0·3 mK (Furukawa *et al* 1971) are not produced when it is simply replaced in the IPTS-68 definition.
(iv) *Triple point of oxygen* (54·361 K). This is a defining point of IPTS-68.

[†] The freezing point of tin is reproduced at IMGC with a ±0·2 mK accuracy; the triple point of water with a ±0·1 mK accuracy.

The accuracy of the IPTS-68 below 0 °C so far ascertained is not better than 1 mK (Tiggleman 1973); the propagation of an error of 1 mK at a fixed point over the scale is between 0·2 and 0·5 mK (Furukawa *et al* 1973, Bedford and Kirby 1969, Bedford and Ma 1970), except in limited intervals, where, essentially, it does not exceed 1 mK.

A 1 mK accuracy appears therefore sufficient to obtain fixed points at low temperatures; in fact, although the stated 'accuracy' is generally far better, the agreement between values obtained even recently in several laboratories is not better than 1 mK for each individual fixed point.

This very fact, however, shows that systematic errors are difficult to overcome, and encourages stricter experimental control leading to unnecessary reproducibility.

On the other hand, some simplified realizations (Furukawa *et al* 1971, Cataland and Plumb 1971, Bonnier 1974) show that it is possible to obtain results whose accuracy is not far from the required millikelvin.

The research undertaken at IMGC was started on a very complex apparatus for checking the greatest possible number of parameters, in order to arrive at a rationally simplified solution, with an apparatus which was inexpensive and easy to handle.

The apparatus is at present made of three independent units: *a thermostat* (the cryostat and its accessories), *a measurement cell* (placed in the cryostat, containing the thermometric gas, where measurements are taken), *a system for manipulating the thermometric gas* (the parts linking the extrapure gas source to the measurement cell).

In the relevant literature, the measurement cell is considered as a part of the cryostat, and the gas-handling system a mere accessory. According to our experience, the causes of the success or failure of an experiment are concentrated in these parts.

At the end of the investigation, the cell has become self-contained, as it holds the thermometric gas permanently and can be easily immersed directly into the coolant Dewar vessel.

A description is given of the various devices used to obtain the triple point of the gases, and results that show their functional characteristics are reported very briefly and schematically.

2. The cryostat

The superiority of an adiabatic cryostat for measuring fixed points is shown by recent studies (Ancsin 1973a,b, Compton 1971, Tiggleman 1973), besides having been known for some years to the specialists who make thermal analyses of purity (Westrum *et al* 1968; McCullough and Waddington 1957).

The cryostat of figure 1 is of this type. Its main characteristic derives from the necessity of making the measurement cell independent and removable from above without having to open the cryostat or warm it up to room temperature.

The cryostat has therefore a re-entrant well, of 25 mm diameter. The two concentric shields inside the 'submarine' are maintained at the same temperature as the measurement cell through automatic regulation, with a sensitivity of approximately 1–2 mK for the two fixed points being investigated.

Vacuum in the cryostat is always better than 10^{-3} Pa (10^{-5} Torr).

Nitrogen is used as coolant, either in its liquid state or, at about 47 K, as the solid;

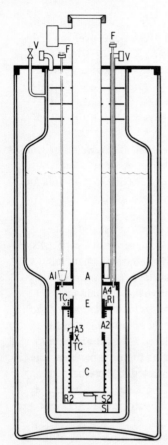
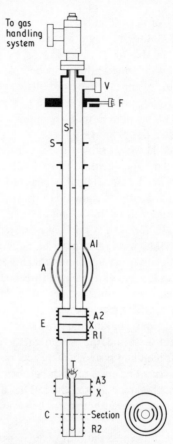

Figure 1. *Triple points cryostat* (not to scale). In the re-entrant well of the cryostat the cell is housed in C, the heat exchanger in E and the thermal anchor in A. V, vacuum fitting; F, electrical feed-through; A1 to A4, thermal tie-down for cryostat leads; on the right-hand side, the leads for thermocouple TC and thermometer T used for automatic temperature regulation enter the cryostat through a vacuum-jacketed tube; on the left side, those for heaters R1, R2; S1 and S2, copper shields.

Figure 2. *Measurement cell* (not to scale). C, copper container for the liquid–solid mixture; for Ar only, it contains three copper shields placed as shown. The diameter reduction allows reduction to 5 cm³ of the liquid amount necessary to fill the can to indicated level. E, copper heat exchanger; A, thermal sliding anchor (cage of wires brazed to two copper rings); S, radiation trap; F, electrical feed-through; V, vacuum fitting; A1 to A3, thermal tie-down for cell leads; R1 and R2, heaters; X, differential thermocouple. The tube above the heat exchanger is vacuum-jacketed to avoid cold spots during cooldown.

an autonomy of 40 h is obtained in the latter case, owing to the particular shape of the inner Dewar vessel.

The temperature of the shields was found not to be critical; its influence on the value of T_{TP} is very slight, as the thermal flow going from the cell to the cryostat is of the order of microwatt.

As a matter of fact, most of the errors of thermal origin are not to be ascribed to the cryostat, but to the measurement cell.

3. The measurement cell

The cell (figure 2) includes all the parts in contact with the thermometric gas that must be immersed into the cryostat well.

The mixture of the three phases is condensed in a OFHC copper can C, weighing 14 g, where one thermometer is immersed via a pass-through tube also made of copper.

Owing to the small thermal mass of the can, the temperature is able to adjust itself rapidly to that of the gas contained in it. Five cm^3 of liquid fill the cell up to the level indicated in the figure (ie 1 cm from the top of the can).

The diameter of the tube above the cell (8 mm, except for the eccentric part, where it is 4 mm; conductance is about $0.11 s^{-1}$) is large, in order to keep a realistic pumping speed, and therefore requires an efficient thermal trap, consisting of a heat exchanger E of massive copper, through which the gas circulates.

The exchanger temperature is regulated to that of can C by means of a triple differential thermocouple with thermoelectric power of 105 or 130 $\mu V K^{-1}$ for O_2 and Ar respectively, reproducible within ±0.15 μV (Pavese et al 1972).

As the cell must be removable, the whole of the measurement wires are fastened thereto. They are spiral-coiled between E and C for a total length of 0.5 m. They come from feed-through F and must be thermally tied-down at the coolant temperature before reaching the heat exchanger. The problem has been solved with a sliding thermal tie-down A consisting of a cage of silver-plated Cu—Be wires (Pavese 1975).

The overall thermal flow reaching can C can be kept within a few tens of microwatts.

Utmost care is given to guarantee that every part in contact with the gas is perfectly clean. In fact, each square metre of effective area, corresponding to 0.2–0.3 m^2 of geometrical area can introduce into the 5 dm^3 of the thermometric gas used about approximately 40 ppm of impurities per layer of gas desorbed. The cell conductance, on the other hand, is too low for effective outgassing through pumping (see also Guildner and Edsinger 1973).

All the metal parts were therefore machined without lubricating oils, cleaned mechanically with microbead blasting and vacuum-brazed at 900 °C without using fluxes[†]; the cell, then, can still be outgassed up to 700 °C. It can be sealed with an ultra-high vacuum all-metal valve; leaks result below 10^{-11} Torr l s^{-1} (in 5 dm^3 at 10^{-7} Torr l s^{-1} leakage would produce a pollution of 2 ppm d^{-1}).

The cell, where the volume of the vapour phase would be approximately 20 cm^3, is fitted with a 1 dm^3 ballast to contain the gas permanently.

4. The gas inlet system

The system is shown in figure 3. To eliminate the contamination produced by gas desorption, it is made up of stainless steel ultra-high vacuum commercial parts, of 16 mm internal diameter, that can be outgassed up to 450 °C; leaks are less than 10^{-10}

[†] These operations were performed in cooperation with Euratom—CCR, at Ispra (Italy).

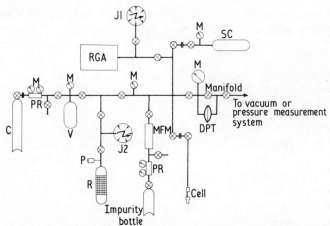

Figure 3. *System for ultra-pure gas inlet.* C, ultra-pure gas cylinder; PR, pressure regulator; V, 1 dm^3 calibrated volume; J1 and J2, ionic pumps; R, rough pump; P, Pirani head; M, manometer; SC, sampling and recovery cylinder; RGA, residual gas analyser; DPT, differential pressure transducer; MFM, mass flow meter.

Torr 1 s^{-1}, including manometers and pressure regulators. To avoid contamination, the pumping system uses ionic getter pumps with zeolite rough pumps.

The system is separable from the cell without having to open it to the atmosphere. The following operations are possible with this system:

(i)   condensation in the cell of a measured amount of gas from the cylinder,
(ii) gas recovery in another container SC by cryopumping,
(iii) linking to pressure gauges through a differential diaphragm transducer,
(iv) gas monitoring.

This last operation is performed with a residual-gas quadrupole mass spectrometer for masses up to 100 amu mounted on-line. Monitoring of the residual atmosphere is thus possible during outgassing cycles (Pavese *et al* 1973), after which a static vacuum better than 10^{-3} Pa (10^{-5} Torr) is obtained in the system with a background of few stable peaks.

Use of this spectrometer was also attempted in direct analysis of impurities, but so far a sensitivity better than 100 ppm has not been obtained with the more favourable peaks, owing to the insufficient ratio between the admissible maximum and minimum pressures.

With oxygen, a method of impurity concentration was devised, using the reaction with yellow phosphorus; the study was done with the cooperation of another laboratory (De Paz *et al* 1974). In all the other cases, the analyses were performed with chromatographic techniques in several laboratories.

5. The measurements

Measurements of thermometer resistance were made with a Kusters thermometric bridge capable of measuring resistance ratios with a relative accuracy of 1×10^{-7}†.

† Guildline Instruments, model 9970, Smith Falls, Canada. The effective precision so far obtained is limited to 1 ppm.

1 Ω and 10 Ω standard resistors were used in measurements, directly traceable to ohm-Italia[†]; calibrations made during the past four years show their reproducibility to be within 1 ppm. The resistors are maintained in a passive thermostat, where temperature variations are always much lower than $0.1\ °C\ h^{-1}$; the mercury thermometers that measure their temperature are also calibrated with an accuracy of $\pm 0.02\ °C$. At both the triple points considered, the inaccuracy introduced by resistance measurements is at present $\pm 50\ \mu K$.

The thermometers are calibrated as well in the NBS–IPTS-68 scale, and one of them was included in the recent international comparison at NPL (Compton and Ward 1975).

Self-heating of thermometers has always been checked, and it was found to be under or equal to 0·1 mK, with 1 mA current.

The melting plateaux were continuously recorded on paper, with a sensitivity of 100 or $100\ \mu K$/division; this makes safer measurement evaluations possible with respect to intermittent measurements.

Intermittent heating of the cell was used to evaluate overheating and the liquidus fraction produced. After each heating phase sufficient time was allowed for the equilibrium state to be restored, so as to avoid temperature variations higher than $20\ \mu K$.

The gas was cooled to a maximum of 7 K for O_2 and about 30 K for Ar below the triple point. With O_2 no evidence was found of the glassy state mentioned by Kemp and Pickup (1971); supercooling is about 1 K for O_2 with very fast recalescence, and 0·1 K maximum for Ar with slower recalescence.

6. Summary of results

6.1. Purity of thermometric gas

In thermal analysis (Westrum et al 1968, Smit 1957, McCullough and Waddington 1957) it is common practice to plot temperature against the inverse of the melted fraction F; the molar fraction of impurities is given by the product of the curve slope and the cryoscopic constant $K = -H_m/RT_{TP}^2$, where H_m is the melting heat, R the gases constant, and T_{TP} the measured temperature. The value of the triple point for the impure substance is obtained for $1/F = 1$ (liquidus point) and that of the pure substance, for $1/F = 0$. As regards the two gases examined, however, the two fundamental hypotheses of the method[‡] are not valid for many of the impurities, as can be seen in figure 4 and 5 ($K_{O_2} = -0.018\ K^{-1}$, $K_{(Ar)} = -0.020\ K^{-1}$). They also explain and confirm some facts observed in recent investigations: for example, Ar in O_2 does not produce a deformation in the plateau (Ancsin 1973a, b), despite a temperature variation, due to the presence of the peritectic and lack of separation of the liquidus and solidus lines. Kr in Ar does not produce appreciable variations when in small concentration. Only Ar in O_2 (and Kr in Ar) are found to produce an increase in the triple point temperature.

It can be said that the representation T vs $1/F$ is at all events much more suitable to

---

[†] Ohm-Italia, maintained at the Istituto Elettrotecnico Nazionale 'G Ferraris', Torino. The difference to the absolute ohm is lower than $1\ \mu\Omega$.

[‡] Hypotheses: that Raoult's law is valid, and that impurities do not form solid solutions with the major component.

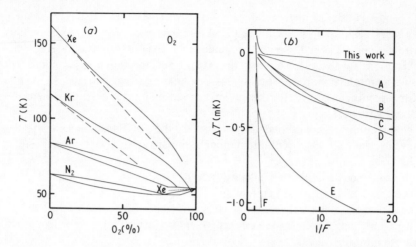

Figure 4. *Oxygen.* (a) Solid–liquid equilibrium diagrams for binary mixtures of O_2 with: Xe (Stackelberg 1934, Ancsin 1970); Kr (Stackelberg 1934); Ar (Din and Goldman 1955); N_2 (Barrett *et al* 1968). (b) Melting plateau (typical slope of 30 plateaux) obtained in present work, compared with other previous studies. F, liquidus fraction formed. The position of the curves with respect to the ΔT axis is arbitrary and does not imply comparison of their values for $1/F = 0$. (A) Ancsin (1973a); (B) Soejima *et al* (1964); (C), (D) Ancsin (1973a); (E) Ancsin (1971); (F) Furukawa and McCoskey (1953).

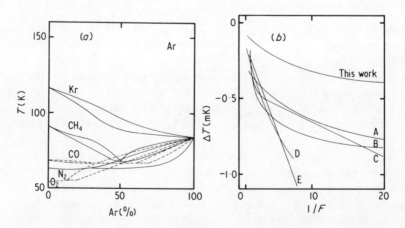

Figure 5. *Argon.* (a) Solid–liquid equilibrium diagrams for binary mixtures of Ar with: Kr (Heastie 1955); CH_4 (Veith and Schröder 1973); CO (Barrett and Meyer 1965); N_2, O_2 (Din and Goldman 1955). (b) Melting plateau (typical slope of 30 plateaux) obtained in present work, compared with other previous studies. F, liquidus fraction formed. The position of the curves with respect to the ΔT axis is arbitrary and does not imply comparison of their values for $1/F = 0$. (A), (B), (C), Ancsin (1973b); (D) Ancsin and Phillips (1969); (E) Flubacher *et al* (1961).

represent the plateau than the traditional T vs F; in fact, it always admits of extrapolation of $1/F = 0$† and shows that this is made possible essentially by the plateau part between $F = 5-10\%$ and $F = 50\%$.

Figures 4(b) and 5(b) show the typical behaviour of the plateau obtained at IMGC for the two gases, as compared to the results of other previous studies. Unfortunately, the data of few of them can be transformed into this form. A much reduced slope was obtained for both gases, a larger one with Ar. The figure of impurities that might be deducted is insignificant; this means that the slope obtained depends completely upon the magnitude of the solid solutions formed, especially in the case of O_2.

The gases used, all of commercial type and contained in 1 dm^3 cylinders, had a nominal impurity content under 20 ppm for O_2 (almost all N_2 and Ar) and 10 ppm for Ar (essentially N_2 and O_2), as certified by a number of analyses.

It was found that the apparatus did not introduce any alteration in the thermometric gas. A sample of O_2 used over three months to make nine plateaux (recovered in SC, figure 3, after each plateau) did not show any systematic variation of T_{TP}. The same gas, kept in SC for over a year, did not show any variation in the N_2 content (13 ppm) in the course of two independent analyses made at about six months' interval. Also the gas of the original cylinder did not exhibit variations in the same period of time.

6.2. Thermal behaviour of the cell

Because of the finite thermal conductivity (of both solid and liquid gas) every thermal flow passing through the cell produces stationary or variable temperature gradients. The former are particularly to be feared, as they are hardly distinguishable from the adiabatic condition and can introduce systematic errors.

The heat exchange with the isothermal shield S2 can be approximately assimilated to the heat supplied by the cell heater and be reduced to a negligible amount with an adequate degree of vacuum.

The heat from the exchanger comes essentially through measurement wires, which are therefore made very long; the power thus transmitted is reduced to about 0·3 mW per kelvin of overtemperature of the exchanger‡, so that it does not produce noticeable influence below 0·1 K of overtemperature.

It must be remarked that this influence is not constant, but increases noticeably with the liquidus fraction increasing, as already observed (eg by Tiggelman 1973), so that it may be the cause of a part of the plateau slope, especially for F higher than 50%.

Nevertheless, the magnitude of the effect produced by these thermal flows on the temperature values of the triple point depends completely upon cell design, as it ultimately depends on the distribution of solid and liquid inside the cell. This allows one to explain how in some cases it was possible to obtain remarkably flat plateaux with the method of continuous heating (Furukawa et al 1971, Bonnier 1974, Soejima et al 1964).

A fact directly or indirectly ascertained by all experimenters is that the liquidus phase does not determine a well-defined temperature, but tends to get overheated. This

† This represents a precise criterion for defining T_{TP}, that otherwise may be left undefined for several tenths of thousands of kelvin.
‡ The cell must be maintained in conditions of heat in-flows, if gas condensation is to be avoided outside it.

depends on the position of the liquid with respect to heat sources and will condition the relationship between the temperature of the solidus–liquidus interface and the temperature measured by the thermometer.

The two models of cell used (figure 2) therefore show a different behaviour; with O_2 a sudden overheating increase during cell heating appears between $F = 50$ and 70%, followed by an imperfect recovery of the temperature that causes the sharp jerk upwards of the plateau shown in figure 4(b). This could be explained by a complete detachment of the solidus inside the thermometer well. In the case of Ar instead, the three internal shields decouple the thermometer from the heater; in this way the overheating curve and the plateau keep regular up to more than $F = 90\%$.

The time for temperature recovery after each heating cycle always resulted in asymptotic behaviour (much higher for Ar than O_2). Overheating was comparable. Pressure measurements with O_2 showed that the recovery time is related to pressure equilibrium, which takes longer to establish itself. Pressure may still not be in equilibrium when temperature drift is reduced to apparently irrelevant values. This fact, if added up cycle after cycle, can bring about a substantial non-equilibrium of pressure with an error in excess of T_{TP}, and contribute to raising the plateau slope and the apparent temperature of the liquidus point.

Recovery time is believed to be related to the fact that overheating of the cell increases the equilibrium pressure above the triple point on the vapour pressure curve, and causes evaporation of a liquidus fraction proportional to dP/dT on this curve. This fraction must then condense again during the non-heating period[†]. This effect is extremely limited with O_2 where P_{TP} is very low. With Ar, it is much greater (it was noted that the rate of temperature change varies as well during cool-down of solid Ar if the cell is sealed or left connected to a volume of about $1 \cdot 7$ dm^3).

The amount of condensation heat involved is incalculable, as the vapour phase is not isothermal; with Ar it is approximately equivalent to a 0·01% variation of F per millikelvin of overheating with respect to T_{PT}, and per cubic decimetre volume of the vapour phase.

7. Conclusions

The experimental apparatus described yielded satisfactory results, whose main points only could be reported here. It is possible with them to maintain with absolute certainty the certified impurity levels in the gas being investigated, even over long periods of time, and therefore to separate thermal from impurity-produced errors.

The reproducibility obtained in experiments similar to those reported in the literature is of about ±0·2 mK.

However, the analysis of the accuracy figure obtained for both gases cannot be said to be finished; in fact, if it is true that most impurities tend to lower T_{TP}, a number of thermal errors might increase it, and they have not been yet made sufficiently clear.

[†] The difference of thermal conductivity between Ar ($K_1 = 1 \cdot 32$ mW cm^{-1} K^{-1}) and O_2 ($K_1 = 1 \cdot 90$ mW cm^{-1} K^{-1}) cannot be said to explain by itself the different behaviour (Ancsin 1973a, b).

References

Ancsin J 1970 *Metrologia* 6 53–6
────── 1971 *TMCSI* 4 part 1 211–16
────── 1973a *Metrologia* 9 26–39
────── 1973b *Metrologia* 9 147–54
Ancsin J and Phillips M J 1969 *Metrologia* 5 77–80
Barrett C S and Meyer L 1965 *J. Chem. Phys.* 43 3502–6
Barrett C S, Meyer L, Greer S C and Wasserman J 1968 *J. Chem. Phys.* 48 2670–3
Bedford R E and Kirby C G M 1969 *Metrologia* 5 83–7
Bedford R E and Ma C K 1970 *Metrologia* 6 89–94
Bonnier G 1974 *Cellule à point triple d'argon CCT* doc 18
Cataland G and Plumb H H 1971 *TMCSI* 4 part 1 183–93
CIPM 1969 *Comité International des Poids et Mesures, IPTS-68, Metrologia* 5 35–44
Compton J P 1971 *TMCSI* 4 part 1 195–209
Compton J P and Ward S D 1975 this volume
De Paz, Pavese F and Pilot A 1974 *Comptes Rendus Journées de Technologie du Vide, Versailles*
Din F and Goldman A G 1955 *9th Congr. Int. Froid, Paris*
Flubacher P, Leadbetter A J and Morrison J A 1961 *Proc. Phys. Soc.* 78 1449–61
Furukawa G T, Bigge W R and Riddle J L 1971 *TMCSI* 4 part 1 231–43
Furukawa G T and McCoskey R E 1953 *NACA Rep.* TN2969
Furukawa G T, Riddle J L and Bigge W R 1973 *J. Res. NBS* 77A 309–32
Guildner L A and Edsinger R E 1973 *J. Res. NBS* 77A 383–9
Heastie R 1955 *Annexe Bull. Inst. Int. Froid* 3 324–8
Kemp W R G and Pickup C P 1971 *TMCSI* 4 part 1 217–23
McCullough J P and Waddington G 1957 *Anal. Chim. Acta* 17 80–96
Pavese F 1975 *J. Phys. E: Sci. Instrum.* 8 508–11
Pavese F, Bonaudo L and Limbarinu S 1972 *La Termotecnica* 26 546–52
Pavese F, Pilot A and Cagna G 1973 *Atti IV Congr. AIV St Vincent* (Milano: AIV) pp65–71
Smit W M 1957 *Anal. Chim. Acta* 17 23–35
Sparks L L and Powell R L 1971 *TMCSI* 4 part 2 1415–21
Soejima T, Takahashi M and Sawada S 1964 *CCT* doc 15 Annexe 2 T29-33
Stackelberg H 1934 *Z. Phys. Chem.* A170 262–72
Tiggelman J L 1973 *Thesis* Leiden 1–162
Veith H and Schröder E 1937 *Z. Phys. Chem.* A179 16–22
Westrum E F Jr, Furukawa G T and McCullough J P *Experimental Thermodynamics* (London: IUPAC, Butterworths) vol 1 pp 133–214

Resistance thermometry

J S Johnston

Rosemount Engineering Company Limited, Durban Road, Bognor Regis, Sussex

Abstract. The author makes reference to resistance thermometry in the definition of the International Practical Temperature Scale and goes on to summarize the characteristics of low-temperature resistance thermometers of the carbon, germanium and thermistor type. The construction, accuracy, reliability, temperature versus resistance characteristic, and application of the platinum resistance thermometer element are next discussed. Comparisons are made of the relative merits of wire-wound and film platinum resistances from the aspect of production and of performance. Descriptions, supported by illustrations, are given of methods of construction and the mechanical support necessary for the temperature-sensing element to endure the rigours of an industrial environment particularly one where high vibrations are encountered. Further consideration is given to the use of elements which are required to have a fast time response while operating in considerable pressures under high vibration conditions.

1. Introduction

In a conference like this the form of resistance thermometer that comes at once to mind is the platinum thermometer used as an interpolation instrument in the definition of the International Practical Temperature Scale.

In fact, of course, platinum resistance thermometer elements are produced and used in industry in numbers of the order of several million per annum; if the thermistor is included, the number of resistance thermometers sold per year probably rises to the order of ten million. It is therefore a fairly common measuring device used in a wide range of industrial and scientific applications.

The greater part of this paper then will deal with thermometer elements for this wide field of practical application although I shall also deal briefly with their use in laboratory measurement.

The very great importance of low temperatures in theoretical and practical physics results in a large proportion of papers on resistance thermometry dealing with cryogenic measurements. In contrast very few industrial devices are used in this temperature range since, so far as I am aware, there is no significant application in industry of temperatures below the normal boiling point of nitrogen.

The industrial applications then range from nitrogen liquefaction through the carriage of liquefied natural gas at sea, the medical applications of thermistors to the juddering steam-lines of a large modern power station.

2. Resistance thermometry and the IPTS

The place occupied by the platinum resistance thermometer in the definition of the IPTS is too well known to require re-stating here. The extension in the 1968 scale of the range covered by this type of thermometer down to the equilibrium point of hydrogen went part way to meet the desire many of us had to see a single type of interpolation instrument used from 13 K to 1300 K.

The upward extension of its use to the gold point, however, was not incorporated into the IPTS although it now seems clear that the resistance thermometer has the necessary repeatability to show a very substantial improvement over the standard thermocouple in this region. At the Washington conference in 1971 a session was set aside for the discussion of this subject (Anderson 1972b, Chattle 1972, Evans 1972, Sawada and Mochizuki 1972); the protagonists clearly felt that an improvement by a factor of 5—10 in reproducibility was case enough for the adoption of the platinum resistance thermometer while others felt that further study of the mechanisms of the remaining uncertainties was of higher priority. Anderson identified the platinum—oxygen reaction as one particular area requiring study and that was reported on by Berry (1975) at this conference. His work is mainly concerned with temperatures up to 630 °C but there is little doubt that it will be important at higher temperatures also. The implication that we should avoid even very low partial pressures of oxygen in precision thermometers is a very strange one to those of us who have always regarded an oxygen-rich atmosphere as quite essential for stable behaviour in the presence of most metals and oxides.

3. Special low-temperature resistance thermometers

Below about 50 K a variety of materials are used for resistance thermometry as the platinum thermometer becomes inconvenient because of its low resistance and rather large size. They have been dealt with in detail by M Durieux (1975) earlier in this conference but for completeness I shall mention them briefly.

3.1. Carbon resistance thermometers

Despite, according to most workers, a poorer stability than the germanium thermometer, carbon thermometers continue to find application in cryogenics because of their ready availability and low magnetoresistance, although the definition of appropriate interpolation equations continues to provide scope for ingenuity (Kes *et al* 1974, Whitehead 1974).

Goer *et al* (1974) have described an interesting technique for calibration in the range up to 2 K using the heat capacity of holmium. Measurements on carbon resistors at helium temperatures in magnetic fields up to 15 T were reported by Sample and coworkers (Sample *et al* 1974, Sample and Neuringer 1974).

A useful review of carbon thermometry was published by Anderson (1972a) and of the properties of Allen—Bradley resistors by Weinstock and Parpia (1972).

A comparison of various cryogenic resistance thermometers, including carbon, and diode thermometers, was given by Swartz and Swartz (1974).

3.2. Germanium thermometers

The germanium thermometer has been used to form the basis of several working temperature scales in the region below 30 K and a comparison of some of these was described by Kemp et al (1972). Some practical aspects of germanium thermometers were reviewed by Halverson and Johns (1972) including some features of construction. At the same conference further detailed calibration work on these devices was reported by workers from the Kamerleigh Onnes laboratory (Van Rijn et al 1972) and curve-fitting techniques were explored by Blakemore (1972) and by Collins and Kemp (1972).

The work mentioned above served to indicate that the germanium thermometer could, with careful calibration, provide temperature measurements to millikelvin accuracy over a very useful range of temperatures but emphasizes the need for care in choosing appropriate curve-fitting techniques. A strong plea for the now fashionable method of fitting using cubic splines is made by Greenfield et al (1974).

The problems of lead and contact resistance variation and self-heating have led to an increasing interest in AC bridge techniques for measurements on these devices while recent improvements in the performance of integrated circuit operational amplifiers have permitted their use in bridge circuits not only as detectors but also as buffers to prevent loading of bridge components (Rubin and Golaling 1972). Some caution is required however, in work of the highest precision as a result of the differences between AC and DC resistances measured by a number of workers (Rusby et al 1972, Swenson and Wolfendale 1973).

4. Thermistors

The problem of what to include and what to exclude is particularly difficult in the field of thermistors: should all papers recording the resistance–temperature relationship of new mixtures of metal oxides be included? I have decided against their inclusion in the full knowledge that ten years from now it will seem incredible that the first disclosure of a revolutionary new thermometer material was not recognized at once. My feeling is, however, that what thermistors require is quiet consolidation rather than further revolutionary new devices.

Some manufacturers have announced devices closely interchangeable with one another at temperatures up to 300 °C; it remains to be seen to what extent these same characteristics will be taken up by other suppliers and hence whether some degree of standardization will become possible.

The highly nonlinear characteristics of the thermistor can be inconvenient in some applications but its high sensitivity makes it the obvious choice for narrow temperature ranges. It is thus not surprising that the majority of papers dealing with thermistors at the 1971 Washington conference were put into the volume dealing with biological and medical applications. Schlosser and Munnings (1972), however, reported remarkably good reproducibility of cryogenic thermistors above 3 K and claimed reproducibility of around 0·5 mK. Over the more usual temperature range of applications Trolander et al (1972) reported good stability and reproducibility for commercial devices over periods up to eight years. They also indicated methods of linearizing the resistance–temperature characteristic using additional resistors and multiple thermistor arrangements. A variety of linearizing circuits using operational amplifiers were described by Broughton (1974) and a further version by Stockert and Nave (1974).

A useful review of thermistors and their applications (not restricted to their use in

Resistance thermometry

temperature measurement) up to about the year 1969, is contained in a book by Hyde (1971).

Perhaps the most significant development in the past few years has been the extension of commercial thermistor capability up to around 1000 °C.

The use of pure boron as a thermistor material for use in the range 200 °C to 700 °C was reported by Prudinziati and Majni (1973) and departs from the more usual mixture of oxides.

E D Macklen (1973 unpublished) reviewed some of the work on high-temperature thermistors and reported on the design of devices based on yttria-doped zirconia for use in the range 500–1000 °C. A similar device for the range 300–500 °C is formulated with praeseodynium oxide and zirconium oxide.

These devices greatly extend the range of application of thermistors but as yet are not available withsthe close interchangeability which would ensure their general acceptance in industry.

5. Other special devices

I have deliberately excluded thermometers based on diodes, transistors and other junction devices from this review since they are not strictly resistance thermometers; apart from such devices however, there are always a number of papers describing special forms of resistance thermometer. A particularly unusual example is contained in a paper by Cussler (1973) entitled 'A liquid thermistor'; the applications would be very limited but it apparently has the advantages of fast response and very high sensitivity over a limited range.

Interest seems to be growing in the properties of carbon-impregnated glass for low-temperature thermometry (Lawless 1972).

There is now some limited commercial application of thermometers using rhodium–iron over a very wide temperature range from below 1 K to ambient temperatures. This type of detector will be particularly useful in monitoring start-up and shut-down operation of cryogenic plant; they have been discussed by Rusby (1975) at this conference.

6. The platinum resistance thermometer in industry

6.1. Standardization

At these conferences the emphasis is naturally on precise thermometry and it was refreshing to see that in the session on thermocouples we had papers concerned with new devices for routine industrial use. The platinum resistance thermometer, however, still has, for many people, an atmosphere of the standards laboratory about it — very accurate but delicate, bulky and difficult to use — despite the fact that millions of them are used each year in industry.

The industrial user requires that the device should offer accuracy and stability at a reasonable price, be small and robust and conform, within defined tolerances, to a standard resistance–temperature relationship.

The last of these requirements is close to full realization with more and more countries falling into line with the relationship which originated in Germany and which has $R_0 = 100\Omega$ and $\alpha = 0.003850$ (BSS 1964, GIS 1964). There is very good reason to expect this to be embodied in IEC recommendations in the near future.

6.2. Thin- and thick-film detectors

Because of the very high price of platinum and the desire to produce small devices, it has for some years been usual to use very fine wire in the construction of resistance thermometer detectors (0.01—0.05 mm).

This has resulted in their construction being rather labour intensive and has led people to consider the development of newer techniques to try to prevent the costs rising directly with labour rates. Various attempts have been made over the years to produce platinum resistance detectors by the deposition of films on to ceramic substrates.

The purity of the vacuum seemed likely to offer the best starting point but in practice it has never proved possible in sputtered or evaporated films to obtain temperature coefficients of resistance approaching those of bulk material. No doubt this would have been possible had much greater thicknesses been deposited, since in thicker films normal electronic conduction becomes more significant than conduction by processes such as tunnelling and thermionic emission which are important in attenuated films. Such thicknesses, however, become unattractive because of the long deposition times involved and the tendency in sputtering and evaporation for the platinum to be deposited all over the apparatus. Finally, the purity of the vacuum proves an illusion in that the chemical and structural purity of the films is in fact frequently rather poor.

Rather surprisingly, the solution has been found in the more cookery-book-like procedures of thick-film techniques and screen printing. As a result, detectors are now being printed as a routine production process to produce devices as shown in figure 1.

These detectors have a performance, over the range $-50\,^{\circ}\mathrm{C}$ to $500\,^{\circ}\mathrm{C}$, comparable with that of wire-wound glazed detectors and can replace these in a wide range of industrial applications.

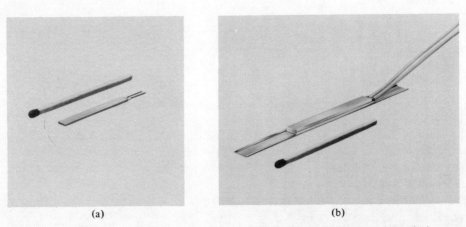

(a) (b)

Figure 1. (a) Thick-film platinum resistance detector. (b) thick-film detector metal-sheathed.

6.3. Wire-wound detectors

For many years, however, more conventional constructions will also continue to be used because of their more convenient shapes for some applications or their wider temperature ranges with good stability.

The fact that the resistance–temperature relationship is internationally defined over a very wide temperature range makes it attractive to mass-produce large numbers of standard detectors of a limited number of different designs. These are then made up into various complete thermometer elements according to the environmental conditions they have to survive. Figure 2 shows a selection of typical sensing elements.

Figure 2. Typical industrial platinum resistance temperature detectors.

The main design problem to be overcome in detector construction is to find a good compromise between the need for a reproducible resistance–temperature relationship with good stability and that for a detector capable of withstanding the vibration and shock associated with an industrial environment.

The first requirement could best be met by a platinum wire with the absolute minimum of support or restraint, as is in fact the case for a laboratory standard thermometer. The vibration resistance would be greatest for a construction in which the platinum wire was fully encapsulated.

Figure 3 shows a compromise which has proved extremely successful and which has been in production for a number of years: the helical sensing coil is anchored down by a glaze to the inside of the bores in a high-alumina ceramic. The result is that the greater part of the coil is free but a portion of each turn is attached. This results in stability of the order of a few hundredths of a degree when used over the range $-200\,^{\circ}\mathrm{C}$ to $+800\,^{\circ}\mathrm{C}$. The elements will survive vibration conditions of up to $350\,\mathrm{m\,s^{-2}}$ and frequencies up to 1 kHz.

Other successful techniques have involved embedding the platinum coil in a powder or a chalky material to reduce the stresses imparted to it by the supports; another more exotic technique used in some aircraft devices involves winding the platinum coil on a platinum former to minimize differential expansion problems.

In some cases the vibration levels may make it essential to use constructions in which the platinum is quite firmly attached to the former. Figure 4 shows an example of

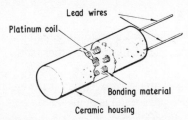 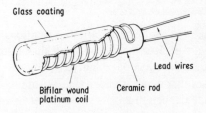

Figure 3. Detector with internal coils.   Figure 4. Externally wound detector.

this construction which is extremely robust but has somewhat poorer stability and hysteresis than that with a partially supported coil.

Figure 5 shows a particular form of detector for surface temperature measurement.

Modern resistance thermometers are capable of excellent long-term stability; changes of less than 1 K over four years on steam-raising plant at 540 °C have been reported. This sort of experience confirms the results of laboratory tests, the results of which are shown in figure 6. This shows the stability of a particular design used in power stations when held for long periods at 600 °C.

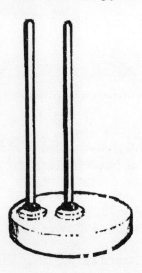

Figure 5. 'Shirt-button' surface temperature detector.

6.4. The complete thermometer element

The design and manufacture of a successful detector is only a part of the total problem of producing a good thermometer element. The design of the sheath and mounting arrangements and particularly the support of the internal leads require careful attention.

This matter is, in fact, the more important since the vast majority of failures of resistance thermometers occur in the leads or sheaths rather than in the detector. The causes of failure are almost always one of the following:

Mechanical damage caused by mishandling, catastrophic failure following an excessive high-temperature excursion, gradual deterioration following long-term exposure to a temperature somewhat higher than the maximum for which the element is designed, and fatigue failure under vibration.

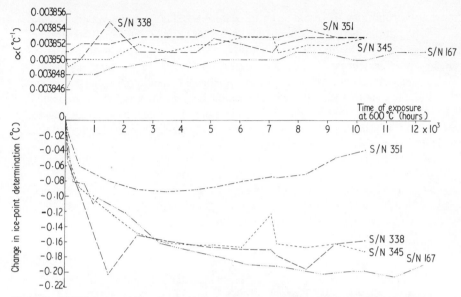

Figure 6. 12 000-hour stability data on industrial thermometer elements at 600 °C.

The first two are obvious and such failures cannot be entirely prevented. Gradual deterioration is generally the result of 'poisoning' of the platinum by its taking into solution or otherwise being contaminated by other materials present in its environment. Iron in particular will poison platinum at temperatures in excess of 550 °C and therefore special sheathing techniques are needed above this temperature. Suitably prepared nickel alloys can be used for sheathing up to say 800 °C; above this temperature very few resistance thermometers are used and those that are must use ceramic sheaths. It is generally essential to ensure an oxygen-rich atmosphere in the sheath to prevent reduction of metal oxides. In particular, lead oxide, which may be present in the glasses used in the detector, may be reduced and the resulting lead can form a low-temperature eutectic with platinum.

Vibration, however, remains the most common cause of failure resulting from a neglect of the principles which should guide the use of this instrument as of any other engineering component.

The typical thermometer element, figure 7, is long and thin and is supported as a cantilever from one end. It is thus a resonant structure with a fairly high Q. The tip

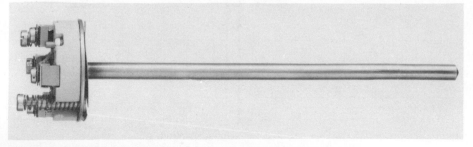

Figure 7. Typical industrial resistance thermometer element.

of the thermometer may therefore reach vibrational accelerations of several thousand m s^{-2} under quite moderate structural vibration. More usually, however, the vibration is the result of vortex shedding in the fluid flow and in any critical installation the wake frequency should be checked against the calculated resonant frequency of the probe.

Vastly greater levels of acceleration can be produced if the thermometer is free to rattle inside the pocket or thermowell. This condition must be avoided by adequate support of the sheath. Figure 8 shows one such mounting system in which the thermometer element is spring-loaded against the end of the pocket while the top is restrained from lateral movement by the spring plate.

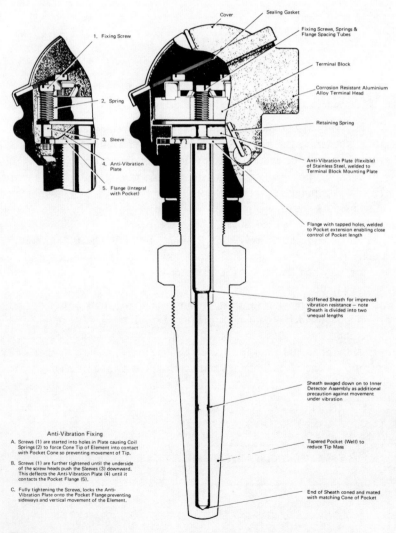

Figure 8. Resistance thermometer element with vibration-resistant mounting.

6.5. Thermometer elements for use under very high vibration

In a few applications very high vibration levels are experienced: aircraft brake-temperature sensors, those used on power-station steam lines and on diesel-engine exhausts. In the last two applications vibration accelerations of 1000 m s^{-2} at frequencies up to 20 kHz have been recorded; these conditions require a much more nearly monolithic construction to avoid all possibility of mutual movement of the constituent parts. Such a structure is shown in figure 9 in which the sensing wire is wound directly onto the insulated surface of a mineral insulated cable.

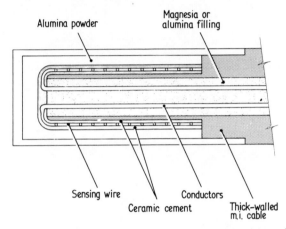

Figure 9. Thermometer element for very high vibration application.

6.6. Robust fast-response elements

Under laboratory conditions a fast-response thermometer element would be produced by making it small and hence of low thermal mass.

This would result in an element too fragile and delicate for use in an industrial environment. In fact the time constant is decided by the ratio of thermal mass to surface area; this permits a low time constant to be produced in a larger element by making it hollow. Figure 10 shows such a construction, used in the primary coolant loops of nuclear reactors, which has a time constant of about 2·5 s but which will survive considerable vibration and withstand pressures up to 31 MPa.

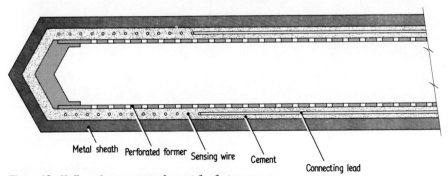

Figure 10. Hollow thermometer element for fast response.

7. Conclusions

Resistance thermometry encompasses a very wide range of devices and applications not all of which can be covered in a single paper — copper thermometers used in car-radiator temperature sensing is an example of a large field not mentioned.

The constructions used and the characteristics produced are just as varied making it difficult to produce a single coherent picture. What I have attempted to do is to show that the field is still active and that these instruments are very widely used in general industry.

References

Anderson A C 1972a *TMCSI* **4** part 2 773–84
Anderson R L 1972b *TMCSI* **4** part 2 927
Berry R J 1975 this volume
Blakemore J S 1972 *TMCSI* **4** part 2 827–33
Broughton M B 1974 *IEEE Trans. Instrum. and Measmt* **IM-23** no 1
BSS 1964 *British Standard Specification* 1904
Chattle M V 1972 *TMCSI* **4** part 2 907
Collins J G and Kemp W R G 1972 *TMCSI* **4** part 2 835–42
Cussler E L 1973 *AIChE Journal* **19** 1111–4
Durieux M 1975 this volume
Evans J P 1972 *TMCSI* **4** part 2 899
GIS 1964 *German Industrial Standard* DIN 43760
Goer D A, Starr E F, Little G R and Erickson R A 1974 *Cryogenics* January p15–20
Greenfield A J, Lieberman D, Zair E and Greenwald S 1974 *Rev. Sci. Instrum.* **45** 1417–22
Halverson G and Johns D A 1972 *TMCSI* **4** part 2 803–13
Hyde F J 1971 *Thermistors* (London: Iliffe Books)
Kemp W R G, Collins J G, Pickup C P and Muijlwijk R 1972 *TMCSI* **4** part 1 85–97
Kes P H, Van der Klein C A M and de Klerk D 1974 *Cryogenics* March p168–9
Lawless W N 1972 *Rev. Sci. Instrum.* **43** 1743–7
Prudinziati M and Majni G 1973 *IEEE Trans. Ind. Electron. and Control Instrum.* **IEC1–20** 30–3
Rubin L G and Golaling Y 1972 *Rev. Sci. Instrum.* **43** 1758–62
Rusby R L 1975 this volume
Rusby R L, Chattle M V and Gilhen D M 1972 *J. Phys. E: Sci. Instrum.* **5** 1102
Sample H H, Neuringer L J and Rubin L G 1974 *Rev. Sci. Instrum.* **45** 64–73
Sample H H and Neuringer L J 1974 *Rev. Sci. Instrum.* **45** 1389–91
Sawada S and Mochizuki T 1972 *TMCSI* **4** part 2 919
Schlosser W F and Munnings R H 1972 *TMCSI* **4** part 2 795–801
Stockert J and Nave E R 1974 *IEEE Trans. Biomed. Eng.*
Swartz D L and Swartz J M 1974 *Cryogenics* February p67–70
Swenson C A and Wolfendale P C F 1973 *Rev. Sci. Instrum.* **44** 339–41
Trolander H W, Case D A and Harruff R W 1972 *TMCSI* **4** part 2 997–1009
Van Rijn C, Nieuwenhuys-Smit M C, Van Dijk J E, Tiggelman J L and Durieux M 1972 *TMCSI* **4** part 2 815–26
Weinstock H and Parpia J 1972 *TMCSI* **4** part 2 785–90
Whitehead N F, Lanchester P C and Scurlock R G 1974 *J. Phys. E: Sci. Instrum.* **7** 36–8

International comparison of low-temperature platinum resistance thermometers

J P Compton and S D Ward
Division of Quantum Metrology, National Physical Laboratory, Teddington, Middlesex, England

Abstract. Standard platinum resistance thermometers from several national standards laboratories have been submitted to the National Physical Laboratory, where they have been systematically intercompared at temperatures between 13·81 and 373·15 K. This work, which has been carried out under the auspices of the Comité Consultatif de Thermométrie, makes it possible to intercompare different laboratories' realizations of IPTS-68 at and between the defining fixed points. It thus allows estimation of the reproducibility of IPTS-68 as it occurs in practice over this temperature range. In addition, it is possible to separate that component of the reproducibility which stems from fixed-point variations from that arising from intrinsic differences between individual platinum resistance thermometers.

The experimental methods will be described together with the results of tests to establish their precision and accuracy. The intercomparison of realizations of IPTS-68 will be presented and discussed.

1. Introduction

Any temperature scale has to meet the twin requirements of reproducibility and thermodynamic accuracy. IPTS-68 is no exception to this, and in the work that preceded its formulation, both aspects were examined in detail. Since then, there have been a number of improvements in the measurement of thermodynamic temperature, and over much of its range the thermodynamic accuracy of IPTS-68 can now be judged far more precisely than was possible in 1968. As a result of these advances in fundamental measurement greater demands will be made upon the reproducibility of IPTS-68 if it is to serve as an adequate vehicle for the dissemination of thermodynamic temperature.

In this paper we give a preliminary account of an experimental study of the reproducibility of IPTS-68 below 273·15 K. The work is being undertaken under the auspices of working group 4 of the Comité Consultatif de Thermométrie, and from it will come a quantitative estimate of the limitations of IPTS-68 and, we hope, guidelines for the scale's improvement.

Standards laboratories have been invited to submit capsule-type platinum resistance thermometers to NPL, where we isothermally compare them with a standard group at 48 temperatures between 13·81 and 273·15 K. Many of the thermometers submitted bear calibrations from their laboratory of origin. The intercomparisons thus make it possible to separate the component of irreproducibility that reflects intrinsic thermometer variation from that due to differences in realization of the defining fixed points. Although we have so far examined only 14 thermometers it is already clear that the fixed points are major sources of the overall irreproducibility of IPTS-68.

Although programmes of thermometer intercomparisons have been reported before (Bedford and Ma 1970, Belyansky *et al* 1969, Preston-Thomas and Bedford 1968, Tiggelman and Durieux 1972) they have always been based upon single sets of fixed-point realizations. They have therefore not yielded information about fixed-point reproducibilities. Furthermore, the lack of individual thermometers common to more than one such intercomparison has prevented the consolidation of all the available data. By bringing together thermometers from all over the world, the present work avoids these difficulties.

2. Equipment and method

The intercomparison equipment, which is shown in figure 1, is straightforward. It consists essentially of an OFHC copper comparison block with locations for the simultaneous intercomparison of nine thermometers. It is surrounded by an adiabatic shield, and except during cool-down, the equipment is operated under high vacuum. Liquid helium, liquid nitrogen and melting ice are used as refrigerants as appropriate. Two simple proportional temperature controllers control the comparison block and the shield, and their performance is such that corrections for temperature drift seldom exceed 0·1 mK.

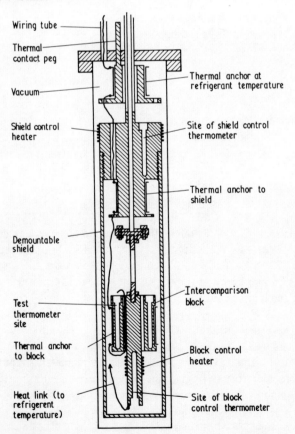

Figure 1. Low-temperature intercomparison equipment (not to scale).

All resistance measurements are made with a Kusters current comparator bridge, which allows a temperature resolution ranging from $16\,\mu\text{K}$ at $13\cdot8$ K (the worst case) to a best figure of $3\,\mu\text{K}$ near 27 K. These temperature resolutions, which reflect the electrical noise in the measuring circuit, are very much better than the accuracy that it is necessary to achieve in this work. The main source of imprecision is found to be temperature instability, and we have therefore paid considerable attention to minimizing this. Once equilibrium has been established, mean drift rates of $0\cdot1$ mK h^{-1} or better are achieved, although this involves waiting periods ranging from 20 minutes to 3 hours following a temperature change. However, occasional transient temperature excursions of the order of $0\cdot1$ mK also occur, probably due to impulsive interference in the temperature control system, and we believe these are the main cause of imprecision.

Because we have good temperature stability and resistance resolution, we are able to adopt a simple measurement sequence. The resistances of the nine thermometers are measured in turn, with repeats of the first at the middle and end of the set. The changes, if any, shown by these repeat measurements are used to calculate linear drift corrections for the intermediate measurements, and a complete set of readings is normally completed in about 20 minutes.

We measure thermometer resistances relative to standards of 1 Ω or 10 Ω, as appropriate. There would have been advantages in comparing each thermometer instead with one thermometer chosen as standard, whose resistance was finally measured absolutely in order to indicate temperature. This method would greatly reduce sensitivity to temperature drift, but we did not adopt it for two reasons. Firstly it would have required modification to the bridge, which is designed to operate with decade values in the 'S' arm. More importantly, however, its main practical advantage, since we have in any case achieved adequate overall precision, would have been to allow measurement before the attainment of thermal equilibrium, and hence give a saving in time. However, this would imply a measurement in a dynamic situation in which temperature gradients were more probable, and we were unwilling to accept this risk. Measurement in terms of a standard resistor thus has little disadvantage in practice.

At each of the defining fixed-point temperatures we make measurements of the self-heating of all the thermometers, from which we obtain effective thermal impedances, $\Delta T/i^2 R(T)$. For each thermometer we then use a least squares procedure to derive a relationship between these thermal impedances and temperature. Quadratic relations are found to be satisfactory, and are then used to calculate self-heating corrections at all temperatures, to permit reduction of all results to zero measuring current.

Finally, before passing to a discussion of results, we set out the normal sequence of measurements to which each thermometer is subject. Following a thermometer's arrival at NPL, we first measure its R_0 and alpha coefficient. It is then assembled in the comparison apparatus together with the three thermometers of the standard group and cooled to $4\cdot2$ K. The standard thermometers, which for this purpose have been calibrated near $4\cdot2$ K, can still be used to indicate temperature, and from the measured resistances we calculate resistances at the helium boiling point.

The low-temperature intercomparisons are now performed, in increasing order of temperature, at each of the defining fixed points and at the following intermediate temperatures: at intervals of $0\cdot5$ K from $14\cdot0$ to $19\cdot5$ K, at intervals of 1 K from 21 to 26 K, at intervals of 2 K from 28 to 40 K, at 45 K, 50 K, and intervals of 5 K from

60 K to 80 K, at 83·7 K, at intervals of 20 K from 100 to 260 K, and at 273·15 K. This gives a total of 48 intercomparison temperatures. To complete the sequence, the thermometers are removed from the comparison equipment and R_0 values re-measured. They are then available for hand carriage back to their laboratories of origin.

3. Performance tests

We have carried out several tests in order to estimate the accuracy of our intercomparisons. The repeatability of the measurements has been examined by repeating some intercomparison sets.

The relative changes between all possible pairs of thermometers observed in the repeated intercomparisons are plotted as a histogram in figure 2. This histogram, which

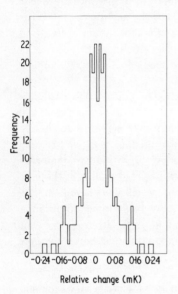

Figure 2. Histogram of relative changes between all possible pairs of thermometers upon repeated intercomparisons at 16, 17, 65 and 240 K.

is necessarily symmetrical about zero relative change, displays the influence of the two sources of irreproducibility that we have already mentioned. The broad, approximately gaussian, distribution represents the effect of temperature excursions caused by interference. Evidently, however, this interference is spasmodic, and when it is absent results conform to the central peak, which reflects the precision of the resistance measurements. If all these deviations are taken together, we find that two-thirds are smaller than 0·07 mK, which we take as representing the precision of this work at the 67% confidence level. The data also indicate that this precision does not vary very much with temperature.

We have sought systematic errors, due to temperature gradients or inadequate thermal anchoring of leads, in two ways. The first of these was to examine relative changes between thermometers when the shield temperature was offset. Figure 3 shows the greatest deviation observed between any pair of thermometers as a result of a 1 K shield offset. The greatest deviation observed at any temperature was 1·7 mK. Since equality of shield and block temperatures was always assured to better than 15 mK, it is clear that thermal anchoring is satisfactory.

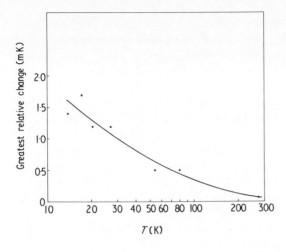

Figure 3. Greatest relative changes induced by a 1 K offset of the shield temperature, plotted against intercomparison temperature.

At certain temperatures, temperature gradient effects were sought by change of refrigerant. Thus, at 80 and 83·7 K both helium and nitrogen refrigerants were used, and at 273 K both nitrogen and melting ice. In addition, at 273 K measurements could be compared with direct measurements of R_0 in a triple-point cell.

The data indicate the presence of small temperature gradient effects at the two lower temperatures. The largest relative changes were 0·16 mK at 80 K and 0·18 mK at 83·7 K, both between the same pair of thermometers. These figures must be treated as representing genuine systematic error, for the relative changes observed at 83·7 K correlate strongly with those at 80 K. In contrast, the deviations found upon repeated intercomparison show weaker correlation between repeats at different temperatures. Two-thirds of the relative changes upon exchanging liquid nitrogen for helium are less than 0·12 mK, and we regard this as the systematic uncertainty at and below 83·7 K, and with a confidence level of 67%.

At 273 K, substitution of melting ice for liquid nitrogen induced a maximum relative change of 2·3 mK. Individual relative changes again correlated well with those observed at 80 and 83·7 K. The observations made using melting ice refrigerant were in closer agreement with the direct triple-point measurements, having a maximum relative change of 0·8 mK. The corresponding figure for nitrogen cooling was 1·9 mK. It is clear that temperature gradients are present, though less when melting ice cooling is used. Because comparisons made below 273 K used liquid nitrogen refrigerant, we have to use the data from the nitrogen-cooled intercomparison at 273 K in our estimate of the uncertainty. In this case, two-thirds of the changes relative to triple-point measurements were less than 0·7 mK, and we take this as the systematic uncertainty, with 67% confidence level, at 273 K. It is not unreasonable to suppose that this uncertainty diminishes linearly with temperature to 0·12 mK at 83·7 K.

As well as testing the performance of the intercomparison equipment, we have examined the repeatability of our measurements of R_0 and alpha. Lack of space forbids more than the briefest account of these tests. In the case of the most carefully studied thermometer, chosen for its stable behaviour, fourteen measurements of R_0, spread over four months, showed a standard deviation from their mean of 14 $\mu\Omega$ or 0·14 mK. The span of the observations was 40 $\mu\Omega$ and four different triple-point cells were used.

There was evidence of systematic differences of up to 0·2 mK between cells, but as we have no grounds for selecting one rather than another we take the overall figure of 0·14 mK as the random uncertainty of this measurement.

Alpha values were measured in two ways: via an absolute realization of the steam point, and also by comparison with standard thermometers that had been calibrated at the freezing points of tin and zinc. Four comparison measurements were made upon the same stable thermometer with a standard deviation in alpha of $14 \times 10^{-9}\,\text{K}^{-1}$, equivalent to 0·36 mK. The two absolute measurements differed from one another by only $7 \times 10^{-9}\,\text{K}^{-1}$, but showed a mean increase of $45 \times 10^{-9}\,\text{K}^{-1}$ over the comparison method. Further absolute measurements are in hand to discover whether this difference is genuine. For the present we are assuming that it is, and measurements made by the comparison method are corrected accordingly.

In table 1 we give the differences between the NPL values for the helium boiling point resistance, R_0 and alpha, and corresponding values measured by submitting laboratories.

Table 1.

Laboratory	Thermometer No	$\Delta R_{4\cdot 2\text{K}}(\mu\Omega)$ (Lab-NPL)	$\Delta R_0\,(\mu\Omega)$ (Lab-NPL)	$\Delta\alpha \times 10^9$ (Lab-NPL)
NML	1731676	−5	−35	−118
	1705628		−114	105
NRC	1722203	−1	14	19
	1158066	−19	21	36
	1158062	−4·4	−92	
KOL	LN 43	0·1	−294	396
	T 4	−2·0	−257	376
IMGC	646		65	−32
ASMW	217990		−136	136
	207278		−87	113
	217997		−267	

A surprising feature of these figures is that ΔR_0, though poorly correlated with $\Delta R_{4\cdot 2\text{K}}$, is strongly correlated with $\Delta\alpha$. At present we have no explanation for this — the dependence of ΔR_0 upon $\Delta\alpha$ is not consistent either with errors in R_0 alone, or the absorption of strain.

We turn now to the results of the low-temperature intercomparisons. Table 2 indicates the differences between the fixed-point realizations at NPL and those elsewhere. Entries are only given where the submitting laboratory has supplied data based upon its own primary realizations. In a number of cases, some fixed-point calibrations are traceable to NBS. We have not yet made low-temperature measurements on the thermometers supplied directly by NBS, which therefore is not represented in table 2.

We come finally to the overall reproducibility of IPTS-68 as realized by the 14 thermometers so far tested. Figure 4 shows the envelope of the magnitudes of the greatest deviations between any pair of thermometers, using for each the calibration supplied with it. Inside this is shown the region within which lie two-thirds of the deviations, and which we propose should be regarded as representing the numerical

Table 2. Comparison of fixed-point realizations. (A positive entry indicates that the NPL realization is cooler.)

Laboratory	Thermometer No	Temperature difference (mK)					
		e–H$_2$ TP	e–H$_2$ '17 K'	e–H$_2$ NBP	Ne NBP	O$_2$ TP	O$_2$ NBP
NML	1731676				−2·5	−0·1	0·2$_5$
	1705628						0·1$_1$
NRC	1722203					−0·4	−0·5
	1158066					−0·6	−1·0
	1158062					0·1	−0·4
KOL	LN 43	−0·83	0·86	0·17	−1·84	−0·24	1·1$_5$
	T 4	−0·70	0·28	0·12	−1·96	0·15	1·1$_0$
IMGC	646					0·31	

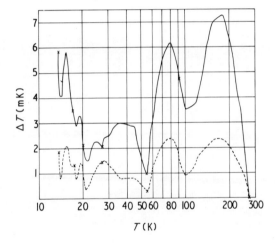

Figure 4. Reproducibility of IPTS-68 deduced from measurements on 14 thermometers, using their original calibrations: greatest deviations (solid curve) and boundary within which lie 2/3 of the deviations (broken curve).

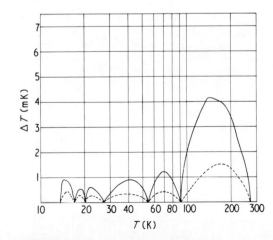

Figure 5. Reproducibility of IPTS-68 deduced from measurements on 14 thermometers, after recalibration in terms of the NPL fixed points: greatest deviations (solid curve) and boundary within which lie 2/3 of the deviations (broken curve).

reproducibility of IPTS-68 as it exists now, and subject to the uncertainty contingent upon the small sample of thermometers.

Figure 5 shows a similar pair of envelopes derived from the intercomparison data, but using instead new calibrations based upon the NPL fixed points. These envelopes thus represent the intrinsic reproducibility of the platinum resistance thermometer used according to the IPTS-68 procedure. The effect of fixed-point variations between laboratories has been eradicated, and the reproducibility is seen to be approximately halved.

Because this is a preliminary report only, and there remain thermometers still to be tested, we confine ourselves to this presentation of results and give no discussion of their implications. When the work is complete, we intend to publish a full account of all the measurements and the conclusions that may be drawn from them.

References

Bedford R E and Ma C K 1970 *Metrologia* 6 89–94
Belyansky L B, Orlova M P, Sharevskaya D I and Astrov D N 1969 *Metrologia* 5 107–11
Preston-Thomas H and Bedford R E 1968 *Metrologia* 4 14–30
Tiggelman J L and Durieux M 1972 *TMCSI* 4 part 2 857–64

Inst. Phys. Conf. Soc. No. 26 © 1975: Chapter 3

Control of oxygen-activated cycling effects in platinum resistance thermometers

Robert J Berry

Division of Physics, National Research Council of Canada, Ottawa, Ontario, Canada

Abstract. It has recently been shown that oxygen-activated thermal cycling effects can occur in platinum resistance thermometers causing significant resistance instability and uncertainty in temperature measurements on the IPTS-68. In this presentation further experiments aimed at understanding and controlling this troublesome effect are reported. A satisfactory solution to the problem is obtained, and the origin of the effect tentatively identified.

1. Introduction

A new type of thermal cycling effect, not related to strains, has been reported (Berry 1974) to occur in nearly all Pt resistance thermometers examined to date that contained oxygen in their gas filling. Since most standard thermometers are filled with dry air. or an inert gas plus oxygen, in accordance with the recommendations of the IPTS-68, this effect is presumably quite widespread.

Basically, it is found that the resistance at the triple point of water, R_{TP}, undergoes small reversible changes when a thermometer is thermally cycled between high and low temperatures in the 0–450 °C range. The general nature of these changes is demonstrated in figure 1 for a thermometer cycled between 450 and 100 °C. Other tests have shown that the presence of oxygen is necessary for the effect to occur. However, a slight trace (0.01 mmHg) is usually sufficient to initiate the process.

The specific properties of this O_2-activated cycling effect, as determined from R_{TP} changes with air-filled (1/2 atm) thermometers (Berry 1974), may be summarized as follows:

(i) The R_{TP} value of a thermometer, cycled between 450 °C and a temperature in the 20–300 °C range, will decrease at 450 °C and increase at the lower temperature in a reversible manner. (A small irreversible change may also occur, but this will not be dealt with here.)
(ii) At 450 °C, R_{TP} reaches equilibrium in about one-half to three hours.
(iii) At temperatures in the 20–250 °C range, R_{TP} is comparatively slow coming to equilibrium, requiring at least 7 days and probably longer.
(iv) The reversible increase in R_{TP} at low temperatures becomes smaller as that temperature is lowered from 200 to 20 °C, and is negligible below −78 °C.
(v) The magnitude of the reversible change in R_{TP} depends on the particular Pt specimen in a thermometer even when the same cycle is used. For example, out of seven

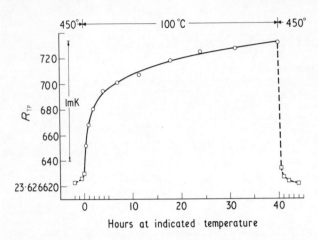

Figure 1. Change in R_{TP} in ohms for thermometer K9 due to thermal cycling between 450 and 100 °C.

air-filled thermometers that were cycled between 200 °C (17–23 hours) and 450 °C (1 hour), three changed by about 1·3 ppm, three by about 3 ppm and one by 4 ppm. (A 4 ppm change in R_{TP} is equivalent to 1 mK at 0 °C.)

(vi) Thermometers that have been heated only to 500 °C tend to have considerably smaller cycling effects.

(vii) Lastly, thermometers having smaller diameter Pt sensor wire tend to give larger cycling effects.

The impact of this cycling effect on precision Pt resistance thermometry is two-fold. Firstly it can degrade the reproducibility of a thermometer's R_{TP} value, and presumably any other $R(t)$ value below at least 300 °C, since these quantities will depend on the immediate thermal history of the thermometer. Secondly, the ambiguity introduced into these resistance values leads to somewhat ambiguous $W(t) = (R(t)/R(0°C))$ ratios and temperature determinations over the entire resistance thermometer range. Under the worst conditions, it is estimated that the uncertainty in temperature measurements is about 7 mK near 630 °C, 3 mK near 420 °C and 1·5 mK near 100 °C, for a thermometer possessing the maximum observed cycling effect.

In this report further experimental investigations of this cycling effect are presented that lead to a better understanding of its origin, and to a procedure for greatly reducing its influence on thermometry.

2. New experimental results

2.1. Dependence on oxygen pressure

It has already been established that the magnitude of the cycling effect depends strongly on the partial pressure of O_2 in a thermometer. Here, in figure 2, we demonstrate how this O_2 pressure also affects the temperature range over which R_{TP} decreases. For each selected O_2 pressure, the 'isochronal-step depression curves' shown were obtained by heating thermometer LN150 overnight at 100 °C to 'maximize' R_{TP}, measuring its initial R_{TP} value, and then measuring R_{TP} after heating for 1 hour at each of the following temperatures 200, 300, 350, 400, and 450 °C in that order. The total decrease

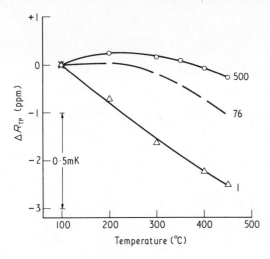

Figure 2. Isochronal-step depression curves for R_{TP} of thermometer LN150 obtained with 3 different gas fillings. The number on each curve gives the partial pressure of O_2 in mmHg in the filling.

in R_{TP} from its initial value is plotted against the heating temperature for each of three different gas fillings: 500 mmHg of O_2, 380 mmHg of dry air, and 1 mmHg of O_2 plus 500 mmHg of N_2.

The results confirm that the total decrease in R_{TP} between the O_2 sensor states corresponding to 100 °C (overnight) and 450 °C (1 hour) becomes smaller as the O_2 pressure is increased from 1 to 500 mmHg (designated property (viii)). However, it can also be seen that the temperature at which R_{TP} starts to decrease becomes much lower as the partial pressure of O_2 is reduced (property (ix)).

Property (viii) indicates that one might greatly reduce cycling effects in thermometers in the 0–450 °C range by filling with O_2 to a pressure higher than normal. Unfortunately this procedure gives an undesirable side effect; viz that the R_{TP} value (corresponding to one O_2 sensor state) drifts with prolonged 450 °C heating much more rapidly than is normal, making measurements in this region more difficult (property (x)). For example the R_{TP} value of thermometer M424 increases by 3 ppm/hour at 450 °C when it is filled with 500 mmHg of O_2 compared to 0·1 ppm/hour with 380 mmHg of air.

Hence the use of high O_2 pressures cannot be recommended, nor can the use of no O_2 since this leads to contamination of the Pt sensor (Berry 1974). It seems, therefore, that an O_2 pressure in the 1 to 76 mmHg range should be used, and that some other means of coping with the cycling effect must be found.

2.2. Effect of thermal cycling on R (100 °C)

The cycling studies reported up to now have all dealt with the reversible changes occurring in R_{TP}. Based on the nature of the experimental results, it seems highly probable that similar cyclic changes would occur in all $R(t)$ values below at least 250 °C. This expectation has been confirmed for the steam-point resistance, $R(100\,°C)$, in the work described below.

The steam-point apparatus used here has been described elsewhere (Berry 1958), and for these particular tests its operating conditions were kept constant to achieve the utmost stability. In figure 3 we show the reproducibility of $W(100\,°C)$ for thermometer

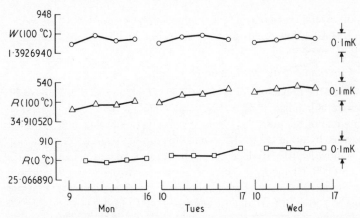

Figure 3. Variation of R (0 °C) and R (100 °C) in ohms, and W (100 °C), with time of day for thermometer M424. The R (0 °C) value measured immediately after R (100 °C) is used to calculate W (100 °C).

M424 after it was first stabilized at 100 °C for four days, and then measured alternately at the steam point and the triple point of water many times over three consecutive days. It can be seen that when little or no cycling effect is present, R (100 °C) and R (0 °C) are usually repeatable to within 3 and $2\,\mu\Omega$ respectively between consecutive measurements, and that W (100 °C) is repeatable to within 2×10^{-7} (equivalent to 0·05 mK) over the whole test period. Similar results have been obtained with two other thermometers, K9 and M163.

Figure 4 shows the cyclic increase in R (100 °C) for thermometer K9 when it is left continuously in the steam-point apparatus after having been exposed to 450 °C for 1 hour. Part of the initial rise is missing because the thermometer was at 100 °C for 7 minutes before the first steam-point reading could be obtained. We see that R (100 °C) increases in much the same manner as R_{TP} in the first 80 hours, and was in fact still drifting up after the thermometer had been exposed six days at the steam point. Other

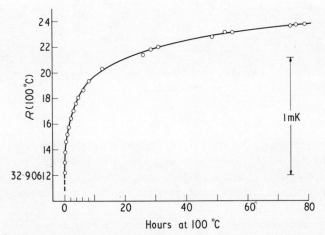

Figure 4. Dependence of R (100 °C) in ohms on time at 100 °C for thermometer K9, following its exposure to 450 °C.

cycling tests on several thermometers, including K9, have confirmed that most of this change in $R\,(100\,°C)$ at low temperatures is recovered by 450°C heating.

2.3. Effect of thermal cycling on $W(100\,°C)$

It follows from the previous results that the $W(100\,°C)$ ratio of a thermometer can take on a considerable range of values if $R\,(100\,°C)$ and R_{TP} are each allowed to correspond to any one of the possible O_2 sensor states. For example $W(100\,°C)$ of thermometer K9 can vary by up to 124×10^{-7}, which is the equivalent of 3·2 mK at 100°C.

To establish how $W(100\,°C)$ depends on the O_2 sensor state when the same state is used for both $R\,(100\,°C)$ and R_{TP}, a special test was performed on several thermometers. This consisted of placing a thermometer in a selected O_2 sensor state by prolonged heating at 100 or 200°C, determining $W(100\,°C)$ in this initial state, then heating the thermometer at 450°C to produce a large change in the O_2 sensor state, and redetermining $W(100\,°C)$ in this final state. In each case $R\,(100\,°C)$ was measured before R_{TP}, and the test measurements were completed in one day.

The results of repeated tests of the above type on five thermometers (described elsewhere by Berry 1974) are summarized in table 1, along with details of the initial (i) and final (f) O_2 sensor states. It can be seen that the change in $W(100\,°C)$ between the two states never exceeds 4×10^{-7}, or the equivalent of 0·1 mK at 100°C, even though a

Table 1. Change in $W(100\,°C)$ and R_{TP} between O_2 sensor states (i) and (f).

Therm. No.	State i hrs	State i t, °C	State f hrs at 450°C	ΔR_{TP}, (i–f) (ppm)	$\Delta W(100\,°C)$ $\times 10^7$ (f–i)
K9	23	200	1	2·5	2
	40	200	1	3·1	0
	89	200	1	3·9	3
	162	200	1	4·2	2
	40	200	0·5	3·2	1
	43	200	2	3·6	3
	114	100	1	3·5	4
	374	100	0·5	3·6	2
M424	21	200	1	2·0	2
	41	200	1	2·2	3
	65	200	1	2·5	1
	90	200	1	2·5	3
	42	200	0·5	2·2	3
	162	100	0·5	1·4	2
K12	20	200	0·5	2·9	0
	118	200	0·5	4·0	2
M327	19	200	0·5	1·0	1
	43	200	0·5	1·0	2
	155	200	0·5	1·6	0
LN150	40	200	0·5	1·5	2
	95	200	0·5	1·6	3
	330	200	0·5	2·1	2

wide variety of sensor states were employed. Furthermore, in 68% of the tests this change in $W(100\,°C)$ is within our experimental uncertainty of 0·05 mK. We conclude, therefore, that for all practical purposes $W(100\,°C)$ can be considered as independent of the O_2 sensor state so long as both $R(t)$ and $R(0°C)$ correspond to the same state (property (xi)). Presumably other $W(t)$ ratios behave the same way.

3. Nature of the O_2–Pt sensor interaction

In this section it will be shown that the reversible resistance changes observed here are most likely due to the first stages of oxidation of the Pt surface in the low-temperature region, and to the subsequent dissociation of this oxide into Pt and O_2 near 450 °C. We cannot be absolutely certain of this explanation since little or no work has been published on the oxidation of Pt under anything close to our conditions. However, as added support for our view, we note that no other plausible mechanism could be found which was entirely consistent with our experimental results.

For oxidation of Pt to take place there must first be chemical adsorption (chemisorption) of oxygen on the metal surface. Recent work (Peng and Dawson 1974, Procop and Volter 1972) has confirmed that this adsorption does occur on clean polycrystalline Pt, and that it proceeds very rapidly at temperatures as low as 100 K. At room temperature the O_2 takes only a few minutes to attain its maximum coverage of about half a monolayer at pressures as low as 10^{-6} mmHg. Thus with an air-filled thermometer one would expect this saturation coverage to occur in a matter of seconds at room temperature, and since the process requires no thermal activation energy the coverage should not increase appreciably as the temperature is increased. Indeed, desorption experiments by Peng and Dawson (1974) indicate that as the temperature approaches 450 °C most, if not all, of this O_2 will desorb from the Pt sensor very rapidly.

The comparatively slow resistance changes that we are concerned with here are found to increase strongly in magnitude as the temperature is increased from about -78 to $+100\,°C$ (property (iv)), and therefore their source must be a process requiring thermal activation energy. Surface oxidation is such a process (Kubaschewski and Hopkins 1962), and will be seen to fit our other results rather well.

To begin with, it is well known that a thin oxide layer may take weeks or years to form on a metal surface, and its growth often depends logarithmically on time, similar to the dependence shown in our figures 1 and 4 (property (iii)). Furthermore, measuring the increase of the electrical resistance of a metal during oxidation is a standard way of following the kinetics of this process. The resistance increase is usually attributed to the effective reduction in cross-sectional area of the specimen caused by the conversion of highly conducting pure metal into a relatively poorly conducting oxide layer. We can use this simple model to show that the increase in resistance, $\Delta R(t)$, is related to the effective decrease in diameter, Δd, of a wire by

$$\frac{\Delta R(t)}{R(t)} \simeq 2\,\frac{\Delta d}{d}. \tag{1}$$

For our maximum observed $\Delta R_{TP}/R_{TP}$ of 4 ppm in an air-filled thermometer with 0·08 mm diameter wire, this change corresponds to about one-quarter monolayer of Pt removed from the conduction process. In addition we see from equation (1) that $\Delta R(t)/R(t)$ should vary inversely with the wire diameter, and that $R(t)/R(0\,°C)$ should not be changed by the oxide growth. Properties (vii) and (xi) of our work agree as well as could be expected with these conclusions.

Assuming that a surface oxide is formed on Pt in the 0–300 °C range, one would expect at some higher temperature that this oxide would start to dissociate fairly rapidly causing the resistance to decrease and ultimately return to its original value. According to recent investigations (Westwood and Bennewitz 1974, Hoekstra et al 1971) on solid Pt oxides this 'dissociation temperature' should be roughly in the 300–600 °C range. Furthermore, it is well known that the dissociation temperature will increase as the ambient O_2 pressure is increased (Ritchie 1971). Thus, our observed properties (i) and (ix) are consistent with this model. It would also be expected that ultimately an O_2 pressure would be reached at which the oxide would not dissociate, at say 450 °C, but would continue to grow causing further increases in resistance. Such a mechanism could account for the results obtained with thermometers filled with 500 mmHg of O_2 (ie property (x)).

The fact that the magnitude of the cycling effect depends on the specimen of Pt (property (v)), and its annealing treatment (property (vi)), can be readily explained by the well known influence of surface contamination (foreign elements or oxides) on chemisorption, and hence oxidation (Hayward 1971). Usually such contamination reduces O_2 chemisorption by blocking adsorption sites, and in some cases has stopped chemisorption completely (Peng and Dawson 1974).

In conclusion, then, surface oxidation of Pt will account for all our observed properties of the O_2 activated thermal cycling effect in thermometers. The known influence of surface contamination on oxidation suggests that the cycling effect might be greatly reduced if less pure Pt, or possibly doped Pt, were used for thermometer sensing elements. This procedure, however, does not look especially promising because it would likely result in worse stability problems due to the migration and oxidation of impurities, and could increase the divergencies between measurements with different thermometers.

4. Improvement of the IPTS-68

Based on the preceding results, it is clear that a substantial reduction in the present ambiguity in $W(t)$ can be accomplished by always measuring $R(t)$ and R_{TP} in the same oxidation state. In practice, for the 20–450 °C range, this simply means that the thermometer should be brought to room temperature after an $R(t)$ measurement, as quickly as normal air cooling allows, and its R_{TP} value measured immediately. For the 450–630 °C range it is recommended that the thermometer first be cooled slowly to 450 °C, in order to avoid the vacancy quenching effect (Berry 1966), and then cooled quickly to room temperature to avoid at least part of the oxidation effect.

The above recommendation could readily be incorporated into a future version of the IPTS-68 by including in the definition of $W(t)$ an explicit statement to the effect that the Pt should be in the same chemical and physical state for both $R(t)$ and R_{TP} measurements, as nearly as is practical, and consistent with the principle that the Pt should contain its equilibrium concentration of thermal vacancies at each temperature.

A general guideline of this type is desirable since it would cover not only the oxidation effect, but also the well known contamination effect (Berry 1962), quenching and recrystallization effects, and presumably any other effect that might be encountered in the future.

Acknowledgments

It is a pleasure to acknowledge the substantial assistance of D G Kearney with all phases of the experimental work. Appreciation is also expressed to Dr G S Kell for the loan of thermometer K9, and to Dr P B Sewell, Dr M Cohen and others for helpful discussions of oxidation phenomena.

References

Berry R J 1958 *Can. J. Phys.* **36** 740
—— 1962 *TMCSI* **3** part 1 301
—— 1966 *Metrologia* **2** 92
—— 1974 *Metrologia* **10** 145
Hayward D O 1971 *Chemisorption and Reactions on Metallic Films* ed J R Anderson (London: Academic Press) vol 1 p227
Hoekstra H R, Siegel S and Gallagher F X 1971 *Adv. Chem.* **1971** 39
Kubaschewski O and Hopkins B E 1962 *Oxidation of Metals and Alloys* (London: Butterworths)
Peng Y K and Dawson P T 1974 *Can. J. Chem.* **52** 3507
Procop M and Volter J 1972 *Z. Phys. Chem. (Leipzig)* **250** 387
Ritchie I M 1971 *Chemisorption and Reactions on Metallic Films* ed J R Anderson (London: Academic Press) vol 2 p258
Westwood W D and C D Bennewitz 1974 *J. Appl. Phys.* **45** 2313

Characteristics of platinum resistance thermometers up to the silver freezing point

P Marcarino and L Crovini
Istituto di Metrologia 'G Colonnetti', Torino, Italy

Abstract. Starting with a review of the proposed equations for the extension of the platinum resistance thermometer characteristic above 630·74 °C, this note deals with the critical examination of the possibility of defining a provisional temperature scale based on the platinum resistance, which should be extended up to the freezing point of silver (t_{68} = 961·93 °C). The scale must agree as far as possible with the IPTS-68, but it should be smooth and unique throughout the entire range 0–961·93 °C.

The behaviour of several thermometers of different construction is examined for the compliance to the aforementioned criteria at both the defining points and the secondary points of IPTS-68.

The trade-off between reproducibility and accuracy on one side and the simplicity of the characteristic equation on the other will determine the best proposal.

1. Introduction

The International Practical Temperature Scale of 1968 (IPTS-68) in the range 630·74–1064·43 °C is defined by means of the EMF versus temperature characteristic of a platinum 10% rhodium–platinum thermocouple (CIPM 1969).

This device exhibits a limited accuracy (not better than ±0·2 °C according to some thorough experiments; see McLaren and Murdock 1972), while the uniqueness of the thermocouple scale may be better than ±0·05 °C in well defined conditions (Bedford 1972). Moreover, there is evidence of the departure of the thermocouple scale from the Thermodynamic Scale (Quinn et al 1973).

Therefore its replacement with a platinum resistance thermometer has been considered many times and in great detail by Berry (1965), Evans and Wood (1971), Chattle (1972) and Anderson (1972).

They considered the sensor design, the annealing techniques and the resistance versus temperature characteristic, taking into account all major sources of systematic errors and aiming at a satisfactory reproducibility of the platinum resistance scale. Regarding the definition of the resistance versus temperature characteristic, two philosophies have been considered so far: either to redefine the IPTS in the range 0–1064·43 °C with a platinum resistance thermometer scale in better agreement with new determinations of the thermodynamic temperatures, or to match as far as practical the thermocouple scale with the platinum resistance thermometer characteristic. Thus an alternative realization of IPTS-68 could be provided with a ten-fold improved reproducibility. The first proposal, however, can be implemented only when more reliable thermodynamic determinations of fixed points are available.

As far as the second proposal is concerned, Evans and Wood (1971) introduced the following equations:

$$t = \theta - 54.84 \left(\frac{\theta}{1064.43\,°C} - 1\right)\left(\frac{\theta}{961.93\,°C} - 1\right)\left(\frac{\theta}{630.74\,°C} - 1\right)$$
$$\times \left(\frac{\theta}{480.081\,°C} - 1\right). \qquad (1)$$

$$\frac{R(\theta)}{R(0\,°C)} = W(\theta) = A + B\theta + C\theta^2. \qquad (2)$$

These relationships cover the range from 630·74 °C to the gold point and require two calibration points (the freezing points of silver and gold) and the value of $W(630.74\,°C)$, which ought to be determined from the calibration according to IPTS-68 in the range 0 to 630·74 °C. The temperature t coincides with t_{68} to better than ±0·07 °C throughout the entire range (Evans 1972).

However, any platinum resistance scale which agrees with the thermocouple scale from 630·74 °C to the gold-point temperature shows a 0·1% discontinuity in the first derivative of $W(t)$ at 630·74 °C, as a consequence of a lack of smoothness in IPTS-68. Moreover, a number of recent optical pyrometer determinations (Quinn et al 1973, Bonhoure 1974, Coslovi et al 1975, Jung 1975) consistently indicate that the temperature interval between the freezing points of silver and gold is smaller than reported by IPTS-68 by 0·14 °C on the average. Then it is evident that a discontinuity in the first derivative is present at 1064·43 °C and the kelvin size will change by approximately 0·14% passing from the thermocouple scale to the radiation scale.

We have investigated the possibility of a platinum resistance scale which follows a somewhat different philosophy from those previously discussed. Our proposed scale covers the interval from the antimony to the silver points, and is smoothly joined to IPTS-68 at 630·74 °C and to the radiation scale at the silver point. Such a scale need not necessarily coincide with IPTS-68 but should not depart from it by more than the standard thermocouple inaccuracy. It should be used as a provisional scale in order to extend platinum resistance thermometry above 630·74 °C until a final replacement of IPTS-68 will be provided.

2. Proposed scale equations

The scale equations have to meet the following requirements:

(i) to match IPTS-68 in the range 0–630·74 °C;
(ii) to require only one additional calibration point — the silver point beside the triple point of water, the tin point and the zinc point;
(iii) to cover by extrapolation a 102·36 °C interval between the silver point and the gold point, in order to match the radiation scale.

Condition (iii) can be fulfilled only if a provisional value of the freezing point of silver is adopted. This should be 962·07 °C according to an authors' average of optical pyrometer determinations. It may be observed that this silver-point shift is still within the accredited accuracy of the thermodynamic determination (CIPM 1969).

We have examined several different equations for this purpose. Simple quadratics or cubics do not meet those requirements even when an exponential term is introduced to account for high-temperature lattice vacancy concentrations in platinum (Quinn et al 1973). A satisfactory result is obtained with two kinds of equation, which are actually closely related. The first scale chosen is given by the combination of equations (3) and (4):

$$t = t' + 0.045 \frac{t'}{100\,°C} \left(\frac{t'}{100\,°C} - 1\right)\left(\frac{t'}{419.58\,°C} - 1\right)\left(\frac{t'}{630.74\,°C} - 1\right)$$
$$\times \left[1 - \epsilon\left(\frac{t'}{1064.43\,°C}\right)^n\right] \tag{3}$$

$$W(t') = 1 + At' + Bt'^2 \tag{4}$$

$$W(t') = (1 + At' + Bt'^2)\left[1 + D\exp\left(-\frac{E}{kT'}\right)\right] \tag{5}$$

and this leads to an extension of IPTS-68 similar to that proposed by Moser (1967). Constants A and B are those in the range 0–630.74 °C and ϵ is derived from the silver-point calibration. The exponent n ought to be selected in such a way that the average characteristic of several comparable thermometers be able to match IPTS-68 at the gold point.

The combination of equations (3) and (5) produces the second scale we are going to consider and which is similar to the previous one. However, in this case t' is derived from a modified quadratic as proposed by Quinn et al (1973). The value of D is 19.8 and that of E is 1.2 eV for all thermometers according to Berry (1966), while k is Boltzmann's constant. These constants take into account the equilibrium concentration of lattice vacancies. As in the first instance, calibrations at tin (or steam) and zinc points, according to IPTS-68, and at the 'modified' silver point, are required to determine constants A, B and ϵ, but a different n value ought to be chosen in order to match IPTS-68 at the gold point.

To provide the average characteristics and compute the best-suited values of n, 21 thermometers were considered: 9 thermometers from the experiment of Evans and Wood (1971), 5 from Chattle's experiment (Chattle 1972) and 7 of IMGC (CIPM 1969); 19 out of 21 are bird-cage-type thermometers of approximately 0.25 Ω at the ice point. The remaining two thermometers, both operated at IMGC, have different design and will be described later in this paper.

Figure 1 shows the deviations of these scales from IPTS-68; t_{68} is obtained by applying equations (1) and (2) to the averaged data from the 21 thermometers. These scales (solid and broken lines) meet IPTS-68 at the gold point within ±0.02 °C and are bound together at 962.07 °C. They do not alter IPTS-68 below 420 °C by any appreciable amount and between 420 °C and 630.74 °C by more than ±4 m °C, which is close to the reproducibility of a standard platinum resistance thermometer at that temperature. The largest deviation from IPTS-68 occurs in the proximity of 840 °C and amounts to approximately 0.23 °C. The two proposed scales are very close and only their reproducibilities will indicate the best one.

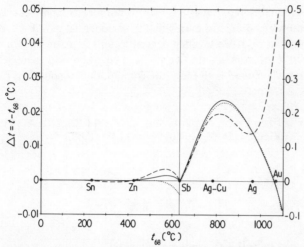

Figure 1. Deviation of extended platinum resistance scales with respect to IPTS-68. The solid line refers to the scale defined by equations (3) and (4) with $n = 3$. The broken line refers to the same equations when $n = 3$. The broken line refers to the scale defined by equations (3) and (5) with $n = 5 \cdot 5$. In all cases the freezing point of silver is set to $962 \cdot 07\,°C$.

3. Experimental results

Seven thermometers were compared in the range $0-1064 \cdot 43\,°C$ in order to test the reproducibility of the proposed scales. The results obtained from them were compared with those from NBS (Evans and Wood 1971) and from NPL (Chattle 1972).

Table 1 reports the more relevant features of IMGC thermometers. Five of them are bird-cage thermometers with minor construction differences. The last two are coiled filament thermometers, but with completely different sensor design. Six out of seven were provided by specialized manufacturers. The seventh one (TP27) is a rather cheap thermometer which had been prepared in the laboratory and kept in use for some years.

Comparisons were made at the triple point of water, the freezing points of tin, zinc, antimony, silver and gold and the melting point of the silver–copper eutectic alloy.

Table 1. Data concerning the seven thermometers compared at IMGC.

Thermometer code	Sensor design	Insulator material	Sensor length (cm)	$R\,(0\,°C)$ (Ω)	As-received α ($°C^{-1}$)	Operation time above $600\,°C$ (h)	Final α ($°C^{-1}$)
R1	Bird-cage	Quartz	4·0	0·2300	0·00392480	75	0·00392465
R2	Bird-cage	Quartz	4·0	0·2320	0·00392558	–	0·00392558
R3	Bird-cage	Quartz	4·0	0·2332	0·00392486	600	0·00392450
R4	Bird-cage	Quartz	4·0	0·2462	0·00392717	100	0·00392698
L1	Bird-cage	Quartz	4·0	0·2364	0·00392694	105	0·00392694
J1	Coiled-filament	Quartz cross	3·0	25·79	0·00392672	155	0·00392656
TP27	Coiled-filament	Quartz capillary	3·0	1·399	0·00392494	205	0·00392474

The fixed point apparatuses were described by Bongiovanni *et al* (1971, 1972, 1975). The gold point was realized by a 1 kg 99·999% gold ingot (Johnson Matthey Metals) which allows the sensor midpoint to be immersed by 14·5 cm below the liquid metal surface, but provides only a 0·5 cm thick metal sleeve around the sensor. Therefore, the resulting plateau is somewhat shorter than reported elsewhere (Evans and Wood 1971, Chattle 1972): for instance, when the temperature uniformity in the liquid phase is accurately trimmed within ±0·15 °C over the entire ingot length, there results a temperature stability better than ±10 m °C for 35 min during the freezing process.

Resistance measurements were taken by means of a G4 Mueller bridge for resistances above 1·3 Ω and with a DC current comparator bridge (Crovini and Kirby 1970) for resistances below 1·3 Ω. Both instruments are able to provide a sensitivity better than 0·2 m °C in terms of temperature. Fixed point reproducibilities and bridge accuracies are summarized in table 2.

Table 2. Fixed-point reproducibilities and bridge accuracies.

Fixed point	Temperature (°C)	Reproducibility (m °C)	Bridge sensitivity (m °C)	Calibration accuracy‖ (m °C)
Triple point of water	0·01	0·1	0·2	±0·2
Freezing point of tin	231·9681	0·2	0·2	±0·5
Freezing point of zinc	419·58	0·2	0·2	±1
Freezing point of antimony	630·755†	1	0·2	±2
Melting point of Ag–Cu eutectic	779·60‡	10	0·2	±10
Freezing point of silver	962·07§	2	0·2	±3
Freezing point of gold	1064·43	30	0·2	±30

† Estimated temperature of IMGC antimony point (Bongiovanni *et al* 1971).
‡ Estimated temperature of IMGC silver–copper eutectic point (Crovini and Marcarino 1974).
§ This value derives from IPTS-68 with a shift of 0·14 °C in order to match the silver-to-gold-point temperature interval as determined on the radiation scale.
‖ It includes the fixed-point reproducibility and the bridge sensitivity and linearity.

In general the thermometers which were used did not show a satisfactory stability above 962 °C and even in the range 800–962 °C the stability of some of them was still questionable. Figure 2 (open circles) shows the behaviour of the water triple-point resistance of R3 during a series of determinations of the silver point; the full circles

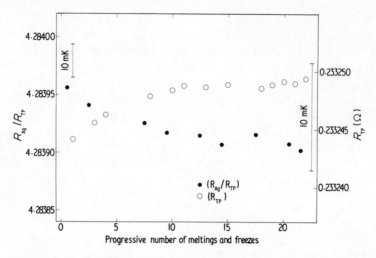

Figure 2. Behaviour of the thermometer R3 when submitted to repeated exposures at silver-point temperature.

represent the behaviour of the silver point reduced resistance and it is evidently correlated to triple-point resistance. All bird-cage thermometers exhibited a comparable behaviour. The maximum instability was found at the gold point where the triple-point resistance change after the high-temperature exposure was equivalent on average to 20 m°C in terms of the gold-point temperature. Coiled filament thermometers required a prolonged treatment at successively higher temperatures (480 °C, 800 °C, 980 °C) to achieve a comparable stability. In consideration of their construction features they were never heated above 1000 °C. TP27 was undoubtedly the best one, its triple point shift after any high-temperature exposure being typically less than 1 m°C, in terms of the triple-point temperature.

Therefore it is not entirely correct to assume that all thermometers were operated in full annealing conditions. Nevertheless, after any high-temperature exposure each thermometer was submitted to a one-hour annealing at 650 °C followed by another one-hour annealing at 480 °C before determining its triple-point resistance. This procedure was very useful for reducing the effect of the unstability to an acceptable level.

The melting point of Ag–Cu eutectic alloy was introduced in order to evaluate the uniqueness of each proposed scale in the temperature region where it deviates more from IPTS-68. A temperature t_{68} = 779·60 °C was assigned to this reference point, according to a previous determination (Crovini and Marcarino 1974). Only R1 and R2 were directly calibrated at this reference point; the other thermometers were compared to them in a copper-block comparator which afforded a comparison error smaller than ±2 m°C. The comparison temperature was adjusted in the close vicinity of the eutectic point.

Final results are reported in table 3. The last four lines of this table refer the average data from NBS, NPL and IMGC and the total average, that is the arithmetic mean of the preceding three lines with the exception of $t_{\text{Ag-Cu}}$.

Table 3. Results of the IMGC comparisons.

Thermometer code	$10^3 A$ (°C^{-1})	$-10^6 B$ (°C^{-2})	ϵ	$\epsilon^\dagger$	t_{Sb} (°C)	t_{Ag-Cu} (°C)	$t_{Ag-Cu}{}^\dagger$ (°C)	t_{Au} (°C)	$t_{Au}{}^\dagger$ (°C)
R1	3·983389	0·587359	0·59470	0·38966	630·745	779·76	779·79	—	—
R2	3·984337	0·587556	0·59101	0·38560	630·753	779·80	779·83	1064·44	1064·45
R3	3·983239	0·587377	0·60310	0·39890	630·743	779·77	779·80	1064·40	1064·41
R4	3·985736	0·587603	0·62000	0·41759	630·759	779·78	779·82	1064·36	1064·36
L1	3·985703	0·587643	0·61705	0·41433	630·757	779·79	779·82	1064·34	1064·34
J1	3·985315	0·587549	0·62512	0·42322	630·751	779·79	779·83	—	—
TP27	3·983483	0·587297	0·62067	0·41829	630·756	779·79	779·82	—	—
Evans and Wood average (1971)	3·984457	0·587459	0·61498	0·41204	630·752	—	—	1064·44	1064·43
Chattle's average (1972)	3·983821	0·587414	0·60928	0·40574	630·740	—	—	1064·45	1064·45
IMGC average	3·984458	0·587484	0·61039	0·40697	630·752	779·78	779·81	1064·39	1064·38
Total average	3·984245	0·587452	0·61155	0·40825	630·748	—	—	1064·43	1064·42

† Data derived with the scale of equations (3) and (5).

4. Discussion of results

On first examination of the experimental results we must conclude that the scale based on equations (3) and (5) does not present any substantial advantage over that based on equations (3) and (4). The exponential term had been introduced with the aim of attaining a better reproducibility by taking into consideration the high-temperature lattice vacancy concentration. However, as this improved reproducibility did not show up, it is not convenient to be stuck to an exponential equation and consequently this scale is no more examined in the following considerations.

The reproducibility and the uniqueness of the proposed scale (equations (3) and (4)) can be evaluated from the results of calibrations at the secondary fixed points. The first seven entries of the sixth column of table 3 exhibit a standard deviation of 6 m°C which is acceptable when the increased measurement difficulty of low-resistance thermometers is taken into account. The results of determinations of the silver–copper eutectic melting point (the seventh column) exhibit a standard deviation of 0·014 °C, which is close to the reproducibility of the fixed point (see table 2).

The reproducibility at the gold point according to IMGC results is not as good in comparison. This fact may be ascribed on the whole to the limited reproducibility of the IMGC gold point. The relatively thin gold sleeve around the thermometer well in the gold-point apparatus may not be sufficient always to provide an extended isothermal region around the platinum sensor during a freeze. A rapid solidification on top of the ingot could eventually reduce the thermometer's effective immersion depth and change the measured temperature. Bird-cage thermometers do in fact exhibit excessive stem-losses, as reported by Evans and Wood (1971). Coiled-filament thermometers are better in this respect as they correctly track the temperature gradient produced by the liquid metal pressure in freezing zinc if an immersion larger than 10 cm is provided. Unfortunately they are not suitable at present for operating at the gold point. The thermometer instability already mentioned contributed the minor but still relevant part of this uncertainty.

Therefore, on the basis of these results it may be estimated that the silver-to-gold temperature interval is reproduced by a standard platinum resistance thermometer with an accuracy of ±0·05 °C.

IMGC results may be better compared to NBS results by considering the eutectic point calibrations. In this case we consider $\Delta t = t - t_{68}$, which is determined by calculating t through equations (3) and (4) and t_{68} by means of equations (1) and (2) with calibration points at 630·74 °C, the eutectic point (779·60 °C) and the silver point (961·93 °C). If $(\Delta t)_{IMGC}$ is calculated from IMGC average data (seven thermometers) it will differ by 0·025 °C at the eutectic point from $(\Delta t)_{NBS}$, which is calculated from average data on nine thermometers (Evans and Wood 1971) with the gold-point calibration. This discrepancy is not surprising since both scales are indirectly based on standard thermocouple measurements.

Therefore, for comparison purposes the IMGC Δt are calculated letting t_{Ag-Cu} = 779·575 °C in order to match NBS results at that temperature. NPL results (with gold-point calibrations) do not differ by more than 5 m°C with respect to NBS. Figure 3 shows the IMGC and NPL Δt with respect to NBS Δt from 750 °C to 1100 °C. Differences below 750 °C are not appreciable; the differences at the gold point do not exceed 14 m°C. It may be observed that the differences of figure 3 closely represent the

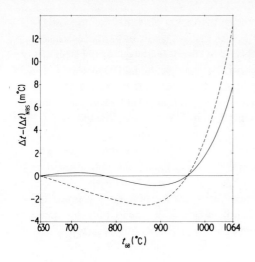

Figure 3. Deviation of IMGC (solid line) and NPL (broken line) scales (equations (3) and (4)) with respect to the NBS one. The deviation is expressed as a difference between Δt and consequently is affected by small differences in the realization of IPTS-68 in the three laboratories. IMGC data for t_{68} were obtained through a calibration at antimony and silver freezing points and the melting point of the Ag–Cu eutectic. This reference point was assumed to be 779·575 °C in order to match the NBS scale.

differences among the three realizations of the proposed scale. They would be exactly the scale differences if the IPTS-68 realizations in the three laboratories were perfectly coincident.

Finally, when Δt at the gold point for each IMGC thermometer are compared to $(\Delta t)_{NBS}$, data reported in table 4 are obtained. The comparison is based on differences between two values of $\Delta t = t - t_{68}$, where t is defined by equations (3) and (4), while t_{68} is obtained from equations (1) and (2). The NBS average data (nine thermometers) are those published as 'first run' by Evans and Wood (1971). Taking the range as the maximum to minimum difference in a set of results, the data from seven thermometers give average of +7 m°C, standard deviation 40 m°C. The worst results are given by thermometers R1 and R3 which in table 3 show values for t_{Sb} that are too low. Their deviation may perhaps be explained by instabilities in the calibration. If these thermometers are not considered the average difference from NBS is −14 m°C (range = 42 m°C) with a standard deviation of 15 m°C which is very promising for the reproducibility of the platinum resistance scale in the silver-to-gold interval.

Table 4. Comparison at the gold point between each IMGC thermometer and the NBS average.

Thermometer code	R1	R3	R2	R4	L1	J1	TP27
$\Delta t - (\Delta t)_{NBS}$ (m°C)	75	46	7	−10	−13	−35	−20

5. Conclusions

On the basis of this investigation it is feasible to affirm that the standard thermocouple scale can be alternatively substituted by a provisional scale which extends the range of the platinum resistance thermometer up to the silver point. This scale is neither exactly coincident with IPTS-68 nor, probably, is it with the Thermodynamic Scale. However, since the maximum difference from IPTS-68 is about the expected thermocouple inaccuracy and the reproducibility of the extended platinum resistance scale is more than ten-fold improved with respect to the thermocouple scale, the proposed alternative is likely to produce more benefits than disadvantages. It has been proved, in fact, that the scale based on equations (3) and (4) is continuous with IPTS-68 at 630·74 °C within 0·005% of the first derivative and matches the radiation scale at the silver point. This last feature is obtained by raising the assigned temperature of the silver point to 962·07 °C. The resulting silver-to-gold-point interval for the platinum resistance scale is in agreement to better than ±0·05 m°C with that obtained by radiation pyrometry.

A quite similar scale (see equations (3) and (5)) with a lattice vacancy concentration correction term has been considered, but it was discarded since it does not present an improved reproducibility.

As far as the reproducibility of the platinum resistance thermometers at present available is concerned, we observed that it is not adequate for repeated measurements at temperatures around the gold point. But also at lower temperature a selection may be needed since the quality of the current production of standards is not conveniently constant. It seems possible, however, to prepare, even in the laboratory, simple and cheap thermometers which are able to operate satisfactorily up to the silver-point temperature.

Acknowledgments

The authors wish to thank Mr G Bongiovanni and Mr G Frassineti for their valuable assistance in all phases of the experimental work.

They are also very indebted to Dr A Coghe of the CSRPE of CNR for the loan of one standard thermometer.

References

Anderson R L 1972 *TMCSI* 4 part 2 927
Bedford R E 1972 *TMCSI* 4 part 1 15
Berry R J 1965 *Metrologia* 2 80
—— 1966 *Metrologia* 2 80
Bongiovanni G, Crovini L and Marcarino P 1971 *CCT 1971* T66
—— 1972 *High Temp. High Press.* 4 573
—— 1975 *Metrologia* in press
Bonhoure J 1974 *CCT 1974* Doc 12
Chattle M V 1972 *TMCSI* 4 part 2 907
Coslovi L, Rosso A and Ruffino G 1975 *Metrologia* 11 85
Crovini L and Kirby C G M 1970 *Rev. Sci. Instrum.* 41 493
Crovini L and Marcarino P 1974 *CCT 1974* Doc 10
Evans J P 1972 *TMCSI* 4 part 2 899
Evans J P and Wood S D 1971 *Metrologia* 7 108
CIPM 1969 *Comité International des Poids et Measures, IPTS-68, Metrologia* 5 35
Jung H J 1975 this volume
McLaren E H and Murdock E G 1972 *TMCSI* 4 part 1 147
Moser H 1967 *CCT 1967* T91
Quinn T J, Chandler T R D and Chattle M V 1973 *Metrologia* 9 44

Thermal drift correction and precision evaluation by data processing of resistance thermometer comparisons

F Pavese and G Cagna
Istituto di Metrologia 'G Colonnetti', Torino, Italy

Abstract. A method is developed for processing data from comparison measurements intended to be accurate within 0·1 mK. Very accurate correction is made possible for thermal drift even with rates and shapes not usually accepted with conventional procedures, but occurring easily at temperatures above 100 K.
   Transcription errors may be easily detected and recovered with a high confidence level.
   Criteria were established, with which the precision of a comparison can be evaluated, without having to refer to previous measurements and resort to calibrated thermometers. Three of these criteria can be used independently so that the consistency of the precision figures obtained can be checked.

1. Introduction

In cryogenics an accuracy of some tenths of thousandths of kelvin is often required in comparisons of resistance thermometers. Accuracy is affected not only by the thermal measurement and control instrumentation, but also — and in no negligible way — by the method with which measurements are organized and the initial data processed.

Despite the copious literature on the mathematical methods more suitable for expressing the $R-T$ characteristics of thermometers, no information is supplied as to the initial processing of data and in particular on the elimination of thermal drift effects of the cryostat: laboratory techniques are often rudimentary as compared to the sophisticated computerized techniques applied in the subsequent calculations.

The thermal stability of a cryostat decreases with increasing temperature, particularly above 50–70 K. A 0·1–0·2 mK stability over 30–60 minutes (the time necessary for taking all the measurements), if obtainable at all, will therefore require increasingly complex regulators and longer regulation time. Consequently, calculation may represent a valid alternative to the elimination of thermal drift. The simplest method is the approximation of drift with a broken line passing through the experimental points. However, apart from obvious limitations, the method does not consider the fact that to the 'signal', namely thermal drift, a 'noise' is superposed, ie, the scatter of experimental data one would have with zero drift.

We therefore developed a method for processing raw measurement data, which, in addition to making the usual corrections, also corrects the thermal drift of the cryostat with a mathematical reconstruction of the curve using all the experimental points. The method includes guarantees against the risk of forcing experimental data and makes data correction possible in the case of transcription errors, should they exceed casual measurement scatter.

Moreover, it is possible to establish a number of criteria for evaluating precision inside a measurement set. These criteria supply far wider information than the commonly adopted methods, do not require comparison with previous measurements and are consequently applicable to uncalibrated thermometers.

2. Measurement pattern and raw data processing

Measuring resistance thermometers requires the correction of measurements to zero current to eliminate the self-heating effect of the transducer that is produced by the measurement current and is supposed to be proportional to the power dissipated in the thermometer.

Generally at the higher temperatures the value of this correction, obtained from measurements made at two different currents, is higher than the measurement accuracy required. Although the relative precision of the correction need not be high, the absolute precision of the resistance measurement at both currents must obviously be the same. This means that the same number of measurements must be made at the two currents, since there is no reason why random errors should have a different incidence in the two cases should the sensitivity of the measurements be the same for the two currents. This method has been applied at our laboratory and we do not consider profitable the practice followed in several laboratories of measuring self-heating only once 'off-line', or worse still, of making no correction at all.

We therefore use the measurement pattern shown in the caption of figure 1, where four thermometers (X) are measured at two currents (Y) for five runs (Z). We will call the whole 'the measurement set'.

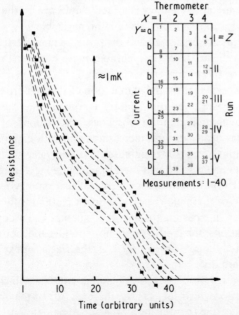

Figure 1. Typical distribution in time of a set of measurements taken in the sequence shown in the table on four thermometers (X), at two currents (Y) for five runs (Z). The broken drift curves have to be reconstructed *a posteriori*: they are exactly superposable along the resistance axis.

The experimental points do not all lie on the same drift curve. The measurement set contains as many drift curves as the product of X and Y (eight in the example), and each drift curve is defined by Z points. A drift curve can be drawn only through the elements of each individual subset (5 out of 40 in the example); this makes the linear interpolation method still less effective. When examining the data layout, it can be observed that most of the differences between the drift curves are due to the systematic differences of resistance between the thermometers.

An assumption is then added: the previously defined drift curves are all exactly superposable through translation along the ordinate axis (resistance). As the temperature drift is itself unique, it follows that the $R-T$ characteristics of the different thermometers must also be the same, except for a constant term. This is justified in the case of platinum resistance thermometers that meet the requirements of the IPTS-68†. In the case of germanium thermometers this may involve some limitations, as will be shown later.

The least-squares method (LSM) is therefore adapted for the case investigated. Each subset is interpolated with a function

$$f'_i(t) = f(t) + \Delta R_i \qquad (i = 1, 7 \text{ in the example}) \tag{1}$$

where $f(t)$ is the function — common to all subsets — that would be used in applying the LSM to each individual subset, and the ΔR_i are the unknown translation parameters.

The normal equations expressed in matrix form are

$$\begin{bmatrix} \mathbf{A}_0 & \mathbf{C}^T \\ \mathbf{C} & \mathbf{D} \end{bmatrix} \begin{bmatrix} \mathbf{a} \\ \mathbf{b} \end{bmatrix} = \begin{bmatrix} \mathbf{c} \\ \mathbf{d} \end{bmatrix} \tag{2}$$

where $\mathbf{A}_0$ is the coefficient matrix relative to function $f(t)$ alone, $\mathbf{C}$ is the translation matrix and $\mathbf{D}$ is a diagonal matrix with equal elements. The vectors $\mathbf{a}$ and $\mathbf{c}$ pertain to the drift equation, $\mathbf{b}$ and $\mathbf{d}$ to the translations ($b_i = \Delta R_i$)‡.

The interpolating function is given by

$$R - R_{1i} = f(t) = a_0 + a_1 \ln(t/t_0 + 1) + a_2 [\ln(t/t_0 + 1)]^2 + \ldots \tag{3}$$

where t is time, t_0 an unknown constant that increases the flexibility of the method and allows fitting to be optimized for each degree of the polynomial (Pavese 1974), and R is the thermometer resistance. R_{1i} is the first value of R in each subset. We use $(R - R_{1i})$ since thermal, and consequently, resistance drift is small with respect to the absolute value. This difference therefore requires a smaller number of significant digits with respect to a direct use of R.

All the measurements are assigned the same weight. This choice does not noticeably affect the quality of fitting, even if the choice is not exact. Verification of the significance of the LSM results is incorporated in the program; the LSM is inserted into a cycle that makes it possible to vary three parameters: n, the polynomial degree, t_0 and M, that of the subsets that is not being translated.

† Strictly speaking it ought to be $W = R/R_0$, but in view of data processing it is the same, and it is simpler, to speak in terms of resistances R.
‡ This result bears a close resemblance to the LSM matrix with constraints, where $\mathbf{C}$ is simply the constraint matrix and $\mathbf{D} \equiv \mathbf{N}$ is a zero matrix.

In the beginning a value of t_0 is established and is then automatically varied inside the cycle in order to minimize an error estimate. This error parameter is the standard deviation (SD) of the differences between the R_C values, as calculated at each measurement time, and the measured values of R.

The maximum data extension is given in figure 1 (with $X = 4$, $Y = 2$, $Z = 5$, whence $M = 8$ maximum). The polynomial degree can vary from $n = 2$ to $n = 6$, a limit supposed to depend initially upon storage capacity but later proving to be connected to the precision of calculation. The minimum extension involves comparison of two thermometers ($X = 2$, $Y = 2$) for three runs ($Z = 3$), whence $M = 4$ maximum.

3. Results of method application

An increase in the order of the polynomial is apt to produce a decrease in the SD that is not necessarily asymptotic but depends on the choice of t_0. For two typical series, to be considered later, the trend followed by the SD in $\mu\Omega$, with varying t_0 and n, is shown in figure 2.

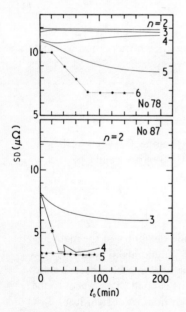

Figure 2. Standard deviation of the differences between measured and calculated resistances with variation of the n polynomial degree, and the t_0 parameter of equation (3). ★ indicates that the parameters for program verification have not allowed calculations to be terminated; ● regular values for which calculations were terminated. With increasing n, a continuous curve can no longer be obtained owing to the presence of 'singularities'.

When we consider the SD behaviour against n, for constant t_0, we observe that a slight improvement may occur when we pass from $n = 2$ to $n = 3$, and a far substantial amelioration is evident when n is increased up to 4 or 5, as happens in measurement set number 78, with $t_0 = 10$. The F test, that is sometimes applied as a significance test for each new coefficient that is found when the order of equation (3) is increased by one (Natrella 1963), gives inconsistent results in this case.

With increasing n singularities will appear. They are values of t_0 that do not pass the significance test established inside the LSM program, so that calculations are interrupted. The singularities will become dominant with increasing n until no more values for t_0 are found to continue calculations. Besides, values of SD optimum do not appreciably improve with increasing n beyond a certain limit. Figure 2 shows that one must stop at $n = 4$ for set number 87, while one can go on as far as $n = 6$ for set number 78. A simple drift curve corresponds in fact to the former case, and an extremely complicated one to the latter (figure 3).

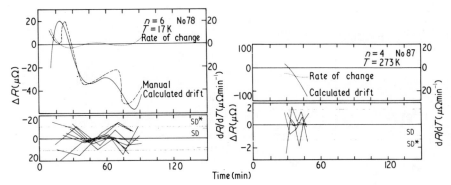

Figure 3. Curves of thermal drift and their derivative for measurement sets number 78 and number 87. The broken line shows the same number 78 drift obtained manually through translation of the experimental points with successive trials. Diagram of deviations between measured and calculated resistances: $SD = [\Sigma \Delta R^2/(n-1)]^{1/2}$; $SD^* = [\Sigma \Delta R^2/(n-k)]^{1/2}$ where k is the number of unknown parameters.

The sudden appearance of singularities for $n = 6$ is an indication that the precision of the calculation will not allow a further increase in n. Even if these facts do not establish a statistical significance, they make possible a detection of an upper limit for the polynomial degree.

Figure 3 shows the capability of the program to represent the more complex drift curves. The accuracy of computation is equivalent to a few μK, ie, to approximately $0 \cdot 1 \, \mu\Omega$ for platinum thermometers.

The program developed also supplies a diagram of the deviation between measured and calculated resistances (figure 3). A coherent behaviour of residuals will suggest an insufficiency of the degree in the interpolating function that was chosen to represent the drift curve, while a random distribution of residuals will indicate that the calculated drift curve is its exact representation. The SD is calculated in the two ways described in figure 3: SD* is nearer to the maximum than to the standard deviation, this being satisfactorily represented by SD.

The deviation charts proved very convenient for revealing errors of measurement transcriptions and for calculating their correction (figure 4). Corrections of one digit only are considered safe and allowable, provided they are higher than the SD.

In the case of germanium thermometers, the fundamental hypothesis of drift curve parallelism can restrict the scope of applicability of the method. In consideration of the very small temperature interval investigated (< 10 mK), the difference between the characteristics of thermometers can be represented by the first derivative dR/dT. If the derivative of two thermometers is different, a deviation from parallelism will occur, the magnitude of which will also depend on the magnitude of the total drift, as appears in figure 5. It is, however, possible to ascertain whether the scatter increase due to this effect keeps within acceptable limits.

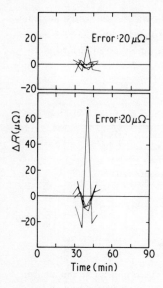

Figure 4. An error of 20 μΩ and 100 μΩ intentionally introduced is clearly visible as to position and magnitude.

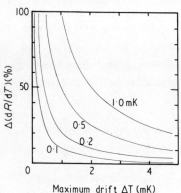

Figure 5. Increase in measurement scatter, produced by a difference between the first derivatives of two thermometers against maximum thermal drift observed. With a 2% difference, a 5 mK maximum drift is necessary to bring about a 0·1 mK scatter.

4. Evaluation of measurement precision

Three mutually independent evaluation criteria can be established. They are based on parameters that are intrinsically stable or unique over each measurement set and whose fluctuations can only be ascribed to imprecision of measurement.

4.1. Curve of cryostat thermal drift

This is intrinsically unique and can be expressed by a mathematical function of sufficiently high degree, within the scatter — characterized by SD — due to measurement imprecision.

4.2. Self-heating of individual thermometers

This depends upon the thermal resistance between the sensitive element and the cryostat serving as a heat sink. Its thermal resistance can be different according to thermometer type, temperature and the way thermometers are mounted in the cryostat. Nevertheless, it remains constant over a measurement set. For each thermometer the scatter of the obtained values is characterized by the maximum deviation from the average.

4.3. Reduction of measurements to the same time

As the temperature of the cryostat varies in time, subsequent measurements can be compared only if they are brought back to the same instant, namely, to the same actual temperature. A meaningful comparison, which is the very purpose of measurements, is therefore connected to the precision with which measurements can be brought back to a common time.

(i) The mean measurement time having been chosen, all the measurements are brought back to this time. Scatter for measurements on each thermometer are characterized as in §4.2.

(ii) Moreover, if the thermometers are calibrated, it is possible to calculate the temperature corresponding to each measurement and thus obtain the calibration difference for each of them with respect to the first, which is taken as a standard. This is possible as well with non-calibrated thermometers using the calibration of another similar thermometer. Obviously, erroneous temperature differences will be obtained, but scatter will still be correct. It is characterized in the same way as in §4.2.

5. Conclusions

The mathematical method adopted for calculating and correcting the drift curve proved to be correct and reliable. With this it is possible, in addition, to evaluate measurement precision.

The evaluation criteria §4.1, 4.2, 4.3 (i) (or 4.3 (ii)) are independent.

A number of criteria are thus available for evaluating the precision of each measurement set, as shown in table 1. Either a whole set or an individual thermometer can be investigated in this way, and the intrinsic consistency of the data obtained can be checked as well.

Table 1. Evaluation of measurement precision obtained with the described measurement data processing.

Measurement set number 80	Criterion			
Thermometer number	§4.1 (mK)	§4.2 (mK)	§4.3 (i) (mK)	§4.3 (ii) (mK)
1	—	0·095	0·15	—
2	—	0·085	0·12	0·085 ‡
3	0·10†	0·22	0·20	0·105
4	—	0·10	0·14	0·145

† SD of the considered measurement set.
‡ Values obtained calculating the calibration differences of the individual thermometers against number 1.

The errors thus analysed are different in the different cases, namely, in §4.1 they concern the individual resistance measurements; in §4.2 and §4.3 (i), the resistance is corrected to zero current: both errors are connected, however differently, to the error produced by the two independent resistance measurements. Finally, in §4.3 (ii), the errors refer to pairs of independent values of resistance corrected to zero current.

The scatter parameter of §4.1 is the standard deviation, while in the other cases the maximum deviation from the average is used, account being taken of the small number of measurements (3 to 5) from which it is calculated.

The measurement pattern applied is confirmed by the fact that repeated measurements of self-heating showed that scatter of the values obtained is not so small as to justify a single measurement (see table 1). This method has been used in all the measurement sets performed at our laboratory and has provided exhaustive information about the precision of comparison measurements, while confirming its steady improvement.

Figure 6 shows the set of SD values obtained with criterion §4.1; on the whole they do not exceed measurement sensitivity. Seventy-five per cent of the SD values obtained for the 1971 measurements remain within 0·2 mK, and 67% within the accuracy limit of the instrumentation. Nearly all the subsequent measurements made with more sensitive and accurate instrumentation remain within 3 μΩ or 0·05 mK or 0·5 ppm.

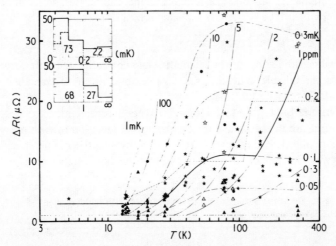

Figure 6. Precision of about 100 measurements as evaluated with criterion §4.1. The curves at constant relative precision (ppm) and at constant precision as to temperature (mK) are also shown. The heavy full curve indicates the present expected accuracy limit of instrumentation. The broken curve indicates sensitivity of measurements ★; that of measurements ▲ and ● is 1 μΩ. In both cases some of the measurements were made with lesser sensitivity: they are indicated with open symbols. The upper histogram indicates the observed frequencies relative to precision in temperature; the lower histogram gives frequencies relative to the expected accuracy of the resistance measurements (1 ≙ limit in bold face). Measurements: ★, 1971, platinum thermometers; ▲, 1972–3, platinum thermometers; ●, 1973, germanium thermometers.

References

Natrella M G 1963 *NBS Handbook* 91
Pavese F 1974 *Cryogenics* 14 425–8

# Resistance thermometry using rhodium–iron, 0·1 K to 273 K

R L Rusby
Division of Quantum Metrology, National Physical Laboratory, Teddington, Middlesex, England

Abstract. A resistance thermometer using an alloy of rhodium with 0·5 atomic per cent of iron has been developed mainly for use in the range 0·4 K to 20 K. At these temperatures the sensitivity of rhodium–iron thermometers is lower than that of germanium thermometers, but the excellent long-term stability of a calibration is a major advantage in standards work. The quasilinear (rather than exponential) variation of resistance with temperature also allows some simplification of calibration procedure, and gives the thermometer its exceptionally wide range of usefulness.

1. Introduction

Anomalies in the properties of dilute magnetic alloys such as the resistance minimum at low temperatures (the Kondo effect) are well known, but an unexpected behaviour was found by Coles (1964) when he investigated the resistivities of alloys of rhodium with less than 1 atomic per cent of iron. The impurity contribution was found to have a large positive temperature coefficient particularly below 30 K so that the total resistivity decreased monotonically on cooling from room temperature. The rate of this decrease led Coles to suggest that rhodium–iron could form the basis of a resistance thermometer whose usefulness would extend from room temperature down to temperatures well below the range of the platinum thermometer.

The rhodium–iron thermometer has been developed at NPL mainly to provide an alternative to the use of germanium thermometers in the range 0·4 K–20 K, which includes the range of practical helium vapour pressure thermometry and overlaps the IPTS-68. It was felt that the resistance–temperature characteristics of rhodium–iron should be more amenable to curve fitting and that a single thermometer could cover the whole range. It was also felt that a wire element, if well mounted, might be more reproducible than a germanium chip in which so much depends on the contacts and the manner in which the chip is suspended. From the standards point of view an improvement in reproducibility is well worth the considerable reduction in sensitivity which the choice of rhodium–iron over germanium thermometers entails.

The results of preliminary testing were encouraging and have been published (Rusby 1972). The alloy chosen contained 0·5 at.% Fe and the resistivity at this composition is plotted as a function of temperature in figure 1. It has since been shown (Rusby 1974) that at higher concentrations interactions between iron atoms cause magnetic ordering and a loss of sensitivity at low temperatures. Other alloy systems have been found to exhibit similar effects but none has the sensitivity of RhFe.

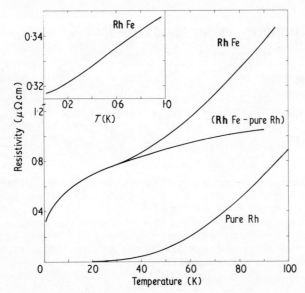

Figure 1. Resistivity of Rh–0·5% Fe alloy, pure Rh (from White and Woods 1959) and of Rh–0·5% Fe after subtraction of the host resistivity.

2. Construction and performance

The alloys used were prepared by Engelhard Industries Ltd by powder metallurgical methods. The major considerations are that the iron should be completely dissolved in an unoxidized state and that the final anneal, which in Engelhard's process is at 1200 °C, should recrystallize the material.

In constructing a thermometer it is desirable to achieve a high value of resistance subject to the necessity of mounting the wire free of strain. The method of encapsulation is similar to that used for platinum thermometers. The wire, of 0·05 mm diameter, is coiled and inserted into four glass tubes (two of which are shown in figure 2) and platinum current and potential leads are flame-welded to each end. The assembly is then given a strain-relieving anneal at 700 °C before being inserted in a platinum sheath and sealed in helium, at a pressure of about 30 kPa by a glass bulb seal. Thermometers of two sizes have been produced: type U in which the sheath length is about 3 cm and the resistance at 4·2 K is about 3·5 Ω (rising to about 6 Ω at 20 K and 50 Ω at 273 K),

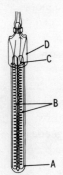

Figure 2. Semi-schematic diagram of a rhodium–iron thermometer: A, platinum sheath approximately 5 mm in diameter, 35 mm (type U) or 50 mm (type W) in length; B, two of four glass tubes containing coils of wire; C, flame-welds to platinum leads; D, glass bulb seal.

and type W in which the sheath is about 5 cm long and the resistance values are about twice those of a type U thermometer.†

Two important quantities in resistance thermometry are fractional sensitivity $(1/R)(dR/dT)$ and voltage sensitivity dV/dT. The former is an intrinsic property of a material and in rhodium–iron it is similar to that of pure Rh (or Pt) at high temperatures but passes through a minimum of about 1%/K near 28 K before rising to more than 10%/K at 1 K due to the impurity effect (figure 3).

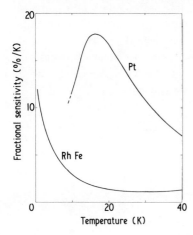

Figure 3. Fractional sensitivity $(1/R)(dR/dT)$ for rhodium–iron and platinum.

For low-resistance thermometers the voltage sensitivity is often limited by the self-heating effect since the Joule dissipation V^2/R tends to be large for any easily measurable voltage. For instance, in platinum the fractional sensitivity reaches its maximum value in the region below 20 K, but the voltage sensitivity of a platinum thermometer is small. The relatively low fractional sensitivity of rhodium–iron is more than compensated by the higher resistance values of the thermometers so that the voltage sensitivities at an acceptable measuring current are superior to those of the standards-type platinum thermometer below 20 K and are comparable at higher temperatures (figure 4). In the case of germanium thermometers the high resistances and fractional sensitivities ensure

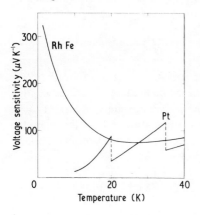

Figure 4. Voltage sensitivity dV/dT of a type W rhodium–iron thermometer at 0·5 mA current, and of a platinum thermometer ($R_0 = 25\,\Omega$) at 5 mA below 20 K, 2 mA from 20 K to 35 K and 1 mA above 35 K.

† These type descriptions are abbreviated forms of those used by the manufacturer, H Tinsley & Co Ltd.

that good voltage sensitivity can be obtained without significant self-heating. However, from figures 3 and 4 it is seen that an accuracy equivalent at 20 K to 0·1 mK is obtainable from a type W rhodium–iron thermometer at 0·5 mA current if the accuracy in resistance measurement is 1·3 ppm and the voltage resolution is 8 nV. This is within the capability of many instruments and at the NPL a calibrated Tinsley 'stabaumatic' potentiometer and, with slightly less accuracy, an ASL 'cryobridge' have been used. The self-heating effect at 0·5 mA, however, is significant in standards work and is acceptable only because it is reproducible. Figure 5 shows that its magnitude, which is about the same in both type U and type W thermometers, increases on cooling, except for the dip produced by the additional conductivity of condensed (superfluid) ^{4}He. Below 1 K thermal contact becomes increasingly poor so that it is essential to reduce the measuring current, and in the region of 0·4 K the lower limit of usefulness of the thermometer is reached. An extension of the range to 0·1 K and below can be achieved in a dilution

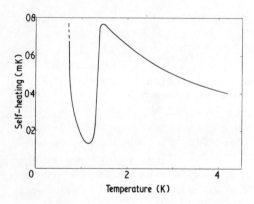

Figure 5. Self-heating of a rhodium–iron thermometer at 0·5 mA below 4·2 K. The value at 20 K is about 0·2 mK.

refrigerator if direct contact is made between the rhodium–iron wire and liquid-helium mixture. In this mode, using a crudely wound sample, we achieved the results shown in figure 1 down to 0·02 K. Here thermal contact is again a problem, and the need to measure resistance to 1 in 10^4 at low current levels in order to achieve a precision of a few per cent in temperature offers little incentive for its solution.

3. Reproducibility

One of the major requirements of a resistance thermometer is that its calibration should be reproducible. It has become increasingly clear that because of their susceptibility to calibration shifts on thermal cycling, germanium thermometers of current manufacture are not suitable for the maintenance of a temperature scale of the highest precision. Tests with rhodium–iron show excellent reproducibility in the short term, partly no doubt because contact effects are unimportant.

A further concern is the long-term stability of a calibration. The resistivity of rhodium–iron may change as a result of induced strain, room-temperature annealing or diffusion of iron, and the sign of the change is not predictable. Repeated calibrations of two of the type U thermometers against T_{58} over a period of 2·5 years indicate that the resistances have decreased at the rate of about 5 ppm/year, equivalent in this range to less than 0·1 mK/year. Three type W thermometers have been tested over 20 months at 20 K. At this temperature the fractional sensitivity of rhodium—iron is near its minimum and small real changes in calibration are most readily observed. Comparisons of the thermometers with each other and with a platinum thermometer show that the calibrations have been consistent to ±0·3 mK (±4 ppm in resistance). These limits are about three standard deviations in the accuracy of the measurements and therefore represent a pessimistic upper bound for the stabilities of the thermometers themselves. It is worth noting that the consequences of a change in calibration could be largely eliminated by the use of resistance ratios (analogous to the Z functions used in platinum thermometry), which are relatively insensitive to the causes of the change.

4. Interpolation properties

The relatively simple variation of the resistivity of rhodium—iron with temperature suggests that the interpolation of values of temperature between calibration points should not be unduly difficult and this is indeed the case. For example, calibration data for three thermometers obtained by precision gas thermometry at 21 temperatures in the range 2·6 K to 20 K could be approximated by eighth-order Chebyshev series, with a standard deviation of $\lesssim 0·2$ mK. Refitting using points at only 11 temperatures gave the same curves to ±0·3 mK, a result due in part to careful spacing of the data. It may be noted that it is not necessary to take logarithms either of resistances or temperatures in fitting rhodium—iron calibrations. The measurements therefore may be more evenly spread over the temperature range than is the case in germanium thermometry. It is also significant that lower-order curves give quite good accuracy, the standard deviations for orders 3 to 8 being 9, 2·5, 1·5, 0·5, 0·2$_3$ and 0·1$_8$ mK respectively. Our experience in fitting less accurate data in the range 0·6 K to 20 K indicates that the downward extension of the range is not accompanied by significant deterioration in the fitting, but we may expect some difficulty in fitting through 28 K where the minimum in dR/dT occurs.

It is tempting to try to use the rhodium—iron thermometer as an interpolation instrument in the manner of the IPTS-68; that is, to select the calibration of one thermometer as a reference function, expressed in terms of resistance ratios against temperature, and subsequently to calibrate other thermometers by interpolating differences from the reference function at a number of fixed points. These would be obtained from helium vapour pressures up to 5·2 K and the IPTS-68 above 13·81 K and perhaps also the superconducting transition temperature of lead near 7·2 K. In practice we have found that resistance ratios for two thermometers, even if made from the same batch of wire, are not in general sufficiently similar for this method to yield either accuracy or uniqueness to better than 1 mK. We have also found that a polynomial fit to the gas thermometry data omitting all points between 5·2 K and 13·81 K did not depart by more than 2 mK from the full-range fit, and with an extra point at 7 K the maximum departure was reduced to 0·6 mK. Neither of these procedures is really satisfactory, however, as the accuracy of interpolation can only be checked by taking more data. While this consideration also applies to the calibration of platinum thermometers using the IPTS-68, it becomes more significant when dealing with an impure material whose batch-to-batch reproducibility is inherently less certain, and it would therefore seem that improved metallurgical reproducibility and if possible the development of a more evenly spaced set of fixed points would be required if the thermometer is to be suitable as the interpolating instrument for a future extension of the IPTS.

Acknowledgments

The author wishes to thank Professor B R Coles of Imperial College, London for his advice and continued interest in the project, and Dr K H Berry of NPL for the use of calibration data based on gas thermometry prior to publication.

References

Coles B R 1964 *Phys. Lett.* 8 243–4
Rusby R L 1972 *TMCSI* 4 part 2 865–9
—— 1974 *J. Phys. F: Metal Phys.* 4 1265–74
White G K and Woods S B 1959 *Phil. Trans. R. Soc.* **A251** 273–302

Resistance thermometers with MOS field effect transistors

I Eisele and G Dorda

Forschungslaboratorien der Siemens AG, München, Germany

Abstract. The temperature-dependent resistance change of n- as well as p-type MOS transistors has been investigated. Implantation of He ions increases the temperature sensitivity of the n-channel devices. The influence of magnetic fields up to 80 kG has been studied.

MOSFET (metal—oxide—semiconductor field effect transistor) devices fabricated on silicon have been investigated for use in resistance thermometry. The basic idea is to develop a sensor which has an extremely small geometry and whose relative resistance change can be controlled. MOS transistors meet both requirements to a large extent.

Transistors according to the schematic presentation in figure 1 have been fabricated on p- as well as n-type 10 Ω cm silicon. A 120 nm silicon oxide was thermally grown on (111) and (100) surface orientations. Two different sample geometries have been measured: (i) channel length $L = 40$ μm, channel width $W = 400$ μm and (ii) $L = 400$ μm, $W = 40$ μm. The depth of the channel depends on the applied gate field normal to the surface and varies at low temperatures between 5 and 20 nm. The carrier transport between source and drain is restricted to an inversion layer as shown in figure 2. The band bending in the z-direction (normal to the surface) is due to the applied electrical surface field and the nearly triangular potential well causes a quantization in energy E denoted by E_i with $i = 0, 1, 2 \ldots$ (Dorda 1974). This effect prohibits a motion of the charge carriers perpendicular to the surface if $kT < (E_i - E_{i-1})$. For a n-type inversion on p-silicon the restriction holds in general even at room temperature whereas the energy splitting for a p-type inversion on n-silicon is lower mainly for a (100) surface (Y Uemura private communication, Bangert and Landwehr 1975). If the magnetic

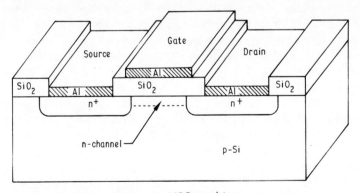

Figure 1. Schematic diagram of a MOS transistor.

field is parallel to the surface the described quantization phenomena in the inversion layer can be utilized to suppress motion due to Lorentz forces. As a consequence the magnetic field dependence of the device is small.

The resistivity of the channel is determined by the gate-field-induced carriers and their mobility. The carrier concentration for a MOS structure is only a function of the applied electric surface field, in other words it is capacitively induced by the gate voltage and therefore temperature independent. Carrier freeze-out does not influence the

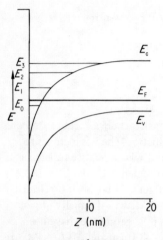

Figure 2. Energy-level diagram representing electrons bound in quantized sub-bands E_i (i = 0, 1, 2, ...) at the semiconductor surface. E_c and E_v are conduction band and valence band edge, respectively. E_F is the Fermi level.

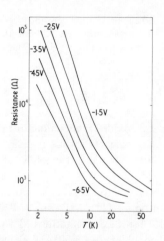

Figure 3. Resistance versus temperature for a p-channel on (111) Si. The parameter is the effective gate voltage $V_G - V_T$. Transistor geometry: width W = 400 μm, length L = 40 μm.

temperature dependence of the resistivity and the mobility μ remains as the only temperature-dependent factor. This is the basis of reliable measurements at low temperatures. The temperature dependence of μ exhibits a maximum which is determined by scattering mechanisms and by the quality of the Si–SiO$_2$ interface. Sensitive temperature detection is possible on the low-temperature branch of the curve. It has been demonstrated that the scattering in this region is mainly dominated by ionic impurities in the channel (Dorda 1974). Evidently the position of the maximum and the slope of the profile can be controlled in a well defined manner by ion implantation of non-doping impurity centres.

Figure 3 presents a plot of the resistance versus temperature for a p-channel MOSFET with (111) surface orientation. The parameter is $V_G - V_T$ where V_G denotes the gate voltage and V_T the threshold voltage of the channel. The transistor was made by the usual standard MOS process without special selection. In the measured temperature range down to 2 K it is obvious that the sensitivity increases sharply with decreasing temperature. For $V_G - V_T$ = −4·5 V the relative resistance change α (4·2 K) = $\Delta R / (R \Delta T)$ appears to be 0·55 K^{-1}. With increasing $V_G - V_T$ a shift of the profile towards lower temperatures is observed and at the same time the resistance decreases by one order of magnitude. This behaviour is due to a change of mobility as well as carrier concentration with gate field. It should be pointed out that this tunable resistance change is advantageous for practical applications and cannot be achieved with other sensors. From our data it can be deduced that the useful detection range can be swept from 80 K down to 2 K and below. Because of our experimental set-up a lower limit has not yet been established.

Measurements of the relative resistance change in magnetic fields B up to 80 kG are shown in figure 4 for various $V_G - V_T$. These results are typical for any direction of B in the plane of the MOS surface. For all other directions the magnetoresistance effect increases and approaches a maximum for B perpendicular to the surface (Eisele and Dorda 1974). At 4·2 K the corresponding temperature inaccuracy for the profile with $V_G - V_T = -4·5$ V amounts to $\Delta T = 0·015$ K at 10 kG and $\Delta T = 0·3$ K at 80 kG. Such low values have only been obtained for (111) planes whereas (100) data yield five times larger effects. This can be understood on the basis of recently calculated energy level diagrams within the framework of the quantization model (Y Uemura private communication, Bangert and Landwehr 1975).

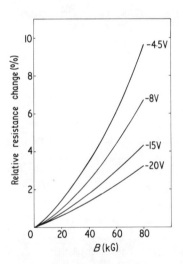
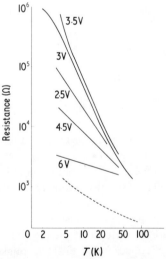

Figure 4. Relative resistance change versus magnetic field B for a p-channel on (111) Si. Magnetic field parallel to the channel current, ie in [110] direction. The parameter is the effective gate voltage $V_G - V_T$.

Figure 5. Resistance versus temperature for a He-implanted n-channel on (111) Si (Solid curves). Transistor geometry: $W = 400$ μm, $L = 40$ μm. For comparison, an equivalent but unimplanted transistor is shown (broken curves). The parameter is the effective gate voltage $V_G - V_T$.

A sensor which exceeds the qualities of the described p-channel device was obtained by implanting ions into a n-channel MOSFET. Best results were observed with a dose of 1×10^{12} He ions/cm^2 implanted at 15 keV, in other words the distribution maximum was located at the Si–SiO$_2$ interface. After implantation a 10 min hydrogen anneal at 400 °C was performed in order to lower radiation damage. Figure 5 demonstrates that the highest sensitivity is obtained only in a narrow region of $V_G - V_T$. It should be mentioned that the decreasing resistance for $V_G - V_T < 3 \cdot 5$ V is a typical anomalous behaviour for implanted samples. At 4·2 K we obtain the sensitivity α (4·2 K) = 0·47 K^{-1} for $V_G - V_T = 3$ V.

The dependence on magnetic field as shown in figure 6 is far below the values for the p-channel and has a smaller gate-field dependence. The existence of a quantized

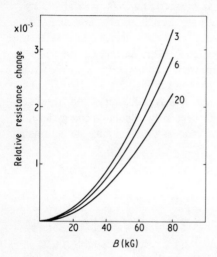

Figure 6. Relative resistance change versus magnetic field B for a He-implanted n-channel on (111) Si. Current is in the [11$\bar{2}$] direction and magnetic field in the [1$\bar{1}$0] direction.

sub-band system hinders the motion of electrons normal to the surface and because for n-channel samples the energy splitting is more pronounced than for p-channel the magnetoresistance effect diminishes. From the obtained data at 4·2 K the following temperature errors are caused by the magnetic field: $\Delta T = 10^{-4}$ K at 10 kG and $\Delta T = 10^{-2}$ K at 80 kG.

Similar small changes of the resistance with parallel magnetic field have been detected with other unimplanted n-channel devices. However, in this case the ion impurity is too small to influence appreciably the scattering mechanism which in turn weakens the temperature dependence of the resistance.

In summary it appears that the MOS transistor could be useful as a temperature sensor in the region from 1 K to about 100 K. Some advantages of such a device are: (i) extremely small geometry due to the planar technology; (ii) tunable resistance change with gate voltage; (iii) high temperature sensitivity for p- as well as implanted n-channel transistors, and (iv) small or negligible influence of magnetic fields parallel to the surface plane.

References

Bangert E and Landwehr G 1975 to be published
Dorda G 1973 *Festkörperprobleme XIII*, ed O Madelung and H J Queisser (Braunschweig: Vieweg)
—— 1974 *Electronica y Fisica Aplicada* 17 203
Eisele I and Dorda G 1974 *Proc. 12th Int. Conf. on Physics of Semiconductors, Stuttgart* ed M Pilkuhn (Stuttgart: B G Teubner Verlag) p704

Inst. Phys. Conf. Ser. No. 26 © 1975: Chapter 3

The 220 Ω Allen–Bradley resistor as a temperature sensor between 2 and 100 K†

B W A Ricketson

Cryogenic Calibrations Ltd, Pitchcott, near Aylesbury, Buckinghamshire, England

Abstract. The Allen–Bradley resistor has excellent resistance–temperature characteristics but the reproducibility is poor, being about ±0·1 K at liquid helium and ±5 K at liquid nitrogen temperatures. An experimental programme was conducted to find a method by which a calibration once obtained could be re-used by measuring again the sensor's resistance at one or two known temperatures. As the effect of water vapour is to alter appreciably the temperature-dependent resistance of the sensor, vacuum-dried and epoxy-potted resistors were calculated.

Using 20 K and 60 K as the reference temperatures, it is shown that over five calibrations, ±10 mK accuracy can be obtained between 2 and 60 K with certain restrictions on the time over which the measurements are made.

1. Introduction

Clement and Quinnel (1952) showed over 20 years ago that certain carbon radio-resistors had very desirable resistance–temperature characteristics, and $\frac{1}{4}$ W and $\frac{1}{8}$ W Allen–Bradley resistors have been used as a cheap and convenient method of temperature measurement ever since. Unfortunately they are not stable under temperature cycling and if used for accurate temperature measurement, must be calibrated each time they are cooled. Kopp and Ashworth (1972) have shown that improved stabilities are obtained if the resistors are held under vacuum and cooled slowly.

In fitting calibrations to carbon resistors, Rose-Innes (1973) and White (1968) give excellent guidance through the very large number of suggested formulae but with the exception of Lounasmaa (1958) none of the authors have indicated how changes in the resistance on recooling can be taken into account.

The cryostat consists of a copper block with an isothermal shield, surrounded by, but thermally isolated from, a can which can be cooled to either liquid nitrogen or liquid helium temperatures. Calibrations appear to be reproducible to ±3 mK on germanium, platinum and rhodium–iron sensors. The temperature standards are three germanium and three platinum sensors; two of each type having been calibrated by the National Physical Laboratory to an accuracy of ±5 mK of the Helium Vapour Pressure Scale (58) and the IPTS-68.

The 220 Ω, $\frac{1}{8}$ W BB resistor was chosen as having a good sensitivity without too large a resistance at low temperatures. The resistance changes from around 3500 Ω (sensitivity 2000 Ω K^{-1}) at 4·2 K to 550 Ω (20 Ω K^{-1}) at 20 K. At 80 K, a resistance of 290 Ω and a sensitivity of just over 1 Ω K^{-1} still allows useful resolution with the simplest of detectors (10 μV K^{-1}).

† Allen–Bradley Company, Milwaukee, Wisconsin (January 1967) Technical Bulletin 5000.

2. Analytical approach

Curves of log $(dR/R\,dT)$ plotted against log T for carbon resistors between 10 and 1000 Ω nominal resistance are approximately straight up to a temperature of 100 K and, what will be shown to be of more significance, the curves appear to be parallel, particularly between 10 and 80 K. A graph for resistors of low nominal resistance and at low temperatures appears in the paper by Clement and Quinnel (1952).

Assuming that the lines are straight, integration leads to an equation of the form

$$\log R - B = A''(T)^{-m} \tag{1}$$

where A'', B and m are constants. This equation is not easily solved. A spline fit of a 220 Ω sensor calibration was made between 2 and 20 K and a 'best fit' obtained by successive approximations of the constants B and m in equation (1). Differences between the temperature derived from the best equation and the true temperature are shown in figure 1. This fit to within 50 mK is satisfactory as indicating that the form of the equation can be used with an accuracy comparable above 4 K with both Clement and Quinnel (1952) and Zimmerman and Hoare's (1960) empirical formulae with three constants (an accuracy of 0·5% of T) but not giving as good a fit as Kes *et al* (1974) whose modification to the above equation has four constants.

To derive an equation suitable for all nominal values of sensor resistance the equation can be normalized by dividing it into the value of the function at 4·215 K.

$$\frac{\log R_{4\cdot 215} - B}{\log R - B} = \left(\frac{T}{4\cdot 215}\right)^m. \tag{2}$$

This equation shows a relationship between resistance and temperature of the form indicated by Clement *et al* (1956). If a scaling constant (A') is introduced and m is assumed to be a function of temperature only, we can define a function.

$$F(T) = \frac{A'(\log R_{4\cdot 2} - B)}{\log R - B} = \frac{A}{\log R - B} \tag{3}$$

where A and B are constants which can be found once the function is known.

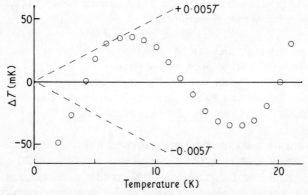

Figure 1. The difference between the temperature derived from the spline fit of a calibration and the function $\log R - 2\cdot 397\,78 = 1\cdot 121\,44 \times T^{-0\cdot 7575}$, for sensor 159. 'Best fit' between 4 and 20 K. All points equally weighted.

This work is concerned with finding experimentally the form of the function, its validity and the effect of resistance changes on the value of the constants.

3. The effect of adsorbed water on resistance

The Allen–Bradley bulletin for hot moulded resistors shows that a typical 5 to 10% increase of room-temperature resistance occurs after a steady state soak of 240 h at 95% humidity. First trials of drying the resistors in an oven were unsuccessful, in that, although the resistance was reduced, the process was not reversible. However, by placing a resistor in a rough vacuum and heating it gently the resistance could be reduced to a value which did not alter with time after 24 to 48 h.

Figure 2 shows for ten resistors from a freshly opened sealed plastic container a plot of the resistance in liquid nitrogen against that at room temperature. Measurements are shown for the resistors, as received from the manufacturer, after drying in the manner described above, and after sealing in a tin containing damp cotton wool for 14 days.

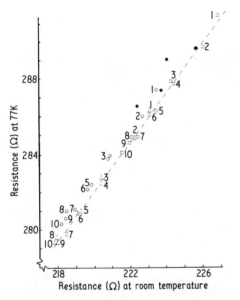

Figure 2. The effect of water vapour on the resistance of a batch of ten 220 Ω resistors. ○, as from sealed case; △, after drying; □, after 15-day soak at 100% humidity; ●, another batch.

There is a close relationship between the room and liquid nitrogen temperature resistances for each of the three sets of measurements. A few measurements from another batch measured over three years ago do not give results lying on any of the lines so that it may be deduced that the correlation within each set of measurements is only peculiar to sensors from a single batch. What is of greater interest is the large rise in resistance after the sensors were stored in a humid atmosphere, the slope of the line and the fact that $\Delta R_{77}/R_{77} \simeq \Delta R_{300}/R_{300}$, indicating that adsorbed water is introducing a temperature-dependent resistance.

Because of this result, all future measurements were made on resistors which had been dried and then potted with an epoxy resin in copper cans of 7 mm length and 3·1 mm OD. They will be referred to as sensors and were treated as any other commercial sensor. Before each calibration the leads were reconnected by soldering.

4. Stability of sensors in liquid nitrogen

Five potted carbon sensors were placed with a platinum sensor in an aluminium block which was immersed in a 25-litre liquid nitrogen container. Figure 3 shows the variation in resistance of the sensors over 12 days. The measured resistances were corrected for variations in the bath temperature. On the eighth day, the block was removed for 24 h to room-temperature conditions and then re-immersed for a further three days. It should be noted that:

(i) Drifts in resistance can be either positive or negative.
(ii) Drift rates equivalent to 50 mK h^{-1} after first being immersed and 25 mK/day after a two-day initial soak period are possible.
(iii) On re-immersion the resistance jumped in the direction in which it had been drifting.
(iv) The resistance of sensor 262, increased on first being immersed, decreased after its second immersion.

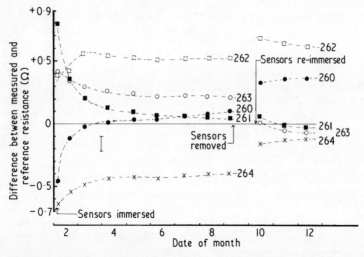

Figure 3. Variation of the resistance of five resistors in liquid nitrogen with time. The reference resistance for each sensor is given in table 1.

Table 1. Reference resistance values to be used with figure 3.

Sensor number	(1)	(2)	(3)
260	274·537	275·0	277·0
261	279·793	279·0	278·0
262	268·383	268·0	268·0
263	273·416	273·0	272·0
264	286·866	287·5	290·0

(1) Resistance 20 min after the first immersion.
(2) Reference resistance from 2–9 July.
(3) Reference resistance from 10–13 July.

The drift of resistance with time is noticeable when calibrating carbon sensors. A normal calibration run consists of cooling down overnight on day 1 to 77 K, cooling to 2 K and calibrating up to 60 or 70 K on day 2, continuing above this temperature on day 3. It will be shown that the drift is probably associated with changes in the constant B in equation (3). Consequently, at low temperatures although the drift is present, the change in resistance is small compared with the transfer of the temperature to ±3 mK. Above 50 K the effect is measurable and often observed over the eight-hour period between day 2 and day 3.

5. Measurements on one sensor

Five runs (25, 26, 27, 28 and 29) covering a period of five months were made on one encapsulated sensor (number 202). On run 29, 80 K was reached on the evening of day 2 and after being repeated the morning after showed a resistance increase of only 0·011 Ω (equivalent to 9 mK). Because of this small change this run was used to generate the function $F(T)$. Smoothed resistance values and corrected values assuming a step change in B for the 90 and 100 K points were obtained. These smoothed values of resistance were inserted into equation (3) with $A' = 1$, by definition, $B = 2·301 030$ an informed guess but shown not to be critical in terms of its absolute value, and $\log R_{4·215} = 3·563\,185$ from the measurements but expressed to six places by definition. A spline fit as described by Ward (1972) of $F(T)$ against T was made with alternate calibration points and the differences with the remaining points noted. Maximum differences of less than 2 mK below 20 K and 6 mK below 80 mK were considered satisfactory and a table of $F(T)$ in terms of T generated using all the points. This table was used to compare all other measurements made on this sensor and as a reference table for sensor 203.

The maximum change in resistance over the five calibrations was equivalent to 13 mK at 4·2 K, 337 mK at 20 K and just over 2 K at 60 K.

In figures 4 and 5 the differences between T, the measured temperature and T_F, the temperature calculated from $F(T)$ are plotted. The values of A and B were found from the 20 K and 60 or 70 K points. Below the upper reference temperature, the values of $T - T_F$ fall within a ±8 mK band except for the 25 and 30 K points of run 28. Table 2 gives the resistance values and constants used in equation (3) for each run. The values of $\log R_{60} - B$, $\log R_{20} - B$ and $\log R_{4·215} - B$ are also given. These latter values remain constant to the equivalent of ±35, ±13 and ±5 mK respectively, suggesting that to a close approximation, but outside experimental error, it is the temperature independent constant B that is changing.

By definition, at a given temperature $dA = F(T)\,d(\log R_T - B)$. Therefore from changes in $(\log R_T - B)$ for each run, it should be possible to calculate the new value of A. This is shown in table 2 in the column headed A_{calc} which has been found from the 60 K values. The close agreement between A and A_{calc} indicates that the form of equation (3) is correct in that A is reflecting the changes in B, eg, A' is a constant associated with a particular sensor. This is confirmed to a lower degree of accuracy by a direct calculation from the 4·215 K resistances.

Figure 5 gives the same information for temperatures above the upper reference point. For run 27 the measurements were obtained 24 h before the low-temperature

Table 2. Selected resistances and constants, sensor 202.

Run no.	$R_{4.215}$	R_{20}	R_{60}	A
25	3626·2	535·93	313·100	1·26209
26	3627·4	538·94	314·825	1·26172
27	3627·4	537·90	314·299	1·26096
28	3623·3	536·91	313·825	1·26091
29	3653·6	542·67	317·030	1·262155
Variation	30·3	6·74	3·930	0·00119
Equivalent (mK)	15	337	2005	—

Run 29 was used to make the reference scale. $A_{\text{calc}} = 1 \cdot 262155 - 6 \cdot 3085 (0 \cdot 200\,072 - (\log R_{60} - B))$.

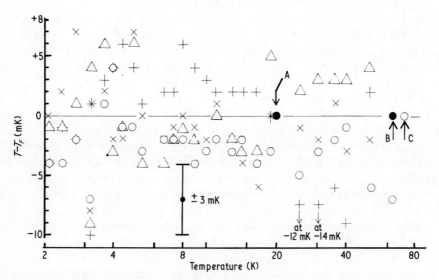

Figure 4. Difference between measured temperature and the temperature derived from $F(T)$. Low-temperature results for sensor 202. See table 2 for constants. ○, run 25; ×, run 26; △, run 27; +, run 28. Point A, reference point all runs; point B, reference point runs 26, 27 and 28; point C, reference point run 25.

points were taken. For run 28, the differences are large reaching nearly 0·2 K at 100 K, but this may be due to a large time-dependent temperature change. Generally, the differences are much larger than those below the reference point. Since the difference between log R and B is quite small in this region, it may be associated with the wrong choice for the value of B. Notwithstanding, the function can still be used for moderately accurate temperature measurements by normalizing the equation at the higher temperatures. This is shown in figure 5 for run 28, the reference temperatures being 80 and 100 K.

6. Measurements with a second sensor

The resistance of sensor 203 was measured on the same runs as sensor 202. Figure 6 shows the difference between the temperature and the temperature calculated from the

The 220 Ω Allen–Bradley resistor as a temperature sensor

B	$\log R_{4.215} - B$	$\log R_{20} - B$	$\log R_{60} - B$	A_{calc}	A'
2·295620	1·2639	0·43350	0·200060	1·26208	0·9986
2·298067	1·2615	0·43337	0·19999	1·26164	1·0002
2·297461	1·2624	0·43311	0·199881	1·26095	0·9995
2·296814	1·2623	0·43309	0·199873	1·26090	0·9989
2·301030	1·2622	0·43351	0·200072	Reference	1·0000
0·005410	0·0024	0·00042	0·000197	–	0·0016
–	10	26	73	–	6

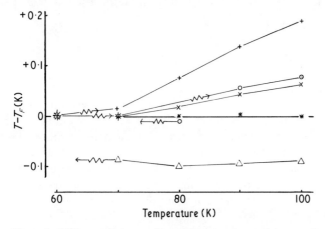

Figure 5. Difference between measured temperature and temperature derived from $F(T)$. Temperatures above the upper fixed point for sensor 202. See table 2 for constants. ~~, time break; ○, run 25, eight-hour break (80 K 36 hours earlier); ×, run 26, eight-hour break; △, run 27, 24 hours earlier than 60 K point; +, run 28, eight-hour break; ∗, run 28 refitted at 80 K and 100 K.

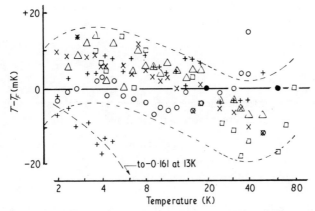

Figure 6. Difference between measured temperature and the temperature calculated from $F(T)$ for temperatures between 2 and 80 K for sensor 203. Reference points at 20 and 60 K. See text concerning low-temperature points of run 28. ○, run 25; ×, run 26; △, run 27; +, run 28; □, run 29.

function derived from the sensor 202. Agreement between runs is to ±10 mK but this could be reduced to ±7 mK if the lower reference point of run 25 was considered wrong and changed in resistance by the equivalent of 6 mK. The differences for temperatures above the fixed point are not shown but are similar to the results of sensor 202 in figure 5.

In run 28, a discontinuity in resistance of 9 Ω occurs between the 13 K and 14 K measurements, which is equivalent to a 0·16 K step change. In over 70 calibrations of Allen–Bradley resistors a step change during a calibration has never been observed and so is an isolated phenomenon in the experience of the writer. Letting $T - T_F = +0.005$ K at 13 K and keeping A constant, it is possible to renormalize the function by attributing the change in resistance to be solely a change in the constant B. The results are plotted in figure 6.

7. Discussion

The average of the points for $T - T_F$ shown for sensor 202 in figure 4 is -0.7 mK and the standard deviation 3·7 mK which are both low figures in relation to the errors that are inherent in the measurements and in the analysis. These are estimated for temperatures below 60 K to be ±3 mK for the reference function, ±3 mK for transferring the calibration from the standards, which must be doubled since it includes the reference points, and ±1 mK in obtaining T_F from the tables. To these uncertainties must be added at the higher temperatures any drift in resistance associated with the time at which the measurements are made, since a time lapse of about six hours passes between taking the 20 K and 60 K points.

Since changes in the calibration can be counteracted by changes in B, $\Delta B = \Delta R/R$ and therefore the mechanism is geometrical. One may postulate a model in which contacts between carbon particles are broken and remade as strains are relieved with thermal cycling or with the removal or addition of water. If macroscopically the mixture is of the same consistency A will remain constant and the conductance will only alter proportionally with the average cross section available for the current. If the mixture does vary in its electrical properties, A will alter as various current paths are used within the resistor.

8. Recommendations

From the information presented in this paper a few recommendations in using Allen–Bradley resistors as temperature sensors between 2 and 100 K are given:

(i) Choose as high a nominal resistance sensor as is compatible with the measuring circuit since under almost any criterion of sensitivity a higher resistance sensor has a better performance, particularly at the higher temperatures.
(ii) Keep the sensor dry.
(iii) The sensor should be left to 'soak' at a low temperature for as long as possible if maximum stability is to be achieved.
(iv) After calibrating the sensor, choose a value of B equal or close to $\log 0.9 R_{\text{nominal}}$ and normalize at some convenient fixed temperature to find A letting $A' = 1$. Fit

$A/(\log R - B)$ against T using a spline or polynomial method to find $F(T)$. On reusing the sensor, measure the resistance at two known temperatures preferably in the higher part of the temperature range. Recalculate A and B and use for calculating the other temperatures with the previously generated $F(T)$. If only one temperature is available for measuring resistance, find a new value for B and change A by $F(T)\,\mathrm{d}(\log R_T - B)$.
(v) Always check the sensor has not drifted or made a step change in resistance during your measurements.

Acknowledgments

The author wishes to thank Mr D Ward for the spline fitting and Dr R W Hill and Professor A C Rose-Innes for reading the draft and for many helpful criticisms.

References

Clement J R, Dolecek R I and Logan J R 1956 *Advances in Cryogenic Engineering 2* (New York: Plenum)
Clement J R and Quinnel E H 1952 *Rev. Sci. Instrum.* **23** 213
Kes P H, van der Klein C A M and de Klerk D 1974 *Cryogenics* **14** 168
Kopp F J and Ashworth T 1972 *Rev. Sci. Instrum.* **43** 327
Lounasmaa O V 1958 *Phil. Mag.* **3** 652
Rose-Innes A C 1973 *Low Temperature Laboratory Techniques* (London: English UP)
Ward D A 1972 *Cryogenics* **12** 209
White G K 1968 *Experimental Techniques in Low Temperature Physics* (London: Oxford UP)
Zimmerman J E and Hoare F E 1960 *Phys. Chem. Solids* **17** 52

Thermocouple thermometry

G W Burns and W S Hurst
Institute for Basic Standards, National Bureau of Standards, Washington, DC 20234, USA

Abstract. A broad overall view of the current status of thermocouple thermometry is given. The salient features and limitations of standard thermocouple types are reviewed, and some of the recent changes in standardization are noted. Some of the non-standardized thermocouple types are discussed, giving particular attention to those employed at cryogenic temperatures and to those intended for use above the temperature limits or under conditions where the standard types are inadequate. Commonly used materials for insulation and protection are described, and some examples are noted of applications where stringent requirements are placed on thermocouple performance.

1. Introduction

Thermocouple thermometry, as it exists today, covers a diverse spectrum of endeavours. In temperature measurement. thermocouples are used from temperatures below 1 K to temperatures above 2800 °C. As measuring instruments, they can be found as devices that are simply constructed at almost no cost or as expensive devices constructed with painstaking care. In physical size, they can be found as fragile wire barely visible to the eye or as robust tubes several centimetres or more in diameter. Their thermoelements may be metallic or non-metallic, and they may be in a variety of configurations. typically as wires, ribbons. rods, tubing or as vapour-deposited thin films. They exist in many different designs which are adapted to a multitude of applications extending into many industries, such as automotive, iron and steel. non-ferrous metals, cement, ceramic, glass, semiconductor, petroleum, aircraft, electrical power, nuclear, aerospace, food, chemical, and biomedical engineering.

There has been a great deal written about thermocouples, especially during the last twenty years. In his recent book Kinzie (1973) has presented a comprehensive collection of data on approximately three hundred different material combinations that have been studied and used, including a thorough treatment of commonly used ones plus a bibliography of over five hundred references. Zysk and Robertson (1972), in their paper at the fifth symposium on temperature, extensively reviewed newer thermocouple materials that have recently become important. The American Society for Testing and Materials (ASTM 1974a) has produced a new and very useful manual which covers the essential aspects of the application and use of thermocouples in the United States. including materials, protection tubes, hardware, fabrication, thermometer designs, extension wires, installation and calibration, and it contains over 150 references and an extensive bibliography. In addition, numerous papers on the use and performance of thermocouples were presented at the fifth symposium on temperature. MacDonald (1962), Pollock (1971) and more recently Barnard (1972) have given comprehensive theoretical accounts of thermoelectricity in metals and alloys. The basic principles or 'fundamental laws' of thermoelectric circuits have been discussed by Roeser (1940) and Finch (1962).

Information pertaining to the calibration and testing of thermocouples and thermocouple materials, principally at temperatures above $0\,°C$, has been presented by Roeser and Lonberger (1958), Jones (1968, 1972), Evans and Wood (1971), Bedford et al (1972), Carter and Wells (1973), and in the ASTM manual mentioned. Hust et al (1972), Sparks et al (1972), and Sparks and Powell (1972a), have described calibration methods employed at cryogenic temperatures.

The major limitation in the use of thermocouples for precise thermometry, of course, arises from the fact that unwanted EMFs are generated in inhomogeneous thermoelements passing through temperature gradients. In practice, no real materials are perfectly homogeneous thermoelectrically, with the exception of materials in the superconducting state. The inhomogeneities, which may be either chemical or physical in nature, may exist in the materials from the start, and they may be introduced during handling and especially during use. Therefore, for the most precise applications, selective testing of the materials, as discussed, for example, by Fenton (1972) and Sparks et al (1972), will usually be required. Furthermore, in precision thermometry it may be necessary to include special heat treatments of the materials beyond that given by the vendor or manufacturer.

The useful temperature range for a particular thermocouple is somewhat arbitrary. One can argue that the upper limit is determined by the melting points of the thermocouple materials, but in practice the useful upper limit is usually well below the melting point and depends upon such factors as time at temperature, environment, thermoelement diameter, thermopower, and the exact conditions of use. The lower limit is also somewhat arbitrary, but it is usually determined by impractically small thermopower.

In this paper we have made a deliberate effort to avoid presenting numerical values for the instability of thermocouples, since thermocouple performance depends very much on the conditions of use. The exact environmental and test conditions are not always well defined, and much conflicting data exist on the performance of the same thermocouple types under apparently similar test conditions. References are cited in this paper that contain information on stability and the reader can make his own judgment.

In this paper, we limit discussion to thermocouples using metallic thermoelements. We shall first briefly review those thermocouple types that have become standardized in the United States, describing some of their features and limitations, and some of the recent changes in standardization will be noted. We shall then discuss some of the non-standardized thermocouple types that are receiving some use or are under development, and following that we shall describe some of the commonly used materials for insulation and thermocouple protection. We shall then conclude with a brief look at some examples of applications where stringent requirements are placed on thermocouple performance.

2. Thermocouple materials

2.1. Standardized thermocouple types

There are seven types of thermocouples that have become highly standardized, primarily because they are the ones most commonly used in industrial pyrometry. In the United States these standardized types are identified by letter designations. This practice was originated by the Instrument Society of America (ISA) and adopted in 1964 as an American Standard, primarily to eliminate the use of proprietary names. The typical nominal compositions†, representative trade names, and corresponding letter designations of the standardized thermocouple materials are given in table 1. The letter designations will be used throughout this paper except when citing the work of others.

Table 1. Compositions, trade names and letter designations for standardized thermocouples.

Type designation	Thermocouple combinations
	Materials
B	*Platinum*–30% rhodium/*platinum*–6% rhodium.
E	*Nickel*–chromium alloy/a *copper*–nickel alloy.
J	Iron/another slightly different *copper*–nickel alloy.
K	*Nickel*–chromium alloy/*nickel*–aluminium alloy.
R	*Platinum*–13% rhodium/platinum.
S	*Platinum*–10% rhodium/platinum.
T	*Copper*/a copper–nickel alloy.
	Single-leg thermoelements
...N	Denotes the negative thermoelement of a given thermocouple type.
...P	Denotes the positive thermoelement of a given thermocouple type.
BN	*Platinum*–nominal 6% rhodium.
BP	*Platinum*–nominal 30% rhodium.
EN or TN	A *copper*–nickel alloy, constantan: Cupron†, Advance§, ThermoKanthal JN‡; nominally 55% Cu, 45% Ni.
EP or KP	A *nickel*–chromium alloy: Chromel‖, Tophel†, T-1§, ThermoKanthal KP‡; nominally 90% Ni, 10% Cr.
JN	A *copper*–nickel alloy similar to, but usually not interchangeable with EN and TN.
JP	Iron: ThermoKanthal JP‡; nominally 99·5% Fe.
KN	A *nickel*–aluminium alloy: Alumel‖, Nial†, T-2§, ThermoKanthal KN‡; nominally 95% Ni, 2% Al, 2% Mn, 1% Si.
RN, SN	High-purity platinum.
RP	*Platinum*–13% rhodium.
SP	*Platinum*–10% rhodium.
TP	Copper, usually Electrolytic Tough Pitch.

Registered trade marks: † Wilbur B Driver Co; ‡ Kanthal Corp; § Driver-Harris Co; ‖ Hoskins Manufacturing Co.
In the above table an italicized word indicates the primary constituent of an alloy and all compositions are expressed in percentages by weight. The use of trade names does not constitute an endorsement of any manufacturer's products. All materials manufactured in compliance with the established thermoelectric voltage standards are equally acceptable.

† Alloy compositions in this paper, unless specifically noted as atomic per cent (at.%), are given as per cent by weight.

These standardized thermocouples are available commercially with thermoelectric properties that comply with reference tables of EMF against temperature to within established tolerances (called limits of error in the US). Tolerances differ in different countries, and those for the US can be found in documents of the American National Standards Institute (ANSI 1973 C96.2) and the ASTM (E230-72). The reference tables in the NBS Monograph 125 (Powell et al 1974) provide the basis for the present ASTM and ANSI standards. The monograph includes the new international tables for types R and S (Bedford et al 1972) which were established to bring into accord UK and US standards for these thermocouples. Tables for types J, K, E, T and B in the monograph are modified and updated versions of previous NBS tables (Shenker et al 1955, Burns and Gallagher 1966, Sparks et al 1972). In addition, EMF tables for the thermoelements against the platinum thermoelectric standard, Pt-67, are included, as are analytical functions for the representation of all the tables. These same tables for thermocouples have been published by the British Standards Institution (BSI) as BS 4937. They have also just recently been proposed as International Electrotechnical Commission (IEC) recommended thermocouple tables by working group 5 of subcommittee 65B, and they have been circulated to national committees for comment.

The use of letter designations for the thermocouple types is primarily a US practice. However, the above mentioned BSI and IEC tables include the letter designations. which may encourage international usage.

ASTM and ANSI standards explicitly state that the letter designations identify only the reference tables and may be applied to any thermocouple with a temperature–EMF relationship that complies with the table within the specified tolerances, regardless of the chemical composition of the thermocouple. Substantial variations in composition for a given letter type do occur, particularly for types J, K and E (Powell et al 1974, Sparks et al 1972, Burley 1972).

2.1.1. Base-metal thermocouple types E, J, K and T. Base-metal thermocouples are employed from moderately high temperatures to cryogenic temperatures. For industrial purposes, the upper temperature limit for continuous use in air is determined by the oxidation resistance of the thermoelements and therefore the limit varies with wire size. Based upon experience with these materials in industrial practices, ASTM has set recommended upper temperature limits for various wire sizes. Values for several wire sizes are given in table 2. All of the base-metal types have been used at cryogenic temperatures, but below 20 K their thermopowers become quite small. Type E has been recommended by Hust et al (1972) as the most suitable of the standardized types for general low-temperature use, since it offers the best overall combination of desirable properties, ie, high thermopower, low thermal conductivity, and reasonably good thermoelectric homogeneity. Typical values for the thermopower of type E at 4, 20 and 50 K are 2·0, 8·5 and 18·7 $\mu V\ K^{-1}$, respectively. Typical data for the thermoelectric inhomogeneity of commercial type E thermocouple materials at cryogenic temperatures, as well as for other standard base-metal types, can be found in NBS Monograph 124 (Sparks et al 1972). Types K and T thermocouples are also often used below 0 °C, but type J is not suitable for general low-temperature use because the positive thermoelement (noted as JP) is composed of iron and thus is subject to rusting and embrittlement in moist atmospheres.

Table 2. Recommended upper temperature limits (°C) for various thermocouples and wire sizes (AWG) (values extracted from ASTM Standard E230-72).

Thermocouple type	AWG 8 (3·25 mm diameter)	AWG 14 (1·63 mm diameter)	AWG 24 (0·51 mm diameter)
B	–	–	1705 °C
E	871 °C	649 °C	427
J	760	593	371
K	1260	1093	871
R and S	–	–	1482
T	–	371	204

This table gives the recommended upper temperature limits for the various thermocouples and wire sizes. These limits apply to protected thermocouples, that is, thermocouples in conventional closed-end protecting tubes. They do not apply to sheathed thermocouples having compacted mineral oxide insulation. In any general recommendation of thermocouple temperature limits, it is not practicable to take into account special cases. In actual operation, there may be instances where the temperature limits recommended can be exceeded. Likewise, there may be applications where satisfactory life will not be obtained at the recommended temperature limits. However, in general, the temperature limits listed are such as to provide satisfactory thermocouple life when the wires are operated continuously at these temperatures.

Above 0 °C, type T thermocouples have a rather narrow range of application, as illustrated by the values in table 2. However, within this range, it possesses stable chemical and physical properties that make it a highly reliable thermocouple. The useful range is limited by oxidation of the TP (copper) thermoelement. As a result of the high thermal conductivity of the TP thermoelement, special care must be exercised to insure that both the measuring and the reference junctions assume the desired temperatures. The thermocouple can be used to somewhat higher temperatures in vacuum and in inert or reducing atmospheres since deterioration of the TP thermoelement is no longer a problem. Its use in hydrogen is limited since embrittlement can occur above 370 °C.

Type J is one of the most commonly used thermocouples in industrial pyrometry due to its relatively high thermopower and low cost. It has been estimated that over 200 metric tons of type J thermocouples are supplied annually to industry in the US. They are used primarily in the range 0 to 760 °C. They can be used in inert, reducing, oxidizing and vacuum environments. The JP (iron) thermoelement undergoes a magnetic transformation at about 769 °C and an $\alpha-\gamma$ crystal transformation near 910 °C (Hansen and Anderko 1958). These transformations, especially the latter, seriously affect the thermoelectric properties of iron. This behaviour and the rapid oxidation rate of the JP thermoelement are primarily responsible for the customary restricted upper limit of temperature. While the new NBS tables do extend to 1200 °C, they are based on limited data above 760 °C. The stability of type J thermocouples in air was reported by Dahl (1941) over 30 years ago and he showed that relatively large inhomogeneities occur with short exposure to temperatures above 500 °C.

Type K is more resistant to oxidation at elevated temperatures than types E, J and T and consequently it finds wide application at temperatures above 500 °C. It can be used in oxidizing or inert atmospheres. Rapid drift can result in reducing atmospheres, while marginally oxidizing conditions can lead to the preferential oxidation of chromium in the KP thermoelement, promoting the well known 'green rot' corrosion that leads to a large decrease in thermal EMF with time. The effect is most serious in the 800 to 1050 °C temperature range. Its use in carbonaceous and sulphurous atmospheres should also be avoided (for example see Kinzie 1973). The thermoelectric stability of type K in air at various temperatures has been studied by Dahl (1941), Potts and McElroy (1962), Burley and Ackland (1967), Burley (1972), Wang et al (1969), Starr and Wang (1975), and Burley and Jones (1975). Reversible thermoelectric effects which occur principally in the KP thermoelement on heating in the 250 to 550 °C range limits the use of this thermocouple for precision thermometry. Fenton (1969) and Burley (1970) have attributed this instability to short-range ordering in the Ni–Cr atomic lattice. The evidence (Burley 1972) is that careful heat treating can stabilize the type K thermocouple for use up to about 550 °C, but instabilities will again develop on subsequent heating to higher temperatures. Sibley et al (1968) have described precautions that can help to minimize the errors from this reversible process during use. One of the strong features of type K is its stability under thermal neutron irradiation. Changes in the composition of the thermoelements under thermal neutron irradiation are small. Carpenter et al (1972), in a well controlled experiment that exposed the thermocouple to a neutron fluence of $1\cdot6 \times 10^{21}$ thermal neutrons/cm^2 and $2\cdot7 \times 10^{21}$ fast ($E > 0\cdot18$ MeV) neutrons/cm^2 at temperatures between 900 and 1100 °C, found no measurable change in thermal EMF. Types E, J, and T exhibit larger compositional changes (Browning and Miller 1962); the copper in their thermoelements is converted to nickel and zinc.

Type E has the highest thermopower above 0 °C of any of the standardized types. It is subject to the same environmental restrictions and thermal cycling behaviour as type K since they commonly employ the same positive thermoelement.

2.1.2. Noble-metal thermocouple types B, R, and S. Thermocouples employing platinum and platinum–rhodium alloys for their thermoelements have been in use for many years and exhibit a number of advantages over base-metal types. They are more resistant to oxidation, their thermoelements have higher melting points, and they have generally been found to be more reproducible at elevated temperatures in air. They are therefore used when higher accuracy and longer life are sought. Compared to the base-metal types, of course, they are considerably more expensive, and they have lower thermopowers. In temperature ranges where both types may be used, the user must weigh cost against performance.

Of the standardized thermocouples, type S is the oldest (Le Chatelier 1886) and perhaps the most important. It has served as a standard instrument on the International Temperature Scale since its inception in 1927 (Burgess 1928). In the IPTS-68 (CIPM 1969), type S thermocouples that meet certain requirements for purity and thermal EMF are the standard instrument for interpolation in the range 630·74 °C to the gold point (1064·43 °C).

The calibrations of standard type S thermocouples have been shown to be dependent upon their thermal histories and therefore upon the annealing procedures for the thermoelements. Following currently recommended practices for annealing (CIPM 1969) and taking care to minimize systematic errors, one generally finds the measurement uncertainty to be ±0·2 °C in the IPTS-68 defining range (Jones 1968). It has been known for some time (Corruccini 1951) that the thermal EMF is affected by rapid cooling from elevated temperatures. This effect has been attributed to quenched-in point defects in the crystal lattice, and it can be minimized by avoiding rapid cooling or by annealing at low temperatures subsequent to high-temperature exposure (Ricolfi and Sartori 1967, McLaren and Murdock 1972). The effect of surface and internal rhodium oxide formation and dissociation on the thermopower of the Pt–Rh alloy thermoelements, as discussed by McLaren and Murdock (1972), constitutes a formidable barrier to the use of Pt–Rh thermocouples in air for highly precise temperature measurements. The limitations of standard type S thermocouples for precise thermometry have been summarized by Quinn and Compton (1975). The 1975 amended edition of the IPTS-68, which is to be published soon, recognizes these difficulties and includes the statement: 'However, in general use, (these) precautions may not ensure an accuracy of better than ±0·2 °C because of continually changing chemical and physical inhomogeneities in the wires in the regions of temperature gradients.' The IPTS-68 criteria for standard type S thermocouples have also been modified in the amended edition. The modifications reflect the recommendations of Bentley (1969) and incorporate the thermal EMF values from the new international reference tables (Bedford *et al* 1972).

Because of these limitations, consideration has been given to replacing the thermocouple as a standard instrument for interpolation on the practical temperature scale. It has already been demonstrated that high-temperature platinum-resistance thermometers are capable of higher precision and stability (Evans and Wood 1971). Another possibility is to extend some form of radiation pyrometry downward from the gold point (Preston-Thomas 1972).

In spite of all these difficulties, relating to scale definition, the long term stability of Pt–Rh/Pt thermocouples in air at elevated temperatures satisfies many industrial requirements. While types S and R have been tested up to the melting point of the Pt thermoelement, in practice they are seldom used above 1500 °C.

Type B was adopted as a standard type in the US in the late 1960s primarily to serve requirements in the 1200 to 1750 °C range. At elevated temperatures, it offers superior mechanical strength and improved stability over types R and S, and it exhibits comparable thermopower. Its thermopower diminishes at lower temperatures and is very small in the room-temperature range, thus eliminating the need for compensating lead wires in many industrial applications. It was selected by ASTM as the most desirable of three thermocouples then in use that employed Pt–Rh alloys for both thermoelements (Burns and Gallagher 1966).

In general, types R, S and B thermocouples are most reliable when used in air. Their long-term stability has been investigated by Walker *et al* (1962), Glawe and Szaniszlo (1972), Selman (1972), and a host of others. The thermocouples have also been evaluated in neutral environments and vacuum by Glawe and Szaniszlo (1972), Hendricks and McElroy (1967), and Walker *et al* (1962, 1965). The effects of atmospheres containing metallic vapours and other contaminants that are reactive with the platinum group metals are discussed by Zysk (1962) and Bennett (1958). Types S, R, and B exhibit changes in composition under thermal neutron irradiation; the rhodium in the thermoelement converts to palladium (Browning and Miller 1962). Recently, however, Bauchede and Haange (1974) have reported surprisingly little change in the thermal EMF of type R thermocouples after an accumulated dose of $1·5 \times 10^{21}$ neutrons/cm^2.

2.2. Non-standardized thermocouples

There are a number of non-standardized thermocouple types that are available for use in those temperature ranges, environments and applications where the standardized types have been found to be unsuitable. For purposes of discussion, these can be separated into those used primarily at cryogenic temperatures and those primarily used above 800 °C in either nuclear or non-nuclear environments. Many of the non-standardized types presently in use were initially developed during the late 1950s and early 1960s as the need arose for reliable thermometry to aid the rapidly advancing technology in the aerospace and nuclear industries. This list is not intended to be exhaustive (for example, see Zysk and Robertson 1972 and Kinzie 1973).

2.2.1. Thermocouples for cryogenic temperatures. At the present time the thermocouples most commonly favoured for use at cryogenic temperatures are those that employ dilute gold–iron alloys in combination with KP, copper or Ag–0·37 at.% Au (called 'normal' silver or 'silver-normal'). The dilute gold–iron alloys form the negative thermoelement, and are composed of 0·02, 0·03 or 0·07 at.% Fe in Au. Zysk and Robertson (1972) have traced the evolution of these alloys. The alloys have replaced an earlier Au–Co alloy which exhibited unstable thermoelectric properties at room temperature (Powell *et al* 1962). The thermoelectric powers of thermocouples formed from combining various positive thermoelements with the gold–iron alloys, as well as those for types E, K, and T, are given in figure 1. As discussed by Berman

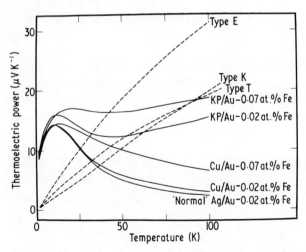

Figure 1. Thermoelectric power of some low-temperature thermocouples.

(1972), factors such as the thermal conductivity, homogeneity, thermoelectric power, and temperature range of use must be considered in selecting the best combination of thermoelements. For general engineering applications, Sparks and Powell (1972a) recommend the KP/Au–0·07 at.% Fe thermocouple for use below 20 K. Reference tables for KP, normal silver and copper/Au–0·02 at.% Fe and Au–0·07 at.% Fe have been prepared by Sparks and Powell (1972b) for the 1 to 280 K range. Rosenbaum

(1968) prepared tabular values for temperatures down to 0·4 K and discussed (Rosenbaum 1970) the feasibility of using the Au–0·02 at.% Fe alloy in combination with a superconducting material such as Nb for measurements down to 0·05 K. Variations in the low-temperature calibrations due to annealing have been described by Rosenbaum (1968) and Sparks and Powell (1972b).

The presence of magnetic fields has a large effect upon these thermocouples, particularly below 25 K. This behaviour has been of particular concern to many investigators, since often low-temperature experiments are conducted in the presence of moderate to strong magnetic fields. These effects have been discussed by Berman (1972) and Neuringer and Rubin (1972). In recent work Sample et al (1974) have found chromel P/constantan to be more suitable than chromel P/Au–0·07 at.% Fe in strong magnetic fields. Chiang (1974) has prepared a series of tables to account for the magnetic field dependence of chromel P and normal silver/Au–0·07 at.% Fe. It has been shown experimentally and theoretically (Richards et al 1969) that the thermal EMF is not affected by a magnetic field if that segment of the thermocouple in the field is also in an isothermal zone.

2.2.2. Thermocouples for high temperatures. There are three thermocouples comprised of W and W–Re alloys that are commercially available and in rather common use for measuring very high temperatures. Three positive thermoelements, W, W–3% Re and W–5% Re, are used in combination with a negative W–Re alloy thermoelement that nominally contains between 25 and 26% Re. The temperature–EMF relationships of these thermocouples are given in figure 2, which includes also the relationships for other non-standardized thermocouples discussed in this section. All of the thermoelements have melting points in excess of 3000 °C (Caldwell 1962). Heating of these thermoelements above their recrystallization temperatures reduces their room-temperature ductility. The positive thermoelements presently sold in the US contain residual impurity dopants that controls their grain growth characteristics. For the dilute alloys, this results in a marked improvement in the room-temperature ductility even after full

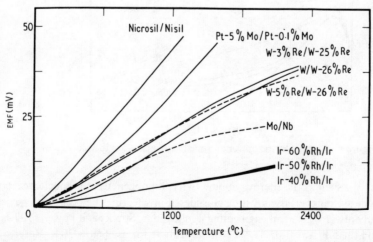

Figure 2. Thermal EMF–temperature relationship of some non-standardized thermocouples; reference junctions at 0 °C.

recrystallization (Burns and Hurst 1972). Doping is not effective in alloys which are rich in rhenium (Pugh et al 1962), such as constitute the negative thermoelement. However, as a result of the high Re content of these thermoelements, they are not as brittle as pure W wire.

The thermocouples are suitable for use in hydrogen and in high-purity inert atmospheres. They may also be used in high vacuum, but for long periods of exposure above 1950 °C substantial changes in their thermal EMF can occur as a result of preferential evaporation of rhenium (Burns and Hurst 1972). Environments containing hydrocarbons, oxygen and oxygen-containing gases such as H_2O, CO and CO_2 can be highly detrimental to the performance of these thermocouples at elevated temperatures. Both W and Re have relatively high neutron-capture cross sections and under thermal neutron irradiation exhibit substantial changes in composition. W is converted to Re and Re to Os.

Reference tables for these types have been developed by the suppliers of the materials. These have recently been modified, smoothed by using computer techniques, and converted to the IPTS-68, and they have been published by the ASTM (1974b) as proposed tables. The tables are published for information only and have no status as ASTM standards. For precise measurements with these thermocouples, the thermoelements must be annealed to stabilize their thermoelectric properties. Burns and Hurst (1972) have determined annealing parameters for this purpose. Further, it has been demonstrated that suitably annealed bare thermoelements exhibit little change in their thermoelectric properties with exposure to temperatures as high as 2125 °C in high-purity environments of argon, helium, hydrogen and nitrogen. Further information on the stability of assembled thermocouple devices will be discussed later.

A thermocouple developed specifically for use in neutron irradiation environments at temperatures above the operating limits of type K and receiving some use is Pt–5% Mo/Pt–0·1% Mo. Reichardt (1963) found this thermocouple to be particularly susceptible to selective oxidation of molybdenum and emphasized the need for a high-purity, non-oxidizing environment for reliable use. Reference tables for the 0 to 1600 °C range as well as experimentally developed annealing procedures have been reported by Tseng et al (1968). Rohne (1974) recently tested the thermocouple under nuclear irradiation and indicated performance superior to that of the tungsten–rhenium alloy thermocouples.

Another thermocouple that has been suggested for nuclear use is Mo/Nb. While Nb and Mo are available from various manufacturers, at this time they are not available as matched thermocouple wires from any supplier in the US. Zysk and Robertson (1972) have given calibration data to 2000 °C, but these are based on a limited amount of material. The thermocouples may not be used under oxidizing conditions. Recently, Schley et al (1974) have reported work on thermocouples formed from Nb, Mo, and Nb–Mo alloys.

For measurements in oxidizing atmospheres at very high temperatures (above the limits of Pt–Rh alloy thermocouples) Ir–Rh alloy thermocouples have been used. Three thermocouples have been developed, employing Ir–Rh alloys with 40, 50 or 60 per cent Rh for the positive thermoelement with a negative thermoelement of unalloyed Ir. Reference tables for these combinations were prepared at NBS by Blackburn and Caldwell (1962, 1964). Limited testing at NBS of recently acquired thermocouples has shown that they do not conform closely with the reference tables. The results suggest that present materials are of higher purity than those employed in establishing the original tables. For this reason the earlier tables have not been updated to the IPTS-68.

Both iridium and rhodium are quite volatile at very high temperatures in oxidizing atmospheres and the useful life of Ir–Rh/Ir thermocouples in air is short, typically about 10 to 20 hours at 2000 °C for 0·8 mm diameter wires (Freeze *et al* 1975, Aleksklin *et al* 1964, Zysk and Robertson 1972). It was postulated some time ago (Carter 1950) that Ir would be lost preferentially from the alloys and that use of an alloy with lower Rh concentration should result in best thermal EMF stability. Recently, Freeze *et al* (1972, 1975) at NBS have determined the stability of Ir–40% Rh/Ir in air at temperatures up to 2000 °C, as well as the effects of catalysis in flowing gas streams. Improved performance is possible in inert or mildly oxidizing atmospheres. The liquidus–solidus temperatures (Caldwell 1962) of the alloys range from about 2150 to 2250 °C, thus defining the maximum upper limit. These thermocouples are very expensive, possess rather poor mechanical properties and have relatively low thermopowers; hence, Pt–Rh alloy or W–Re alloy types are usually preferred when environmental conditions permit.

It has long been recognized that in many applications the performance of types J or K thermocouples is either marginal or inadequate. A number of proprietary thermocouples have therefore been developed, but in most cases they have been burdened with the constraint that their EMF–temperature relationship should approximate types J or K. This constraint permits the analogue instrumentation for types J and K, which is a considerable investment for industry, to be retained. The thermocouples that have been developed are both nickel-base alloy and noble-metal alloy combinations. Zysk and Robertson (1972) have discussed some of these in their recent review. No definitive study, including reference tables, has been accomplished for the base-metal types, but reference tables for some of the noble-metal types have been prepared.

At the fifth symposium on temperature, Burley (1972) reported on some new nickel-base alloys for thermocouples named Nicrosil and Nisil that he was then developing. The primary consideration in their formulation was to minimize two kinds of thermal EMF instabilities that are inherent in type K thermocouples, namely, the gradual and generally cumulative drift that occurs with long-term exposure in air at elevated temperatures and the short-term changes (mentioned earlier) that occur with thermal cycling. Subsequent work since the symposium has led to the establishment of final compositions for the alloys: Ni–14·2% Cr–1·4% Si for Nicrosil and Ni–4·4% Si–0·1% Mg for Nisil. In a joint programme between the Australian Materials Research Laboratories and the National Bureau of Standards, reference tables for the above combination have been prepared and are to be published soon (Burley *et al* 1975b). This joint programme, as well as recent performance data on these new alloys, is discussed by Burley *et al* (1975a) and Burley and Jones (1975).

3. Thermocouple insulation and protection

The reliable use and performance of thermocouples for thermometry require the proper selection of materials for insulation and protection. Rarely in practice are there situations where the thermoelements can be kept physically and electrically isolated without introducing some form of electrical insulation. The choice of insulating materials for thermocouples will, of course, depend upon the application. Today thermocouple wire insulation is available in many different varieties, and the materials used will depend upon a number of different factors, including the temperature range, thermocouple type, environment, flexibility and thermocouple size.

For temperatures below 0 °C, compatibility of the insulation with the environment and the thermocouple wire is usually not a problem. Typical insulation materials employed at subzero temperatures are enamels, polyethylene, polytetrafluorethylene, polyimide, and spun glass. For the intermediate temperature range, a variety of commonly used types of non-ceramic insulants are discussed in the ASTM manual (1974a), which rates them for their moisture and abrasion resistance and gives their useful temperature range.

At high temperatures, ceramic insulators are usually required. Base-metal type thermocouples typically employ mullite, steatite, porcelain and other ceramics in the form of single or multi-bore tubing and beads. For platinum metal thermocouples high-purity sintered Al_2O_3 multi-bore tubing has traditionally been used, particularly at temperatures greater than 1000 °C. Chemical contamination of the thermocouple by impurities present in the insulator can occur, and it can proceed at an accelerated rate in non-oxidizing atmospheres. Recently, Darling and Selman (1972) have demonstrated that, when the oxidizing potential of the surrounding atmosphere is low, MgO is more suitable than Al_2O_3. At very high temperatures, the chemical compatibility of the insulator with the thermoelements and the surrounding environment must be considered. The thermodynamics of the chemical compatibility of some refractory metals with refractory oxides has been considered by Droege et al (1972). The electrical resistivity of thermocouple insulators becomes low at temperatures greater than 2000 to 2200 °C, resulting in considerable shunting errors on the thermocouple EMF (Shepard et al 1972).

With W–Re alloy thermocouples above about 1700 °C, sintered BeO insulators are often employed because of their high resistivity and because they are readily available in high purity. For some applications, the anomalous thermal expansion at the α–β transition temperature, which results in decrepitation of sintered BeO, may limit the temperature of use to below 2050 °C (Ryshkewitch 1966). However, sintered BeO has been used successfully by Asamoto and Novak (1967) at temperatures up to near its melting point (about 2550 °C according to Ryshkewitch 1966). Other refractory oxides, such as HfO_2 and ThO_2, have also been used for high-temperature applications. The properties of these high-temperature insulators, as well as their limitations when used with W–Re alloy thermocouples, have been summarized by Anderson and Bliss (1972) and Zysk and Robertson (1972).

If it is necessary to protect the thermocouple from contamination by the environment in which it is used, protection tubes constructed from metals, ceramics or cermets are available. They are essential in industrial applications where corrosive chemical environments are encountered, and the ASTM manual (1974a) gives an extensive list of suggested protection tube materials. Important factors in the choice of a suitable protection tube can be its thermal shock resistance and mechanical strength. At high temperatures, the protection tube must be compatible with the thermocouple insulator as well as the environment. In addition, it may serve as a containment for gaseous reaction products within the sheath, thereby preventing deterioration of the thermoelements (Droege et al 1972). Typical metal sheath materials for use at high temperatures include Pt and Pt—Rh alloys which are normally used in oxidizing atmospheres, and refractory metals such as Mo, Ta, Nb, and alloys of W—Re and Mo—Re that are used in non-oxidizing atmospheres. In special cases, the performance of refractory metal protection tubes in hostile environments can be enhanced by applying various types of protective coatings or by encasing the tube within another protection tube that is more compatible with the environment.

Thermocouple assemblies can be constructed with sintered insulators placed into protection tubes, or they can be a compacted insulation assembly. This latter form is constructed by surrounding the thermoelements with non-compacted insulating material held within a metal tube. The insulation is typically a crushable performed ceramic and is compacted around the thermoelements by employing a mechanical reduction process such as swaging or drawing to reduce and elongate the tube. In general, any of the commonly used thermoelements and any metal tubing that can be successfully worked to smaller sizes can be employed.

Compacted insulated thermocouples are in widespread use and offer many advantages, such as their ease of production, the ability to be manufactured in very long lengths and small diameters (down to 0·5 mm or less), their flexibility, which allows them to be bent during installation to reach difficult locations, and their ability to withstand thermal shock.

For base-metal thermocouples, satisfactory long-term performance with smaller-sized thermoelements than those listed in table 2 can be obtained with the compacted design, since the thermoelements are protected by the insulation and sheath. Stainless steel and nickel—chrome alloys are the most widely used sheathing materials with the base-metal thermocouple types. Other compacted thermocouple assemblies frequently employ Pt—Rh alloy thermoelements and sheaths, and W—Re alloy thermoelements with refractory metal sheaths. MgO is the most widely used insulant because of its superior compaction properties, because it does a minimum of abrasion damage to the thermoelements and the sheath during the reduction process, and because of its compatibility with commonly used thermocouple materials (see ASTM manual 1974a). It is seldom used above 1500 °C, however, because of its relatively low electrical resistivity. Other insulators, such as Al_2O_3, BeO and ThO_2, are used, depending upon the application. The insulation must be kept dry both during the thermocouple fabrication process and in use. This is particularly critical with MgO since it is hygroscopic. MgO insulated thermocouple assemblies should be hermetically sealed if the insulation resistance is to be kept high (Horton 1971, 1972) and the thermocouple is to give reliable performance. Information on the design of thermocouple measuring junctions can be found in the ASTM manual (1974a). The manual includes information on terminations as well.

Some cold working of the thermoelements from the compaction process necessarily occurs, and heat treatments are usually performed by the manufacturers to relieve these effects. Fenton (1972) has shown that the homogeneity of the mineral—oxide insulated thermocouples is not as good as that of typical bare thermoelements in an 'as-received' condition. Techniques and apparatus for checking the homogeneity have been given by Carr (1972) and Fenton (1972).

4. Selected thermocouple applications

There are numerous applications of thermocouples in science and technology. Some of these, having engendered their own core of investigators and experts, are worthy of an entire review paper by themselves. We wish to draw attention to a few applications here to illustrate the severe requirements on thermocouple materials and performance demanded by today's technology.

One of the most active fields in recent years has been concerned with thermocouples for nuclear environments. The requirements for temperature measurement in nuclear reactors varies widely, as evidenced by recent symposia on this subject (conferences at Petten, Holland (1974) and Sandusky, Ohio (1969) and the fifth symposium on temperature (1972)). In many applications, such as the determination of coolant and cladding temperatures, mineral–oxide insulated type K thermocouples sheathed in stainless steel or nickel–chrome alloys are typically employed. Nuclear fuel development studies usually require temperature measurements above the capabilities of type K thermocouples, and refractory-metal-sheathed W–Re alloy thermocouples have been used almost universally for this purpose. This is a difficult problem: not only must the effects of neutron irradiation upon the thermocouples be endured, but the thermocouples must often operate in hostile material environments, at high temperatures, and for periods as long as a year or more.

Experiments to determine the in-pile thermal EMF drift of high-temperature thermocouples typically have been plagued by incompatibilities in the thermocouple materials. It is for this reason that many studies have been directed at the out-of-pile performance. Glawe and Szaniszlo (1972), Burns and Hurst (1972, 1974), Heckelman and Kozar (1972), and Kuhlman and Baxter (1969) have reported on the long-term thermoelectric stability of insulated W–Re alloy thermocouples in argon or helium atmospheres at elevated temperatures, while Schwarzer (1969), Droege *et al* (1972), and Glawe and Szaniszlo (1972) have reported on their stability in vacuum. In addition, protective coatings on the sheath have been shown to enhance material compatibility (Haas *et al* 1974).

The reliable evaluation of the thermal EMF drift of thermocouples installed in a reactor core is difficult, but some rather unique techniques have been developed to accomplish this. Carpenter *et al* (1972) have developed a procedure to determine the thermal EMF drift during irradiation using an internal calibration system that employs the melting point of copper. Heckelman (1969) devised a comparison technique using a reference thermocouple which can be extracted from the reactor core and replaced with another recently calibrated thermocouple as often as required throughout the experiment. Johnson *et al* (1974) have developed irradiation equipment for long-term testing of thermocouples at high temperatures under neutron irradiation. Recently, Shepard and Kollie (1974) have shown that the calibration drift during irradiation can be determined quantitatively by measuring the thermocouple EMF and the electrical resistance of the thermocouple loop and then applying Matthiesen's rule to interpret the results.

In general, experience shows that W–Re alloy thermocouples exhibit an overall decrease in thermal EMF with extensive neutron exposure. It has long been recognized that both transmutation and fast neutron damage affect the thermopower. Just recently, Williams *et al* (1974) have reported unexpected precipitation reactions in

W—Re alloy samples containing as little as 5% Re when irradiated at temperatures as high as 1500 °C. It is very likely that the change in thermocouple calibration is at least partly associated with the precipitation phenomena.

In addition, other instabilities may be present even when the thermocouple is heated out-of-pile. Recent studies at NBS, yet unpublished, indicate that in commercial W—25% Re thermocouple wire, precipitates of σ phase are formed with long-term exposure in the 800 to 1300 °C range. These precipitates result in substantial increases in the thermopower of this alloy. Such precipitates do not readily form in the wires if they have previously been annealed for a short time at about 2100 °C.

There are other new and interesting applications, besides the nuclear field, where W—Re alloy thermocouples are being used today. In the petroleum industry, oil refineries are using Claus sulphur recovery unit thermal reactors to reduce sulphur-containing gas emissions. Temperature sensors are needed to provide an alarm against overtemperatures that are disastrous to the reactor. W—Re alloy thermocouples are being evaluated for this purpose. The thermocouples operate in an environment containing gases such as H_2S, NH_3, H_2O, CO_2, SO_2 and CH_4 and at temperatures in the 1200 to 1500 °C range. They are protected by a nitride-bonded silicon carbide outer sheath and an inner molybdenum sheath which has a molybdenum disilicide coating. Thermocouple assemblies of this design are also being installed in a pilot coal gasification plant, and are employed in incinerators as well. The measurement of burner exhaust temperatures for jet engines requires thermocouple probes to operate at temperatures as high as 2100 °C with a measurement accuracy of ±1% of gas-stream absolute temperatures (Anderson and Bliss 1972). Future engines may require operation for 10 000 hours at temperatures of 1600 to 2000 °C (Baas and Mai 1972).

While a multitude of special applications could be cited for any of the standardized types of thermocouples described, type B perhaps warrants special consideration because it is the newest of the standardized types. It was only after standardization of type B thermocouples by ANSI and ASTM that US industry began to accept them for use. They are now employed in favour of types S or R in a number of applications. For example, modern glass plants now being built are installing type B with computerized process controls for monitoring the temperature in the 'crown' area of glass tanks. Greenberg (1974) has indicated that crown temperatures can reach 1570 °C and has described typical thermocouple designs used for such measurements. Type B thermocouples are also being used for the process control of high-temperature tunnel furnaces employed in the production of ferrites and ceramics (D B Sharp, private communication). Tauras (1972) has described an aspirating probe using type B thermocouples for measuring gas-stream temperatures as high as 1750 °C in the development of jet engines. Type B thermocouples are also finding application in expendable thermocouple devices for instantaneous molten-metal temperature measurements in the steel industry.

5. Concluding remarks

We have described the principal materials that are used in thermocouple thermometry as it is practised today. While the standard letter-designated thermocouple types have been satisfactory for a great number of the industrial and engineering applications over the range 20 K to 1700 °C, other non-standardized types, some of which we have described, are available for use in special applications both within and outside of this range. The simplicity, versatility and reliability of the thermocouple as a temperature sensor assures its continued and expanding use in science and technology. We can therefore expect continuing improvement in thermocouple materials, development of new materials and techniques as needs arise, and an ever growing array of routine and special applications. We can also expect an increasing effort toward international standardization of all aspects of thermocouple thermometry.

References

Aleksaklin I A, Lepin I R and Bragin B K 1964 *Issl. Splav. Dyla Termop.* **22** 143–58 (1967 *Wright Technology Div., Wright–Patterson Air Force Base, Ohio* AD-663573)
Anderson T M and Bliss P 1972 *TMCSI* **4** part 3 1735–46
Asamoto R R and Novak P E 1967 *Rev. Sci. Instrum.* **38** 1047–52
ASTM 1974a *Manual on the Use of Thermocouples in Temperature Measurement* STP 470A
—— 1974b *Annual Book of ASTM Standards, part 44* pp 730–6
Baas P B R and Mai K 1972 *TMCSI* **4** part 3 1811–22
Barnard R D 1972 *Thermoelectricity in Metals and Alloys* (London: Taylor and Francis)
Bauchede J and Haange R 1974 *Proc. Int. Colloq. on High Temperature In-Pile Thermometry, December 1974, Petten, Holland* to be published
Bedford R E, Ma C K, Barber C R, Chandler T R, Quinn T J, Burns G W and Scroger M G 1972 *TMCSI* **4** part 3 1585–602
Bennett H E 1958 *Platin. Metals Rev.* **2** 120
Bentley R E 1969 *Metrologia* **5** 26–8
Berman R 1972 *TMCSI* **4** part 3 1537–42
Blackburn G F and Caldwell F R 1962 *J. Res. NBS* **66C** 1–11
—— 1964 *J. Res. NBS* **68C** 41–59
Browning W E Jr and Miller C E Jr 1962 *TMCSI* **3** part 2 271
Burgess G K 1928 *J. Res.* **1** 635
Burley N A 1970 *Australian Defence Stds Lab. Rep.* pp 353
—— 1972 *TMCSI* **4** part 3 1677–96
Burley N A and Ackland R G 1967 *J. Aust. Inst. Metals* **12** 23–31
Burley N A, Burns G W and Powell R L 1975a this volume
Burley N A and Jones T P 1975 this volume
Burley N A, Powell R L, Burns G W and Scroger M G 1975b *NBS Monograph* to be published
Burns G W and Gallagher J S 1966 *J. Res. NBS* **70C** 89
Burns G W and Hurst W S 1972 *TMCSI* **4** part 3 1751–66
—— 1974 *Proc. Int. Colloq. on High Temperature In-Pile Thermometry, December 1974, Petten, Holland* to be published
Burns G W, Hurst W S and Scroger M G 1974 *NBSIR* 74–447, *NASA* CR-134549
Caldwell F R 1962 *NBS Monograph* 40
Carpenter F D, Sandefur N L, Grenda R J and Steibel J S 1972 *TMCSI* **4** part 3 1927–34
Carr K R 1972 *TMCSI* **4** part 3 1855–68
Carter D F and Wells P H 1973 *Platin. Metals Rev.* **17** 52–6
Carter F E 1950 *Trans. Am. Soc. Metals* **42** 1151
Chiang C K 1974 *Rev. Sci. Instrum.* **45** 985–9
CIPM 1969 *Comité International des Poids et Mésures* 1968 *Metrologia* **5** 35–44
Corruccini R J 1951 *J. Res. NBS* **47** 94
Dahl A I 1941 *TMCSI* **1** 1238
Darling A S and Selman G L 1972 *TMCSI* **4** part 3 1767–80
Droege J W, Schimek M M and Ward J J 1972 *TMCSI* **4** part 3 1767–80
Evans J P and Wood S D 1971 *Metrologia* **7** 108

Fenton A W 1969 *Proc. IEE* **116** 1277
—— 1972 *TMCSI* **4** part 3 1973–90
Finch D I 1962 *TMCSI* **3** part 2 3
Freeze P, Thomas D, Edelman S and Stern J 1972 *NASA* CR-120990
Freeze P, Thomas D and Stern J 1975 *NASA* CR to be published
Glawe G E and Szaniszlo A. J. 1972 *TMCSI* **4** part 3 1645–62
Greenberg H J 1974 *ISA Trans.* **13** 40–5
Haas H, Gerken W, Delis K and Schmidt F 1974 *Proc. Int. Colloq. on High Temperature In-Pile Thermometry, December 1974, Petten, Holland* to be published
Hansen M and Anderko K 1958 *Constitution of Binary Alloys* (New York:McGraw-Hill)
Heckelman J D 1969 *National Symposium on Developments in Irradiation Testing Technology, Sandusky, Ohio* AEC CONF-690910
Heckelman J D and Kozar R P 1972 *TMCSI* **4** part 3 1935–50
Hendricks J W and McElroy D L 1967 *Environ. Q.* 34–8
Horton J L 1971 *Instrumentation and Controls Div. Ann. Prog. Rep.* ORNL-4822 p70
—— 1972 *Instrumentation and Controls Div. Ann. Prog. Rep.* ORNL-4822 p66
Hust J G, Powell R L and Sparks L L 1972 *TMCSI* **4** part 3 1525–36
Johnson F A, Walter A J and Brooks R H 1974 *Proc. Int. Colloq. on High Temperature In-Pile Thermometry, December 1974, Petten, Holland* to be published
Jones T P 1968 *Metrologia* **4** 80
—— 1972 *TMCSI* **4** part 3 1561–8
Kinzie P A 1973 *Thermocouple Temperature Measurement* (New York:Wiley)
Kuhlman W and Baxter W 1969 *General Electric Company* GEMP-738
—— 1969 *Trans. Am. Nucl. Soc.* **12** 319
Le Chatelier 1886 *CR Acad. Sci. Paris* **102** 819
MacDonald D K C 1962 *Thermoelectricity: An Introduction to the Principles* (New York: Wiley)
McLaren E H and Murdock E G 1972 *TMCSI* **4** part 3 1543–60
Neuringer L J and Rubin L G 1972 *TMCSI* **4** part 3 1085–96
Pollock D D 1971 *ASTM* STP 492
Potts J F Jr and McElroy D L 1962 *TMCSI* **3** part 2 243
Powell R L, Caywood L P Jr and Bunch M D 1962 *TMCSI* **3** part 2 65–77
Powell R L, Hall W J, Hyinck C H Jr, Sparks L L, Burns G W, Scroger M G and Plumb H H 1974 *NBS Monograph* 125
Preston-Thomas H 1972 *TMCSI* **4** part 1 3–14
Pugh J W, Amra L H and Hurd D T 1962 *Trans. Am. Soc. Metals* **55** 451–61
Quinn T J and Compton J P 1975 *Rep. Prog. Phys.* **38** 151–239
Reichardt F A 1963 *Platin. Metals Rev.* **7** 122–5
Richards D B, Edwards L R and Legvold S 1969 *J. Appl. Phys.* **40** 3836–7
Ricolfi T and Sartori S 1967 *La Termotecnica* **21** 29
Roeser W F 1940 *J. Appl. Phys.* **11** 388
Roeser W F and Lonberger S T 1958 *NBS Circular* 590
Rohne B 1974 *Proc. Int. Colloq. on High Temperature In-Pile Thermometry, December 1974, Petten, Holland* to be published
Rosenbaum R L 1968 *Rev. Sci. Instrum.* **39** 890–9
—— 1970 *Rev. Sci. Instrum.* **41** 37–40
Ryshkewitch E 1966 *Air Force Materials Lab. Tech. Rep. Wright–Patterson Air Force Base, Ohio* AFML-TR-65-378
Sample H H, Neuringer L J and Rubin L G 1974 *Rev. Sci. Instrum.* **45** 64–73
Schley R, Liermann J, Mataver G and Gentil J 1974 *Proc. Int. Colloq. on High Temperature In-Pile Thermometry, December 1974, Petten, Holland* to be published
Schwarzer D E 1969 *National Symposium on Developments in Irradiation Testing Technology, Sandusky, Ohio* AEC CONF-690910 pp307–29
Selman G L 1972 *TMCSI* **4** part 3 1833–40
Shenker H, Lauritzen J I, Corruccini R J and Lonberger S T 1955 *NBS Circular* 561
Shepard R L, Hyland R F, Googe J M and McDearman J R 1972 *TMCSI* **4** part 3 1841–54

Shepard R L and Kollie T G 1974 *Proc. Int. Colloq. on High Temperature In-Pile Thermometry, December 1974, Petten, Holland* to be published
Sibley F S, Spooner N F and Hall B F 1968 *Instrum. Tech.* 53
Sparks L and Powell R L 1972a *TMCSI* 4 part 3 1569–77
—— 1972b *J. Res. NBS* 76A 263–83
Sparks L L, Powell R L and Hall W J 1972 *NBS Monograph* 124
Starr C D and Wang T P 1975 to be published
Taurus J A 1972 *TMCSI* 4 part 3 1805–10
Tseng Y A, Robertson A and Zysk E D 1968 *Engelhard Industries Tech. Bull.* 9 77–84
Walker B E, Ewing C T and Miller R R 1962 *Rev. Sci. Instrum.* 33 1029–40
—— 1965 *Rev. Sci. Instrum.* 36 101
Wang T P, Gottlieb A J and Starr C D 1969 *Society of Automotive Engineers, New York* SAE-690426
Williams R K, Stiegler J O and Wiffen F W 1974 *Oak Ridge Natn. Lab. Prog. Rep.* ORNL-TM-4500 §4·3
Zysk E D 1962 *TMCSI* 3 part 2 135
Zysk E D and Robertson A R 1972 *TMCSI* 4 part 3 1696–34

Nicrosil and Nisil: their development and standardization

N A Burley[†], G W Burns[‡] and R L Powell[§]

[†] Metallurgy Division, Materials Research Laboratories, Australian Defence Scientific Service, Department of Defence, Melbourne, Victoria 3032, Australia

[‡] Heat Division, Institute for Basic Standards, National Bureau of Standards, Washington DC 20234, USA

[§] Cryogenics Division, Institute for Basic Standards, National Bureau of Standards, Boulder, Colorado 80302, USA

Abstract. This paper reviews the development of the new nickel-base thermocouple alloys Nicrosil and Nisil by the Australian Defence Standards Laboratories (now the Materials Research Laboratories of the Australian Government Department of Defence), and their standardization by the US National Bureau of Standards.

The relevant properties of the new alloys are described, and they are shown to have much higher environmental, structural and thermoelectrical stabilities, and to be more suitable for use at the higher operating temperatures, than existing Type K nickel-base thermocouple materials.

The standardization procedures are summarized, including the derivation of reference tables. Calibration data were obtained, over the range 5 K to 1575 K, from prototype alloys specially fabricated by five major manufacturers of base-metal thermocouple alloys in the UK, the USA and Sweden.

1. Introduction

Of the base-metal thermocouples commonly used for temperature measurements up to about 1000 °C, the nickel-base alloy varieties presently designated Type K[∥] by the American National Standards Institute (ANSI 1964) are the most versatile. Indeed, their use at the higher end of the temperature range is almost universal because they possess the best combination of such desirable properties as calibration accuracy and stability, oxidation resistance, high thermoelectromotive force or thermal EMF, and reasonable cost. The accuracy of thermocouples of Type K materials, however, can be significantly impaired by certain characteristic types of change (Burley 1972) which can occur in their temperature—thermoelectromotive force characteristics, principally: (i) a gradual and generally cumulative drift in thermal EMF on long exposure at high temperatures; and (ii) a short-term change in thermal EMF on heating in the temperature range ~250 °C to 550 °C.

Burley (1969, 1972) has demonstrated that the long-term EMF drifts are caused by the development of compositional inhomogeneities as reactive solutes are depleted chiefly by oxidation, in particular by internal oxidation. Fenton (1969) and Burley (1970) independently have adduced much circumstantial evidence in support of a hypothesis that the short-term EMF changes are due to inhomogeneous short-range ordering in the Ni—Cr atomic lattice of the Type K$^+$ alloy.

[∥] The compositions of typical examples of various Type K alloys at present available are given in table 1. All alloy compositions in this paper are expressed as percentages by weight.

Table 1. Typical analyses of various Type K alloys presently available (after Burley 1972).

Alloy	Composition (wt%)†							Traces of other elements	Variety
	Cr	Mn	Al	Si	Co	Nb	Fe	Ni		
Positive	9.2_0	T	ST	0.2_5	P		T	bal	Mg, Mo, Zn, Sn	conventional
	9.3_4	T	T	0.2_6	0.2_0	ST	T	bal	Mg, Mo, Cu, Ca	(lower Si)
	9.3_5	T	FT	0.4_5	0.1_5	ST	ST	bal	Mg, Cu, Ca, Zr	conventional
	9.3_4	T	FT	0.4_6	ST		0.1_7	bal	Mg, Mo, Cu, Ca, Zr	(higher Si)
	9.3_1	ST	FT	0.3_5	ST	0.2_2	0.3_5	bal	Mg, Mo, Cu, Ca, Zr	special conventional (Nb bearing)
Negative	ST	2.8_7	1.9_6	1.1_5	0.4_9		T	bal	Mg, Cu, Ti, Pb	conventional
	T	2.7_8	1.8_0	1.0_2	P		T	bal	Mg, Cu, Zn, Pb	conventional
	ST	1.6_7	1.2_5	1.5_6	0.7_2		T	bal	Mg, Mo, Cu, Ca, Pb	modified conventional
	ST	0.3_7	T	2.3_9	0.3_1		ST	bal	Mg, Cu, Ca, Pb	(Mn and Al decreased, Si and Co increased)
	ST	ST	FT	2.5_0	1.0_0		0.2_3	bal	(Cu-2.2_2)Mg, Ca	special conventional (Mn and Al eliminated, Si and Co increased)
	ST	FT		0.1_3	2.5_8	ST		ST	bal	Mg, Ba, Cu, Ca, Pb	special conventional (Mn and Co eliminated, Al reduced, Si increased)

† The numerals refer to chemical analysis and the symbols to spectrographic analysis as follows:
P = 0·1–0·5%; ST = 0·05–0·1%; T = 0·01–0·05%; FT ⩽ 0·01%.

Burley (1972) has shown that the thermoelectric stability of nickel-base thermocouple alloys can be significantly enhanced, particularly at temperatures above 1100 °C, by increasing alloy solute levels above those required to cause a transition from internal to external modes of oxidation, and by selecting solutes which preferentially oxidize to form impervious diffusion barrier films. Furthermore, the short-term EMF change can be virtually eliminated by the choice of higher solute levels at which this structure-dependent effect is not evident. Based upon these considerations, and following an extensive programme of research at the Australian Defence Standards Laboratories (now the Materials Research Laboratories of the Australian Government Department of Defence) two new nickel-base alloys for thermocouples have been developed. These alloys, at present called Nicrosil (Ni–14·2Cr–1·4Si) and Nisil (Ni–4·4Si–0·1Mg) are found to be more resistant to air oxidation, to be usable at higher maximum temperatures, to be substantially free of the effects of structural ordering and, as a consequence, to have much higher thermal EMF stability than existing nickel-base alloys of Type K.

The basic thermoelectric properties of Nicrosil and Nisil have recently been the subject of a joint research programme between the Australian Materials Research Laboratories and the US National Bureau of Standards. The chief aim of this project, which was conducted at NBS under the auspices of the 1968 US/Australia Agreement on Scientific and Technological Cooperation, was to make possible the formulation of basic reference data on the thermoelectric and other properties of the new alloys which could be recognized by various standards bodies around the world.

Thermoelectric calibration data were obtained, over the range −268 °C to 1300 °C using a number of prototype alloys specially fabricated to close chemical tolerances by five major manufacturers of base-metal thermocouple alloys in the UK, the USA and Sweden. The standardization procedures including the derivation of standard reference tables, summarized in this paper, are described in full in Burley et al (1975).

2. The formulation of highly stable nickel-base thermocouple alloys

Burley (1972) and Burley et al (1975) have set down in detail the conceptual and theoretical rationale for the optimal formulations of Nicrosil and Nisil. This is summarized in the following sections.

2.1. Positive alloy (Nicrosil)

There are sound reasons for retaining both chromium and silicon as the major solute elements in a preferred nickel-base positive thermoalloy. Nickel and nickel–chromium alloys, in addition to their economic and metallurgical advantages, have most desirable thermoelectrical characteristics. The ability of silicon to form stable, continuous and impermeable oxide layers at the metal/scale interface in oxidizing nickel–chromium alloys is of considerable significance to the enhancement of their environmental and thermoelectrical stability.

There are equally sound reasons, however, for asserting that the conventional positive thermoalloys can be greatly enhanced in stability by substantially increasing their chromium and silicon contents. It is of considerable significance that the differences between thermal EMF outputs corresponding to disordered and ordered states in Ni–10Cr and Ni–20Cr are of opposite algebraic sign (Burley 1970); an increase in the chromium content of Ni–Cr from 10 to 20 per cent reverses the direction of EMF change due to short-range order from positive to negative. This suggests that in this compositional range there is an alloy which is immune from EMF change effects induced by structural ordering. It is fortuitous that an increase in chromium content to, say, 15 per cent would also significantly enhance the oxidation resistance of the alloy. Recent studies of the composition dependence of the parabolic rate constants in the oxidation of Ni–Cr (Wood and Hodgkiess 1966, Giggins and Pettit 1969) have shown that in the temperature range 800 °C to 1100 °C the addition of chromium to nickel up to about two per cent increases the rate constant, that this rate remains substantially constant with further chromium additions up to about ten per cent, but that with further additions the rate is substantially reduced. Reference to the nickel-rich zone of the

1000 °C isothermal section of the Ni–Cr–O equilibrium diagram (Croll and Wallwork 1969) suggests that about 10Cr is the transition composition at which the spinel $NiCr_2O_4$ gives way to Cr_2O_3 as the stable oxide in Ni–Cr alloys. In alloys of 15Cr, the tendency at these temperatures for Cr_2O_3 to form as a continuous passivating layer at the metal/scale interface, instead of as an internal oxide precipitate as in the lower-chromium alloys, has been proposed (Wood and Hodgkiess 1966) as the major reason for their increased oxidation resistance. Since, as a rule, high-temperature parabolic oxidation signifies that a thermal diffusion process is rate-determining, the effectiveness of the Cr_2O_3 layer is presumably due to the low rate at which cations diffuse through a low concentration of chromium vacancies in the defected cation sublattice of this oxide, which is nearly stoichiometric.

There are several features of the oxidation mechanisms in Ni–15Cr, however, which require further consideration. First, as the chromium concentration at the solute-depleting metal/scale interface falls during protracted heating at 1000 °C to 1200 °C, the tendency for Cr_2O_3 to form as a non-protective internal precipitate rather than as a protective external film will be increased. Secondly, it is known that when Cr_2O_3 is heated above 1000 °C it tends to oxidize to CrO_3 which is volatile at such temperatures. The evaporation of the higher oxide from any exposed regions of the healing layer will tend to produce paralinear kinetics, and hence continuous chromium depletion and thermal EMF drift. Thirdly, whilst the Cr_2O_3 film will inhibit outward cation diffusion to the extent that this process becomes the rate-limiting step in the oxidation process, the film could still thicken slowly by oxygen reactions at the metal/film interface. A second oxidation-inhibiting mechanism is thus to be desired in the preferred positive alloy. The formation of a continuous silica film, in the form of alpha cristobalite, which occurs at the metal/scale interface in oxidizing Ni–Cr alloys when a small quantity of silicon is present, appears to be such a mechanism. Since the solubility of most elements in SiO_2 is virtually nil, there will be very small chemical potential gradients across the film and hence very small driving forces for the diffusion of oxidation reactants such as nickel and chromium through it. Provided the SiO_2 film remained continuous, this very low rate of diffusion would be the oxidation rate-controlling factor rather than the diffusion of chromium ions in Cr_2O_3.

Standard free energy data for the formation of the various oxides produced by a Ni–Cr–Si alloy at high temperatures suggest that SiO_2 will tend to form preferentially because it is the oxide with the largest negative free energy value. This factor alone, however, will not guarantee the formation of a complete healing layer of SiO_2; the composition of the bulk alloy must also be taken into account in determining whether this thermodynamically favoured oxide will eventually form a complete external layer or merely appear as an internal precipitate. The presently available Type K[+] alloys have silicon contents which do not exceed about half a per cent. There is strong evidence to suggest that this amount is significantly less than that required to produce a SiO_2 film of optimum diffusion inhibiting propensity. It has been shown (Gil'dengorn and Rogel'berg 1964), for example, that with oxidation in air in the range 1000 °C to 1200 °C, silicon additions reduce the oxidation rate of Ni–10Cr by an order of magnitude or more, the drop being most marked after the addition of from one to two per cent. Of considerable significance to the present study is the result that for such alloys an increase in temperature of exposure up to 1200 °C had an almost negligible effect in increasing the weight increment. It seems (Burley et al 1975) that in Ni–Cr–Si alloys containing (14 to 15) Cr, maximum thermal EMF stability in air at temperatures up to 1250 °C occurs in alloys containing 1·4% Si. Also it has been shown (Burley et al 1975) that minimal structure-related EMF variations occur in Ni–Cr–1·4Si alloys when the chromium content reaches 14%.

From the above considerations, the optimum prototype composition of Nicrosil appears to be Ni–14·2Cr–1·4Si.

2.2. Negative alloy (Nisil)

Burley (1972) has shown that the high-temperature air oxidation of the conventional negative thermoalloy (Ni–3Mn–2Al–1Si–$\frac{1}{2}$Co) produces substantial depletion of the reactive solutes manganese and aluminium in the vicinity of the internal oxidation reaction front, and that this process causes substantial thermal EMF drift. It can be argued (Burley 1972) that the readily oxidizable elements manganese and aluminium can be deleted from the composition of a preferred negative alloy for use at high temperatures in air. Since cobalt is not required as an EMF-modifying element, at least *ab initio*, the formulation of such a preferred alloy can be developed from 'first principles'. The virtue of nickel as the base for such an alloy has been established in earlier discussion, and there are sound reasons for retaining silicon as the major solute.

The preceding discussion proposes that silicon can suppress solute depletion and thermal EMF drift in oxidizing Ni–Cr alloys by forming a stable, continuous, and impermeable film of its oxide alpha cristobalite at the metal/scale interface. There is evidence (Gil'dengorn and Rogel'berg 1964) which suggests that silicon can perform a similar role in binary Ni–Si alloys, but prior to the development of Nisil there were no published data on the stability of the temperature–thermoelectromotive force characteristics of Ni–Si in air at temperatures above 1000 °C. For silicon to have optimum effect as an EMF stabilizer in Ni–Si its oxide would have to form as a continuous and impermeable layer, exclusively on the surface of the metal, which would persist indefinitely at high temperatures not only under isothermal conditions in air but also under conditions of very low oxygen pressure and/or rapid thermal cycling. Using classical Wagner theory (1959, 1965), Burley (1972) has calculated the theoretical critical concentrations of silicon, in a binary Ni–Si alloy, above which its oxide forms exclusively on the surface as a compact film. Examples of his typical values, which are temperature and oxygen pressure dependent, are 4·20 and 4·25 per cent silicon at 1100 °C and 1200 °C, respectively, at ambient oxygen pressure. These values are consistent with the results of an empirical investigation (Wolf et al 1965) in which the depth of internal oxide penetration in these alloys was measured as a function of silicon content. After heating for 200 h in air at 980 °C, for example, the depth of oxide penetration was found to be inversely proportional to solute content, the amount of internal precipitate being negligible when the concentration reached 4·0 per cent silicon. The optimum silicon content of a preferred negative thermoalloy should thus be between $4\frac{1}{4}$ and 5 per cent which is the limit of binary solid solubility of silicon in nickel at room temperature. There are sound thermoelectric reasons (Burley et al 1975) for a choice of 4·4 per cent silicon in this range.

A complete layer of SiO_2 on the surface of an oxidizing Ni–Si alloy should greatly reduce diffusion and hence oxidation rates. The long-term persistence of such a film will be governed by a number of factors, in particular its volatility and reactivity. These factors should not be a serious hindrance to the retention of SiO_2 as a protective diffusion barrier. It has been shown (Gulbransen and Jansson 1970), for instance that a SiO_2 scale at 1225 °C would exhibit negligible vapour losses in air and that SiO volatilization would be appreciable only in highly reducing gases. The reactivity of SiO_2 with nickel in the alloy to form Ni_2SiO_4 should likewise present little problem (Burley 1972, Burley et al 1975).

A compact and continuous scale-layer of silica on the surface of alpha Ni–Si is not assumed to be a perfect diffusion barrier; hence, small quantities of oxygen may dissolve in the solid solution substrate at high temperatures. It would be desirable, therefore, to incorporate in a preferred negative thermoalloy a small amount of a highly reactive solute metal which would preferentially 'getter' any such oxygen in forming its own oxide. In particular, this would suppress any tendency for silicon to oxidize internally, and perhaps reduce any NiO which might form concurrently with SiO_2 in the early stages of oxidation. Magnesium, at a concentration of about 0·1 per cent, has been found (Burley 1972) a most suitable element for this role. The mechanism of its beneficiation is at present under study at Materials Research Laboratories.

From the above considerations, the optimum prototype composition of Nisil appears to be Ni–4·4Si–0·1Mg.

3. The environmental, structural, and thermoelectrical stability of Nicrosil/Nisil

Recently the thermoelectric properties of Nicrosil/Nisil have been the subject of exhaustive studies at three different national laboratories. In addition to facilitating the establishment of basic thermoelectric reference data, the research at the US National Bureau of Standards has determined the sensitivity of thermoelectromotive force to solute concentration, internal stresses residual from plastic deformation and annealing, structural ordering, and magnetic transformations. This work is presented in Burley et al 1975 referred to above.

The thermal EMF stability of the new alloys has been extensively studied in Australia. Some of the results of this work are summarized in a separate conference paper (Burley and Jones 1975). The initial EMF stability has been determined at the National Measurement Laboratory, by successive calibrations of prototype samples of Nicrosil and Nisil having differing thermal histories, using techniques also described at this conference (Jones and Egan 1975). The long-term EMF stability has been assessed at the Defence Materials Research Laboratories by measuring thermal EMF drifts in similar samples on prolonged isothermal heating in air at temperatures up to 1300 °C.

4. The development of the thermoelectric reference data

4.1. Acquisition and preliminary testing of prototype alloys

For the establishment of the thermoelectric reference data, prototype alloy melt batches were acquired from five major base-metal thermocouple manufacturers (one in the UK, three in the USA, and one in Sweden). It was specified that the alloys should be manufactured in accordance with close tolerance limits on chemical composition, namely: Ni–(14·2 ± 0·15)Cr–(1·4 ± 0·1)Si for the Nicrosils and Ni–(4·3 ± 0·1)Si–(0·1 ± 0·05)Mg for the Nisils. Further, it was recommended that the concentrations of carbon and iron should not exceed 0·02 and 0·1, respectively, and that the total content of other elements should not exceed 0·1. Later, sound scientific reasons (Burley *et al* 1975) ultimately led to the adoption of a slightly modified nominal composition for Nisil, namely Ni–(4·4 ± 0·2)Si–(0·1 ± 0·05)Mg–(0·1 ± 0·03)Fe.

Samples of Nicrosil and Nisil, in the form of both 1·63 and 0·32 (or 0·25) mm diameter wires (AWG 14 and 28 (or 30), respectively), were supplied from at least two melt batches by each manufacturer. Altogether, samples from 15 melt batches of Nicrosil and 16 melt batches of Nisil were procured.

All the prototype alloys were received at NBS by January 1974. The thermoelectric and chemical differences between the various melt batches were determined in preliminary tests to decide which of them were most suitable for detailed calibration. The thermoelectric differences were established by taking samples from each melt batch and testing them against the platinum thermoelectric reference standard, designated Pt-67 (Powell *et al* 1974), at about $-196\,°C$ (near the boiling point of nitrogen) and at 100°C intervals from 0 to 1100°C. The chemical differences were ascertained from compositional data furnished by the manufacturers and from independent chemical analyses carried out by the Australian Defence Materials Research Laboratories. First, melt batches not in compliance with the above-mentioned compositional tolerances were excluded from further consideration; then, as discussed more fully in Burley *et al* 1975, six melt batches of Nicrosil and seven melt batches of Nisil were specially chosen for more extensive final calibration. The differences between the thermal EMFs of these batches at 1000°C amounted to about 160 μV for the Nicrosils and about 410 μV for the Nisils.

Samples for final calibration were taken from widely separated locations on each spool or coil of wire in order also to obtain thermoelectric information related to material inhomogeneity. So as to stabilize their thermoelectric properties prior to calibration these samples were isothermally annealed in air, using time–temperature parameters (Burley *et al* 1975) derived experimentally earlier in the project.

4.2. Experimental calibration methods

Detailed calibrations were carried out in the range $-268\,°C$ to $1300\,°C$ using calibration equipment located in laboratories of NBS at Boulder, Colorado and Gaithersburg, Maryland. Calibrations of the smaller diameter wires were made in the range $-268\,°C$ to $+7\,°C$ in the Cryogenics Division at Boulder. In the Heat Division at Gaithersburg, the larger diameter wires were calibrated in the range $100\,°C$ to $1300\,°C$; in addition, calibrations of the smaller diameter wires were made in an overlapping temperature range $-75\,°C$ to $400\,°C$. The calibration methods employed for these three temperature ranges are summarized in the following subsections.

4.2.1. *Cryogenic temperature range ($-268\,°C$ to $+7\,°C$).*
The basic apparatus and methodology of the cryogenic measurements have been described in Sparks *et al* 1972. The

main modifications for the present series of measurements were the installation of new temperature controllers and the use of a calibrated digital multimeter in place of the previously used six-dial potentiometer and Mueller bridge. The cryostat contained 22 different thermoelement wires (four reference wires and nine each Nicrosil and Nisil wires, mostly AWG 28 in size) which were measured in 37 different thermocouple combinations. Using an advantageous experimental design based on graph theoretical concepts (Sparks *et al* 1972), a total of 49 Nicrosil, Nisil, and reference-metal combinations were numerically analysed and studied statistically for errors. The average standard deviations varied between 0·1 and 0·2 μV.

4.2.2. High-temperature range (100 °C to 1300 °C). In this range the thermal EMFs of 1·63 mm diameter Nicrosil and Nisil wires were measured at 50 °C intervals against platinum reference wires whose values of thermal EMF were known relative to Pt-67. The platinum thermoelement of a standard Pt–10Rh/Pt thermocouple served as the reference wire. Testing was done in laboratory tube furnaces; the nickel–chromium tube furnace described in Roeser *et al* 1958 was used between 100 °C and 1000 °C and a furnace with a tubular silicon carbide heater (Burns and Gallagher 1966) was used between 800 °C and 1300 °C. The thermal EMF of the standard thermocouple and that of the nickel-base alloy wires against the platinum reference wire were measured simultaneously by the two-potentiometer method (Roeser *et al* 1958). All calibration runs were taken with increasing temperature and the reference junctions of the wires were maintained at 0 °C in ice baths. In this manner, corresponding values of EMF and temperature were obtained for 18 Nicrosil wires and 20 Nisil wires.

4.2.3. Overlap temperature range (−75 °C to 400 °C). The thermal EMFs of 0·32 (or 0·25) mm diameter Nicrosil and Nisil wires against platinum reference wires were measured at 25 °C intervals from −75 °C to 100 °C and then at 50 °C intervals from 100 °C to 400 °C. For these measurements, a standard platinum resistance thermometer (SPRT) was used to determine the temperatures of the measuring junctions. A cryostat (below 0 °C), together with a series of stirred liquid baths (above 0 °C), provided the controlled temperature media (water, oil and molten tin) for this purpose. Swindells (1965) has described the salient features and temperature ranges of these calibration baths.

The immersion depths, mounting techniques, and configurations of the wires and the SPRT in the various baths were similar to those employed in a previously reported investigation (Bedford *et al* 1972). The calibration points were taken in order of increasing temperature and the reference junctions of the wires were maintained at 0 °C in ice baths. The values of thermal EMF were measured with a calibrated six-dial laboratory potentiometer and all SPRT resistances were measured with G-2 Mueller bridges. By this calibration method, corresponding values of EMF and temperature were obtained for 14 Nicrosil wires and 16 Nisil wires.

4.3. Mathematical analyses and results

Two major steps of data analysis were involved in transforming the raw experimental data into its final presentable form. The first involved minor adjustments to allow for known systematic errors of measurement, and the formation of desired thermocouple combinations using graph theory considerations. The second involved fitting the adjusted data, using orthonormal polynomial equations, in order to provide continuous $E = f(T)$ relationships for appropriate thermocouple combinations. The equations representing the thermal voltages varied in order from 5 to 10.

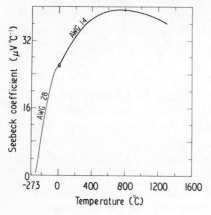

Figure 1. Seebeck coefficient for Nicrosil against Nisil thermocouples as a function of temperature. The curve above 0 °C applies for 1·63 mm diam (AWG 14) wire while the curve below 0 °C applies for 0·32 mm diam (AWG 28) wire.

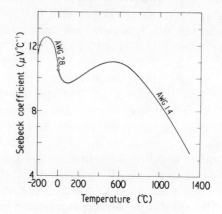

Figure 2. Seebeck coefficient for platinum, Pt-67, against Nisil thermoelements as a function of temperature. The curve above 0 °C applies for 1·63 mm diameter (AWG 14) wire while the curve below 0 °C applies for 0·32 mm diameter (AWG 28) wire.

Table 2. Thermal voltages (E) and Seebeck coefficients (S) of Nicrosil against Nisil thermocouples at various temperatures (T): reference junctions at 0 °C.

T (°C)	E (mV)	S ($\mu V\,°C^{-1}$)
−200	−3·99	9·93
−100	−2·41	20·93
0	0·0	26·15
100	2·77	29·63
200	5·91	32·99
300	9·34	35·43
400	12·97	37·11
500	16·74	38·26
600	20·61	38·97
700	24·53	39·29
800	28·46	39·26
900	32·37	38·99
1000	36·25	38·55
1100	40·08	37·98
1200	43·84	37·17
1300	47·50	36·15

A full treatment of the results for the thermal voltages, Seebeck coefficients, and their derivatives, along with the representing functions and error analyses are given in Burley et al 1975. Some salient information, however, is presented here. The Seebeck coefficients, for example, for Nicrosil/Nisil and Pt-67/Nisil are shown in figures 1 and 2, respectively, for two sizes of wire. Results on the smaller diameter wires (AWG 28) between $-268\,°C$ and $400\,°C$ were analysed separately from those on the larger (AWG 14) wires between $0\,°C$ and $1300\,°C$. Because of different volume/surface area ratios the relative effects of the skin layers in the two wire sizes are different, and therefore the Seebeck coefficients are slightly different, for example by about 0·1% at $0\,°C$. Figure 2 clearly shows the effect of the magnetic transformation that occurs in Nisil near room temperature. A representative set of values of thermal voltages and Seebeck coefficients for Nicrosil/Nisil is also given in table 2.

References

ANSI 1964 *American National Standards Institute* C96.1
Bedford R E et al 1972 *TMCSI* 4 part 3 1585
Burley N A 1969 *J. Inst. Metals* 97 252
—— 1970 *Australian Defence Standards Laboratories Report* 353
—— 1972 *TMCSI* 4 part 3 1677
Burley N A and Jones T P 1975 this volume
Burley N A, Powell R L, Burns G W and Scroger M G 1975 *NBS Monograph* to be published
Burns G W and Gallagher J S 1966 *J. Res. NBS* 70C 89
Croll J E and Wallwork G R 1969 *Oxidation of Metals* 1 55
Fenton A W 1969 *Proc. IEE* 116 1277
Giggins C S and Pettit F S 1969 *Trans. Metall. Soc. Am. Inst. Metall. Engrs.* 245 2495
Gil'dengorn I S and Rogel'berg I L 1964 *Fiz. Metal. Metalloved* 18 955
Gulbransen E A and Jansson S A 1970 *Heterogeneous Kinetics at Elevated Temperatures* (New York: Plenum) p181
Jones T P and Egan T M 1975 this volume
Powell R L et al 1974 *NBS Monograph* 125
Roeser W F and Londberger S T 1958 *NBS Circular* 590
Sparks L L, Powell R L and Hall W J 1972 *NBS Monograph* 124
Swindells J F 1965 *NBS Monograph* 90
Wagner C 1959 *Z. Electrochem.* 63 772
—— 1965 *Corros. Sci.* 5 751
Wolf J S, Weeton J W and Freche J C 1965 *NASA Tech. Note* TN D-2813
Wood G C and Hodgkiess T 1966 *J. Electrochem. Soc.* 113 319

Practical performance of Nicrosil–Nisil thermocouples

N A Burley[†] and T P Jones[‡]

[†] Materials Research Laboratories, Australian Defence Scientific Service, Department of Defence, Melbourne, Victoria 3032, Australia

[‡] National Measurement Laboratory, Commonwealth Scientific and Industrial Research Organization, Sydney, New South Wales 2008, Australia

Abstract. The practical performance of the new nickel-base thermocouple system Nicrosil–Nisil has recently been assessed in laboratory investigations in Australia, at the Defence Materials Research Laboratories (MRL) and the National Measurement Laboratory (NML).

The initial thermoelectric stability has been determined at NML by successive calibrations, in differing temperature gradients, of prototype samples having different thermal histories. The long-term stability has been investigated at MRL by measuring thermal EMF drifts in similar samples on long exposure in air at constant temperatures up to 1275 °C. The salient results of this work are summarized.

The thermoelectric instability of Nicrosil–Nisil, and hence the uncertainty in its calibration and use, is shown to be markedly less than with existing nickel-base thermocouples of Type K. A comparison is also made with rare-metal thermocouples.

1. Introduction

It has been shown (eg Burley 1972) that the accuracy of existing nickel-base thermocouples of Type K (ANSI 1964) can be affected deleteriously by inconstancy of thermoelectromotive force (thermal EMF) caused by environmental and structural instabilities in the component alloys. It has also been shown (Burley 1972, 1975) that these instabilities can be greatly attenuated by changing the solute concentration of certain of the component elements of the nickel-base alloys involved. The same author has derived formulations for a pair of highly stable nickel-base thermocouple alloys called Nicrosil and Nisil.

Another paper in this conference (Burley *et al* 1975a) reviews the development of the new alloys by the Materials Research Laboratories of the Australian Department of Defence and their standardization by the National Bureau of Standards of the US Department of Commerce.

The thermal EMF stabilities of prototype samples of Nicrosil and Nisil are being studied by various scientific and industrial authorities around the world. In Australia, the initial EMF stability has been studied at the National Measurement Laboratory by successive calibrations of prototype samples having differing thermal histories, whilst the long-term EMF stability is being assessed at the Materials Research Laboratories by measuring thermal EMF drifts in similar samples on long exposure in air at various constant temperatures up to 1275 °C. This paper summarizes a representative selection of the results of the Australian work which demonstrates that Nicrosil–Nisil thermocouples have much higher thermoelectric stabilities in air than Type K thermocouples.

2. Materials, equipment and experimental procedures

2.1. Materials tested

The Nicrosil–Nisil thermocouple alloy samples used in these experiments were taken from the same prototype batches used in the NBS standardization exercise, and are fully characterized by Burley *et al* (1975b). The nominal prototype compositions (percentages by weight) are

Nicrosil: Ni–14·2Cr–1·4Si
Nisil : Ni–4·4Si–0·1Mg.

The Type K thermocouples used were of the conventional variety also described in the same paper (Burley *et al* 1975b).

All the wires were annealed by their manufacturers in a protective atmosphere, typically for a few minutes at 1100 °C, prior to delivery. Test samples were re-annealed in air for 30 min at 850 °C immediately prior to the experiments.

2.2. Initial thermal EMF stability

The investigation of thermoelectric changes occurring during initial heating of the test thermocouples utilized two separate facilities, a rapid computerized calibration system (Jones and Egan 1975) and a special three-zone ageing furnace. In the automatic calibration system the test thermocouples are compared with a standard thermocouple in a programmed furnace whose temperature is raised to 1100 °C and lowered to ambient at rates which allow the thermocouple measuring junctions to remain within the range 700 to 1100 °C for only two hours. In addition, the system is programmed to establish different temperature gradients along the thermocouples during the cooling and heating portions of the calibration cycle to afford a sensitive means of detecting changes in the inhomogeneity of the thermocouples. Successive calibrations in this equipment were used to show changes in the thermal EMF outputs of the test thermocouples during the initial hours of their use. The ageing furnace consists of a Mullite tube (32 mm ID) on to which a Kanthal wire (1·3 mm diameter) heater is wound so that there are three turns per centimetre in the two 75 mm end zones and two turns per centimetre in the 560 mm central zone. The temperature distribution along the tube is adjusted by changing the resistances of rheostats connected in parallel with each end zone. A test region, uniform in temperature to ±10 °C over 650 mm was achieved at the two desired test temperatures of 1100 °C and 1250 °C. The ageing furnace was used for short-term thermal ageing of test thermocouples as an integral part of the initial EMF stability investigation. Identical heating of test assemblies comprising Nicrosil–Nisil and Type K thermocouples was ensured by welding their measuring junctions into a common bead, into which was peened a standard rare-metal thermocouple, and inserting the assembly 520 mm into the ageing furnace. At this common immersion, the portion of the wires within 460 mm of the common measuring junction lay within the uniform temperature zone.

Two separate experiments involving ageing were carried out. In the first experiment base-metal test thermocouples of 3·3 mm diameter (8 AWG) were aged at 1250 °C, as measured by the standard thermocouples, and their EMF outputs measured as a function of time. The overall uncertainty of the EMF measurement system in this experiment, including connections, selector switch, and potentiometer, was less than ±0·2 μV for rare-metal thermocouples and ±2 μV for base-metal thermocouples. In the second experiment similar thermocouples were calibrated in the automatic facility before and after ageing for various periods of time up to 100 h, with an uncertainty in the EMF measuring system of less than ±1 μV for rare-metal thermocouples and ±4 μV for base-metal thermocouples. The immersion in the calibration furnace was selected so that the length that had received uniform heating in the ageing furnace extended out into ambient temperature.

To obtain further information on the effects of changes in the inhomogeneity of the test thermocouples during short-term ageing, measurements were repeated with the thermocouples withdrawn 75 mm after the first ageing experiment and also during successive calibrations in the second ageing experiment. The thermal EMF outputs of the base-metal thermocouples and the EMF of each of their thermoelements with respect to the platinum of the standard thermocouple were determined in each measurement. Experiments, similar to those above, were also performed on 1·6 mm diameter (14 AWG) Nicrosil–Nisil and Type K thermocouples. For these the ageing temperature was reduced to 1100 °C.

As a comparison, similar experiments were conducted with rare-metal thermocouples (Types S and R, 0·5 mm diameter). The two different ageing experiments were conducted at 1100 °C for up to 200 h and at least 10 successive calibrations were performed on each rare-metal thermocouple.

2.3. Long-term thermal EMF stability

The investigation of long-term drifts in the thermal EMF outputs involved prolonged exposure of 3·3 mm diameter test thermocouples in air at constant high temperatures in a specially constructed furnace which consists essentially of an alumina tube (40 mm ID) over which a Kanthal spiral heater is wound with uniform spacing. The temperature distribution along the tube is adjusted by varying the output voltages of stabilized manual autotransformers connected one to each of a series of tapped segments of the heater. A central test region, uniform in temperature to ±2 °C over 400 mm, was achieved at a test temperature of 1250 °C. This control arrangement also permits the temperature gradients along the colder portions of the test thermocouples to be held constant throughout the tests, a condition essential to this type of experimentation (Burley 1972). The test assemblies, comprising Nicrosil–Nisil, Type K, and calibrated rare-metal reference thermocouples were similar to those used in the initial stability investigation described above. The composite thermocouples were inserted horizontally into the test furnace and clamped so that the portion of the wires within 300 mm of the common measuring junction lay within the uniform zone.

The thermal EMFs of all the thermocouples, and of their individual positive and negative thermoelements against platinum, were measured initially and at various times during the prolonged high-temperature exposure. Two principal ageing temperatures are considered here, namely 1000 °C and 1250 °C. The Type K thermocouples were found to be quite unstable at 1250 °C, so results for them are presented at an upper test temperature of 1200 °C. A six-dial precision potentiometer was used to measure the thermal EMFs. Particular care was taken in the detection and compensation of spurious thermal EMFs occurring in the lead-wire, reference-junction, and switching portions of the measuring circuitry. The calibrations of the rare-metal reference thermocouples were checked initially and at the conclusion of each series of tests by a special technique which is described elsewhere (Burley 1975). Any changes in the reference calibrations were allowed for at the various ageing intervals on the basis of linear interpolation. The overall uncertainty of the EMF measurements in this experiment was less than ±2 μV.

3. Results

3.1. Initial thermal EMF stability

The short-term variations with time in the thermal EMF outputs of 3·3 mm Nicrosil–Nisil and Type K thermocouples located in a fixed position in the ageing furnace maintained at 1250 °C are summarized graphically in figure 1. Similar curves were obtained for the 1·6 mm wires aged at 1100 °C where, after 180 h heating, the difference in the EMF from that at zero time was stable at 20 μV ($\simeq$0·5 °C) for the Nicrosil–Nisil thermocouple, whilst the difference for the Type K thermocouple was 450 μV ($\simeq$11 °C) and increasing at the rate of 2·5 μV h^{-1}. At the conclusion of the ageing test from which figure 1 was derived, the immersion of all thermocouples was decreased by 75 mm and this caused an increase in the output of the 3·3 mm type K thermocouple of 1000 μV and in the Nicrosil–Nisil thermocouple of 280 μV.

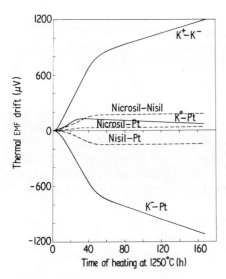

Figure 1. Short-term changes in the thermal EMF output of 3·3 mm diameter Nicrosil–Nisil and Type K thermocouples, and their individual thermoelements against platinum, in a fixed installation at 1250 °C.

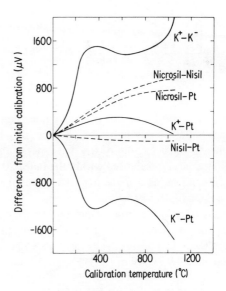

Figure 2. Variation in the calibration of 3·3 mm diameter Nicrosil–Nisil and Type K thermocouples after 50 h uniform ageing at 1250°C.

The differences between calibrations of the larger diameter thermocouples in the second experiment before and after uniform heating at 1250 °C for 50 h are given in figure 2. The smaller diameter wires again gave similar curves; after 50 h at an ageing temperature of 1100 °C, the 1·6 mm Type K thermocouples had changed by +900 μV and Nicrosil–Nisil by +750 μV at a calibration temperature of 1000 °C.

The reproducibility of the calibrations of both sizes of each type of base-metal thermocouple, during successive calibration runs at 1000 °C, are shown in figure 3. The datum points of figure 3 are the mean calibrations of the heating and cooling cycles of the thermocouples. The variations in the calibrations of the base-metal thermocouples

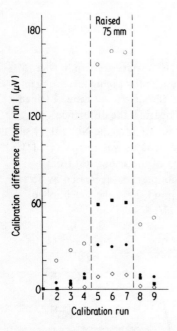

Figure 3. Variation of thermocouple calibrations at 1000 °C for successive calibrations in a fixed installation. The datum points are the mean calibrations of the heating and cooling cycles. ○, Type K (1·6 mm); ■, Type K (3·3 mm); ●, Nicrosil–Nisil (1·6 mm); ◇, Nicrosil–Nisil (3·3 mm).

at the same measuring-junction temperature during the heating and cooling cycles were similar for both types of thermocouple. During all calibration runs the maximum difference in EMF between heating and cooling occurred in the mid-range temperatures and, whereas the behaviour of the two types of thermocouple was different, the maximum differences typically fell within the range 20 to 70 μV. The differences which occurred between heating and cooling calibrations at 1000 °C were as high as 35 μV for either thermocouple type.

By comparison, the total change which occurred in the calibration of both the Types R and S thermocouples during all experiments and calibrations was always less than 0·3 °C. In most cases the changes were much smaller and no systematic trends could be associated with any of the heating schedules.

3.2. Long-term thermal EMF stability

Typical long-term drifts in the thermal EMF outputs of 3·3 mm Nicrosil–Nisil and Type K thermocouples on long exposure at constant high temperatures in air are presented graphically in figure 4 (1000 °C) and figure 5 (1200 °C and 1250 °C). The results

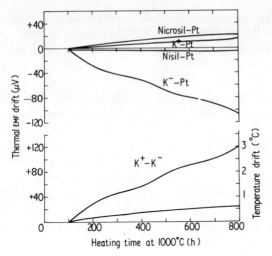

Figure 4. Long-term thermal EMF drifts in Nicrosil–Nisil and Type K thermocouples and their individual thermoelements against platinum. The drifts are changes from EMF output values existent after 100 h constant temperature (1000 °C) exposure in air.

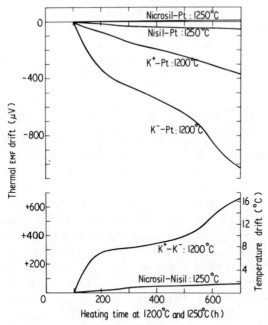

Figure 5. Long-term thermal EMF drifts as in figure 4, but on exposure at 1200 °C (Type K) and 1250 °C (Nicrosil–Nisil).

are given in terms of deviations of thermal EMFs from 'original values' as functions of time at a particular temperature. The original value is taken as the thermal EMF output after 100 h heating so as to effectively separate long-term EMF drifts from short-term calibration changes.

It will be seen that, on long-term exposure at 1000 °C in these tests, Nicrosil–Nisil thermocouples were about five times as stable as Type K thermocouples. The typical deviation of the Nicrosil–Nisil system at 700 h was about $+25\,\mu V$ ($\simeq 0.6$ °C) whilst that of the Type K thermocouples was about $+120\,\mu V$ ($\simeq 3$ °C). Whilst both the Nicrosil and Nisil thermoelements were stable at this temperature, the relative instability of the Type K thermocouples was predominantly due to drift in their negative thermoelements.

On exposure at 1250 °C, the EMF outputs of the Type K thermocouples were quite unstable, drifting continuously at the rate of about $2\,\mu V$ per hour. On the other hand, at 1250 °C the deviation of the Nicrosil–Nisil thermocouples at 700 h was about $+60\,\mu V$ ($\simeq 1.5$ °C). By way of further comparison, at 1200 °C the EMF of the Type K thermocouples had drifted by about $+650\,\mu V$ ($\simeq 16$ °C) at 700 h. It is also to be noted that whereas the positive thermoelements of the type K thermocouples drifted by more than $-350\,\mu V$ at 1200 °C, the Nicrosil elements at 1250 °C were virtually inert and showed negligible EMF changes. Were it not for the fact that the EMF drifts in the individual thermoelements of the Type K thermocouples were of the same polarity against platinum at 1200 °C, the net thermocouple drift would have been even greater, eg, at 700 h a $-360\,\mu V$ drift in K(+) and a $-1020\,\mu V$ drift in K(–) resulted in a net $+660\,\mu V$ drift in the combination.

It is of considerable significance that, at 1250 °C, the calibration of the Type R rare-metal reference thermocouples, as monitored during the test by the special technique referred to above, changed by about 1 °C in the 700 h heating period. Since the thermal EMF drift in Nicrosil–Nisil in the same test was equivalent to about 1.5 °C, it seems that the new thermocouples are about as stable in the long-term as platinum-base thermocouples at 1250 °C in air.

4. Discussion

In the formulation of Nicrosil and Nisil alloys (Burley 1972, 1975) the aim was to develop a new nickel-base thermocouple system having much higher thermoelectric stabilities, and usable at higher maximum temperatures, than existing nickel-base thermocouples of the ANSI Type K. In the main, this aim has been fulfilled. Not only are Nicrosil–Nisil thermocouples much more stable than those of Type K, particularly at the higher temperatures, but their long-term EMF stability in air is similar to that of rare-metal thermocouples in the vicinity of 1250 °C, a temperature at which Type K materials are quite unstable. Furthermore, Nicrosil–Nisil thermocouples are virtually free of the structural ordering effects which further attenuate the thermoelectric stability of Type K thermocouples (Burley et al 1975a, b) and are much more stable under cyclic conditions of heating (Burley 1972). The present results suggest that the long-term calibration of Nicrosil–Nisil will remain quite constant at high temperatures in air, in a fixed location, for at least up to about 1000 h of exposure.

It is in the initial calibration of Nicrosil–Nisil that some problems arise. The short-term changes in the thermal EMFs of both Nicrosil–Nisil and Type K thermocouples in the initial hours of heating clearly indicate limitations in the manner in which they can be calibrated and used. It can be seen, however, that in all cases the short-term performance of Nicrosil–Nisil is markedly better than that of Type K. To complement the long-term data, the results in figure 1 show that 3.3 mm diameter Type K thermocouples are quite unstable, *ab initio,* at 1250 °C. Similarly, 1·6 mm diameter Type K thermocouples are unstable at 1100 °C. In contrast, Nicrosil–Nisil thermocouples are completely satisfactory for fixed location measurements at these temperatures within a stable furnace. However, the results also indicate that if these thermocouples are moved or the furnace temperature gradients change so that a part of the Nicrosil–Nisil thermocouple which was previously at temperatures above 1000 °C falls in a steep temperature gradient, large systematic changes can occur.

The short-term results show that the calibration of both types of base-metal thermocouple tested after extended use at elevated temperatures should not be performed in a different furnace. If an application calls for extended use in a fixed location up to about 1250 °C, a satisfactory calibration can be performed on Nicrosil–Nisil before it is used. It can be seen in figure 1 that the EMF output may increase by the equivalent of 4 °C before settling down to a stable value. Since EMF differences between the heating and cooling curves of as much as 70 μV have been noted for both types of thermocouple during initial calibrations, the least possible uncertainty that can be achieved in a separate calibration furnace is approximately ±2 °C. It would be possible to calibrate a Nicrosil–Nisil thermocouple with a lesser uncertainty, say ±1 °C, if the calibration was performed after approximately 40 h in the location where it was to be used.

The relative lack of reproducibility in the initial calibration of the two types of base-metal thermocouple investigated is due to the production of compositional inhomogeneities as reactive solute elements are non-uniformly depleted by oxidation reactions (Burley 1972). When these inhomogeneities react with steep temperature gradients after movement of the thermocouple or on recalibration in a remote location, large systematic changes in EMF output can occur. All the present results suggest that these compositional inhomogeneities continue to be produced to a considerably greater degree and over considerably longer periods of time in Type K thermocouple alloys than in Nicrosil and Nisil.

Nevertheless the uncertainty in the initial calibration of Nicrosil–Nisil thermocouples is considered to be undesirable and steps are being taken to reduce it. The high environmental stability of the new alloys in the long-term is due to the development at high temperature of highly efficacious diffusion barriers in the form of concentric oxide films located at the metal/scale interface. These oxide layers, once established, virtually inhibit oxidation, solute depletion, and hence long-term thermal EMF drift. The early instability in Nicrosil–Nisil is due to the fact that the passivating oxide layers require a finite time to attain optimum diffusion inhibiting propensity when forming in air. To overcome this a process called 'thermal passivation' is in course of development at MRL. This process, which involves a simple initial heat-treatment of Nicrosil and Nisil wires in a special low-pressure atmosphere, promotes a more rapid and efficacious formation of the diffusion-barrier oxides than is possible by heating in air. It is expected that this process will passivate the alloys prior to initial usage, with the result that the short-term stability of the new alloys will be greatly enhanced and their initial calibration established with minimal uncertainty.

References

ANSI 1964 *American National Standards Institute* C96.1

Burley N A 1972 *TMCSI* **4** part 3 1677
────── 1975 *Australian Defence Scientific Service Materials Research Laboratories Report* to be published
Burley N A, Burns G W and Powell R L 1975a this volume
Burley N A, Powell R L, Burns G W and Scroger M G 1975b *NBS Monograph* to be published
Jones T P and Egan T 1975 this volume

Inst. Phys. Conf. Ser. No. 26 © 1975: Chapter 4

Improved compensating lead systems for platinum-base thermocouples

W G Bugden†, J A Tomlinson‡ and G L Selman

Johnson, Matthey & Company Limited, Research Laboratories, Wembley, London

Abstract. A new compensating lead system has been developed for rhodium–platinum thermocouples. Accuracies within ±2 °C when the junction between lead and thermocouple is varied between 20 and 500 °C can be obtained and more accurate compensation is possible over narrower working ranges. The wide working range and close accuracy of this system represents a considerable advance over those which are currently provided. Precision is obtained by using a 'three-leg' compensating lead, the negative limb of which is produced from two stainless steels of similar composition. The positive limb is a 20% Cr, 70% Ni, 10% Fe alloy. By combining the two negative wires as a braided cable high accuracy is possible without the necessity for close compositional control.

The principle involved allows the thermoelectric characteristics of a wide range of thermocouples to be rapidly and accurately simulated, and should find wide application.

1. Introduction

The compensating leads for rhodium–platinum thermocouples have hitherto consisted of a positive limb of pure copper, the negative limb being a dilute solution of nickel in copper. This combination produces a thermal EMF which is comparable to that of the platinum-metal thermocouple, and can if necessary be made to coincide with it at one particular temperature. The shape of the temperature–EMF curve of the copper–nickel compensating lead is different, however, to that of the rhodium–platinum thermocouple so that accurate compensation is only achievable with the junction between base and platinum-metal wires at one particular temperature.

In recent years the requirements for accurate compensating leads have increased rapidly because of an increasing tendency to situate the junction between base and noble metals in areas which might sometimes rise to 500 °C. This situation occurs with expendable thermocouples and with other more conventional sensing elements used for measuring the temperature of oxygen-blown steels.

The present work was undertaken primarily, therefore, to develop a compensating lead suitable for use with the platinum–13% rhodium–platinum thermocouple, and which could be safely used at temperatures up to about 500 °C without the introduction of errors in temperature measurement significantly higher than those inherent in the thermocouple itself.

† Present address: British Rail Technical Centre, Derby.
‡ 25, Lower Paddock Road, Oxhey, Watford, Herts.

2. Compensating leads and the three-leg system

The ideal compensating lead is composed of two base-metal wires, the integral EMFs of which coincide over a wide range of temperature with the integral EMFs of the platinum thermocouple wires to which they are joined. The practical conditions which must be satisfied are shown graphically in figure 1.

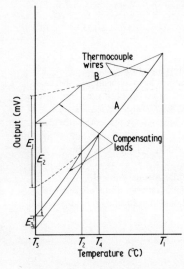

Figure 1. Graphical representation of the outputs of a thermocouple compensating lead system.

Here we have a thermocouple limb A, which, under the action of a temperature gradient, generates the integral thermal EMF indicated on the curve. The hot junction with limb B is made at temperature T_1. At this junction both wires have the same potential, the EMF measured by a potentiometer at T_2 being that generated by the difference between the two integral EMF curves.

The function of the compensating lead is to allow the potentiometer to be displaced from a hot and unfavourable environment at T_2 to more equable conditions at the lower temperature T_3. For reasons of economy the leads employed must be base metals.

If this displacement is to be made without error, the EMF measured at the cool end of the compensating lead must be the same as the EMF which would have been measured if the noble-metal thermocouple wires had been extended to the cold junction terminals. In other words, E_2 on the diagram must be equal to E_1. This could be achieved by using compensating lead wires with the thermoelectric characteristics shown, neither of which coincide with those of the thermocouple wires. This arrangement would only work, however, at the particular temperature T_2 which is illustrated. If one of the junction terminals were displaced, as shown, to temperature T_4, an EMF error equal to E_3 would be introduced into the system and even if both terminals were kept at the same temperature gross inaccuracies would be introduced as soon as the junction temperature changed from T_2.

The only leads which would provide adequate temperature compensation over a wide range of temperature would be those having integral EMF values coincident with those of the thermocouples illustrated by the two broken lines. The accurate simulation within a new alloy of EMF values produced within an already existing alloy is a very difficult problem. An attempt to achieve this simulation has been made with Thermolede where a copper positive limb is coupled with a negative limb consisting of a dilute alloy of nickel in copper. Here fairly reasonable compensation with the 13% rhodium—platinum thermocouple is obtained when the wires are joined at 100 °C, but at higher and lower temperatures errors equivalent to several degrees are introduced into the measuring circuit.

Industrial temperature measurements are now characterized by higher accuracy requirements and by operating conditions which tend to heat the junctions between platinum metal and compensating lead wires to increasingly higher temperatures. When measuring the temperature of liquid steel with disposable thermocouples, the junction between base and noble metals frequently rises to 500 °C and the need for accurate compensating leads for this purpose alone is very considerable.

A comprehensive survey of published information on the thermoelectric characteristics of base-metal alloys showed that the integral EMF curves of certain compositions in the ternary system iron—nickel—chromium approximated closely to those of platinum and rhodium—platinum alloys. The match obtainable by judicious selection from these alloys was infinitely better than that provided by copper and copper—nickel alloys, and it seemed appropriate therefore that these iron—nickel—chromium alloys should be used to provide accurate thermocouple compensation by the three-leg system.

3. Construction and operational characteristics of a three-leg system

The principle of a three-leg compensating lead has been previously described[†]. The basic construction is shown in figure 2, the criteria for its successful operation being: (i) that the temperature—EMF relationships of A—C and B—C straddle that of platinum—13% rhodium—platinum; and (ii) that the resistance ratio R_A/R_B is equal to the ratio of the thermoelectric deviations of the two pairs, from the thermocouple at the temperature T_J. If the alloys A, B and C are such that the ratio of these deviations varies little with temperature then the utility of this construction is considerably extended.

Figure 3 shows that the alloy compositions 14%Cr—12%Ni—74%Fe, 22%Cr—12%Ni—66%Fe, and 20%Cr—70%Ni—10%Fe, can be combined to satisfy the first criterion stated above. The ratio of the thermoelectric deviations from the platinum thermocouple X/Y varies between 1·4 and 2·5 over the temperature range 0—500 °C. Table 1 provides the results obtained from a thermocouple having a 20%Cr—70%Ni—10%Fe positive limb and a negative limb comprising 15 strands of 22%Cr—12%Ni—66%Fe and

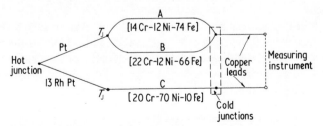

Figure 2. Diagram of a thermocouple three-leg compensating lead system.

† British Patent No 1 379 546.

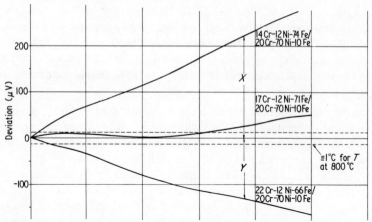

Figure 3. Deviations of Fe–Ni–Cr alloy combinations from Pt–13RhPt.

Table 1. Deviation of the 'stranded' compensating cable AB–C (14%Cr–12%Ni–74%Fe/22%Cr–12%Ni–66%Fe/20%Cr–70%Ni–10%Fe).

Temperature (°C)	Predicted deviations from Pt–13% RhPt			Actual deviation
	A–C (μV)	B–C (μV)	AB–C (μV)	AB–C (μV)
116·6	+77	−45	+1	−2
138·6	+87	−54	−1	−4
211·8	+120	−86	−9	−12
310·5	+179	−116	−5	−12
400·6	+234	−136	−3	−5
496·1	+284	−162	+5	+6

9 strands of 14%Cr–12%Ni–74%Fe wire 0·2 mm in diameter. This thermocouple was compared directly against a calibrated 13% rhodium–platinum thermocouple at temperatures up to 500 °C. The measured deviations compare remarkably well with those predicted from the initial thermoelectric data on the assumption that the specific resistances of the two iron-rich alloys are equal; and they indicate that compensation to within ±1 °C would be achieved at lead junction temperatures (T_J) within this range. Even better agreement could clearly be obtained over more limited temperature ranges by suitable adjustment.

A major advantage of this compensating technique is that it allows one to take account of the unavoidable batch-to-batch variations in the compositions of the alloys involved. From a knowledge of the thermoelectric characteristics of each batch of alloys, it is possible to compute the combination of wires required in the negative limb to achieve accurate compensation.

4. Thermoelectric characteristics of Fe–Ni–Cr alloys

The thermoelectric characteristics of a large number of the ternary alloys from the γ region of the system were evaluated. Curves illustrating the behaviour of three alloy

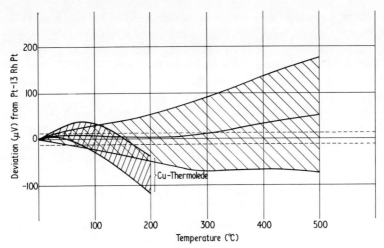

Figure 4. Predicted tolerance range of 17Cr–12Ni–71Fe/20Cr–70Ni–10Fe compared with that of Cu–0·56NiCu (Thermolede).

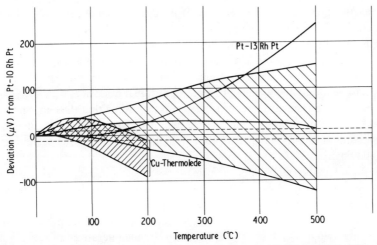

Figure 5. Predicted tolerance range of 19·5Cr–20Ni–60·5Fe/20Cr–70Ni–10Fe.

combinations of interest within the present context are shown in figure 3 where the deviations of the base metal pairs from the temperature–EMF relationship of the 13% rhodium–platinum thermocouple are plotted. The broken lines on the graph represent a deviation of ±1 °C for thermocouple hot junction temperatures of 800 °C and above.

The alloy combination 17%Cr–12%Ni–71%Fe/20%Cr–70%Ni–10%Fe is seen to give a very similar signal to that of the 13%rhodium–platinum thermocouple at temperatures up to 300 °C, and the error introduced by such a compensating lead in this range would be less than 1 °C. However, when the effects of the inevitable batch-to-batch variations in composition are considered, the picture becomes less bright. Figure 4 predicts the errors which are likely to accrue within a compositional tolerance of

±0·5% on the minor, and ±1% on the major constituents (>20%) of the alloys involved. Whilst the ternary alloy combination continues to show advantages over Thermolede in the lower temperature ranges, its performance at higher temperatures leaves much to be desired. Figure 5 presents similar information for the experimental alloy pair which most closely matched the thermoelectric characteristics of the 10% rhodium−platinum thermocouple.

It is probable that better performances than those predicted could be obtained by batch selection. The general conclusion to be drawn from these results, however, is that the three-leg compensating lead system offers a far more practical solution to the problem of batch-to-batch variations in composition.

5. Practical approaches to the three-leg system

The lightweight form of compensating cable, which has 24 roped strands of wire in each limb, would appear to provide an ideal basis for the three-leg system. The thermoelectric potential generated by the negative limb of such a device can be readily adjusted by varying the relative number of fixed diameter strands of the two iron-rich alloys within the stranded cable, and the resistance ratio thus achieved would be independent of cable length. This would not be the case if the division was accomplished by means of an external resistor or potential divider.

The nominal compositions of lead alloys which were selected for the 13 and 10% rhodium−platinum thermocouple are given in table 2. It will be noted that these alloys are slightly modified versions of the original experimental compositions, which were considered to be unsuitable in view of the compositional tolerances which have to be accepted in practice. On the limits of tolerance it is possible that the experimental alloy 14% Cr−12% Ni−74% Fe might become two phase ($\alpha + \gamma$). Also the increase in chromium content to 15% will give a better ratio balance between the deviations of the two alloy pairs.

Figure 6 depicts the variation in performance of the recommended alloy pairs with composition assuming the tolerances in table 2 and predicts the limits of accuracy which could be obtained using a stranded three-leg compensating lead system. The natural variations in alloy composition are accommodated by selecting the optimum number of wires, of the two iron-rich compositions, for a stranded negative limb having 22 to 26 strands in total. It is reasonable to expect on this basis that compensation to within

Table 2. Recommended nominal compositions for compensating lead alloys.

Thermocouple	Limb	Compensating lead alloys		
		wt% Cr	wt% Ni	wt% Fe
Pt−13% RhPt	−ve	15·0 ± 0·5	12·0 ± 0·5	balance
	(Pt)	22·0 ± 1·0	12·0 ± 0·5	balance
Pt−10% RhPt	−ve	16·5 ± 0·5	20 ± 1·0	balance
	(Pt)	23·5 ± 1·0	20 ± 1·0	balance
Pt−13% RhPt Pt−10% RhPt	+ve	20·0 ± 1·0	balance	10 ± 0·5

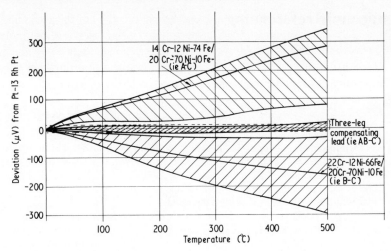

Figure 6. Predicted tolerance range for 15Cr–12Ni–73Fe and 22Cr–12Ni–66Fe/20Cr–70Ni–10Fe and accommodation of these by stranding the Fe-rich alloys together to form a three-leg compensating lead.

2 °C could be achieved over the temperature range 0–500 °C. Over shorter temperature intervals suitable adjustments would clearly give even greater accuracy if desired.

6. Conclusions

The remarkable feature of the compensating lead described above is its extreme flexibility. The Fe–Ni–Cr alloys, selected for their construction, have two basic attributes. Firstly, their EMF–temperature characteristics are very similar to those of the 10 and 13% rhodium–platinum couples, and compensation by means of the three-leg technique can therefore be achieved over wide temperature ranges. Secondly, they have good oxidation resistance and therefore their thermoelectric stability over long periods, at compensating lead temperatures is excellent. Experimental compensating lead alloys of this type have been operating in air at lead junction temperatures within the range 400–550 °C for over 750 h and no significant instability has yet been detected.

The preparation and fabrication of these alloys should present little difficulty.

The general applicability of the new compensating lead in industrial temperature measurement is obvious, since it represents such a considerable technical improvement over existing devices. More specifically, it is tempting to consider its use, possibly in simplified form, as a lead wire system for the disposable forms of thermocouple, employed for process control in the steel industry.

Acknowledgments

Acknowledgments are due to the Directors of Johnson, Matthey and Co Ltd for permission to publish this paper.

Inst. Phys. Conf. Ser. No. 26 © 1975: Chapter 4

Thermocouple referencing

G R Sutton

Royal Aircraft Establishment, Farnborough, Hants

Abstract. Various techniques used to establish the temperature of the reference junction of a thermocouple are considered. Recent work to obtain an accuracy of 0·01 °C with an ice–water reference chamber is described. The replacement of a large number of individual reference chambers by a single chamber, either by using a pumped system circulating water as a uniform temperature zone, or by using a very fine thermocouple wire to extend the thermocouple to a central point is discussed.

1. Introduction

Since the output of a thermocouple depends on the temperatures of both the measuring and reference junctions, its accuracy will depend both on its sensitivity at the two temperatures and on the reliability and accuracy of the referencing technique. In most laboratory installations it is possible to obtain reproducibilities for base-metal and rare-metal thermocouples of ±0·25 °C and ±0·1 °C respectively. Hence if the reference junction temperature is known to within a few hundredths of a degree Celsius the full potential accuracy can be achieved.

The primary fixed point, the triple point of water, is the most accurate form of reference, with a reproducibility of ±0·0001 °C. However this degree of accuracy is not necessary and the somewhat complicated method of use is a disadvantage. If sufficient care is taken the melting point of ice can provide more than adequate accuracy for thermocouple referencing; it is comparatively easy to prepare and has the advantage of operating at 0 °C (standard thermocouple tables are based on the ice point). This method has therefore been investigated in detail.

For permanent installations an automatic ice-point chamber can be used with accuracies similar to the ice–water mixture. When multiple thermocouple installations have to be referenced the cost of providing reference chambers can be reduced significantly if the reference junctions are centralized in as few chambers as possible. With this in mind two remote thermocouple referencing techniques have been investigated. Both methods and the ice-point reference are discussed in detail in the following sections.

There are other forms of referencing that may be used depending on the accuracies and characteristics required; for example, with many installations it may not be necessary to reference at 0 °C and a crystal oven can be used which controls at a phase change temperature other than 0 °C (eg benzophenone 53 °C). However, care must be taken to ensure that an adequate length of wire is at the oven temperature since the available volume is generally small.

When the accuracy required is not more than ±0·1 °C a controlled temperature chamber can be used requiring only supervision of the controlled temperature. Bimetal control can give a sudden change in controlled temperature due to friction in the mechanism which can be detected by comparing the temperatures of similar units.

2. Ice point

The compensation technique of producing an EMF equal to that of the thermocouple at the reference-junction temperature has been available for many years, particularly on potentiometric recorders and indicators. Devices using the temperature coefficient of resistance are available commercially, offering compensation to ±0·1 °C over an ambient range of ±10 °C.

2. Ice point

Melting ice was, until 1960, a fundamental standard of temperature, assigned 0 °C and its ease of preparation has ensured its common use as a reference for thermocouples. It is usual to avoid shorting and electrolytic potentials by enclosing the metal wires in glass tubes closed at one end, and to obtain good thermal contact with the ice by putting a liquid, such as dry transformer oil, in the tubes. To extend the life of the chamber the container is usually a vacuum flask which is closed by a fairly thick cork through which the tubes are passed. The accuracy that can be achieved will depend on many parameters, such as purity of water, chamber preparation, immersion depth and thermal load, which are, to some extent, inter-related. To obtain a useful measure of the contribution of each parameter they have been considered individually. Impurities in the water depress its freezing point. However the impurities in even the hardest tap water depress it by only a few hundredths of a degree Celsius; the actual depression may be calculated from analysis of the water and its cryoscopic constant. RAE tap water having about 300 ppm total impurity gave a depression, varying with actual analyses, of between 0·014 and 0·019 °C. For most cases this is sufficiently accurate and only for the highest precision is it necessary to use distilled water. Ice prepared from distilled water in a refrigerator is cooled to between −10 and −18 °C. Crushing the ice blocks usually left lumps big enough to cause local cold spots but shaving with a carpenter's plane or a commercial rotary shaving machine gave ice flakes which, in water, rapidly attained 0 °C. If however the shaved ice was used directly without wetting, it was found that it could still be well below the freezing point and the temperature at the thermocouple junction in the chamber dropped during the first hour because ice melted by the insertion of the tubes was refrozen; after this the temperature increased, and once the freezing point was reached rapid deterioration of the chamber occurred as the ice melted away from the tubes. This was prevented by thoroughly wetting the ice with distilled water, thus ensuring it was at 0 °C. By packing the ice fairly hard and adding distilled water so that it was visible in the ice at the top the useful life of the chamber was extended considerably. The inserts in figure 1 give some indication of the way in which the ice melted when packed hard. Initially the ice tended to recede from the top of the flask and to a lesser extent from the walls. This recession continued until a core of ice was held down by the tube. The bottom of the tube was maintained at the ice point by the core during melting until the core finally broke up and floated to the surface; consequently once the temperature started to rise it did so quickly. This method of deterioration was also apparent under heavy thermal load conditions with additional melting of ice around the tube causing the temperature of the thermocouple to rise continuously. A vacuum flask of about one litre capacity with sufficient depth (about 40 mm greater than the immersion depth) to house the tubes was required. Caldwell (1965) had shown that the wall thickness of the tubes caused insignificant errors and

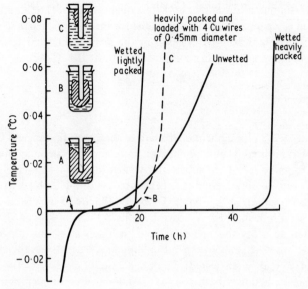

Figure 1. Effect of preparation on chamber temperature. Immersion depth = 140 mm.

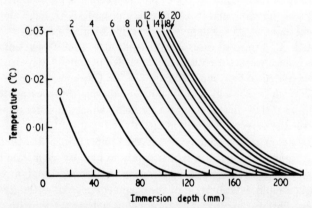

Figure 2. Effect of loading. Ambient temperature = 20 °C. Readings taken within 15 minutes of preparation. Each curve is labelled with the number of 0·45 mm diameter copper wires in one tube.

tubes of about 5 mm inside diameter with a wall thickness of 1 mm are a convenient size. The depth of oil required was investigated and it was found that best results were obtained when it was within 2 mm of the ice–water level. The depth of immersion required will depend upon the thermal load introduced by the thermocouples as this will affect the initial temperature at the junctions and the deterioration with time. Figure 2 shows the effect of thermal loading by copper wires on the accuracy of a freshly prepared reference chamber. By considering the thermal conductivities and cross sectional areas the loading effect of thermocouple wires can be estimated (table 1). The temperature was measured with a glass-dipped bead thermistor which was calibrated from −2 to +2 °C. Spot checks on the thermistor, using a triple-point cell, were made

Thermocouple referencing

Table 1. Comparison of different forms of loading.

Type of wire	No. of 0·45 mm diam Cu for same load
0·45 mm diameter constantan	0·062
0·45 mm diameter iron	0·200
0·45 mm diameter chromel	0·050
0·45 mm diameter alumel	0·077
7/0·2 insulated Cu	1·7
16/0·2 insulated Cu	3·5
1/0·6 insulated Cu	2·8

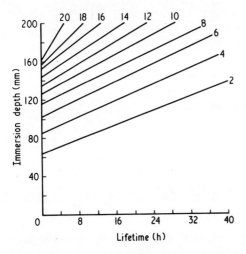

Figure 3. Effect of load and immersion depth on life. Lifetime = time to reach 0·01 °C, ambient temperature = 20 °C, flask depth = 240 mm. Each line is labelled with the number of 0·45 mm diameter copper wires in one tube.

throughout the work enabling an accuracy of 0·0017 °C to be achieved. Fairly high loads can be used without much loss of accuracy provided the immersion depth is adequate. These results do not take into account deterioration with time. Figure 3 shows the time for the junctions to reach 0·01 °C at various depths and for various loadings. Thus the minimum immersion depth required will depend upon the thermal load and lifetime requirements. Generally for a life of 24 hours and two junctions per tube an immersion depth of 140 mm or greater is satisfactory. The end of the chamber life is easily detected from its rapid increase of temperature.

3. Automatic ice reference chambers

These have water enclosed in a metal chamber with re-entrant metal tubes for the thermocouple reference junctions, a metal bellows to sense the internal volume and semiconductor cooling modules fixed to two sides. On cooling ice forms adjacent to the coolers and grows towards the tubes. The bellows movement is used to control the cooling modules.

It was found that, providing oil was present in the tubes to the depth of the chamber only and not the insulation above it, maximum efficiency was achieved. Temperature variations at the point of measurement in the tube did not exceed 0·03 °C unloaded and were less than 0·01 °C when loaded with up to 8 × 0·45 mm copper wires and at about 3 mm from the bottom of the tube. Control was lost only when the ice grew until it touched the tube when a sudden drop in temperature occurred by conduction to the cooler through the ice. This was however outside the normal range of internal volume controlled by the bellows.

4. Remote thermocouple referencing

Remote thermocouple referencing enables a dispersed installation to be referenced at a central point, eliminating many chambers with considerable cost savings in labour and equipment. Such a system can however involve either long lengths of thermocouple wire or the use of thermocouple extension wire. Both methods are expensive and some loss of accuracy must be expected with extension wire. In order to overcome these drawbacks two possible methods have been investigated. Firstly, the use of a water circulation system to provide a uniform temperature zone around the installation has been evaluated; this would enable the thermocouple extension to be made in copper. Secondly, the effectiveness of using very fine thermocouple wire to extend the expensive rare-metal thermocouples has been determined.

5. Uniform temperature zone

Uniform temperature zones are well known for compact installations (Billing 1964). To extend such zones around widely dispersed thermocouple installations the use of a piped water circulation system was considered (Sutton 1970). The extended zone, figure 4, is the pipe surface which must remain within the required accuracy of measure-

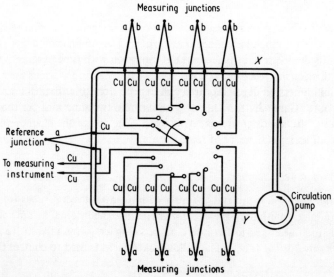

Figure 4. Schematic diagram of circulation system. The zone junctions are electrically insulated from the pipe.

ment (this has been taken as 0·1 °C) between the first and last junctions at X and Y. This requires that the system operates with minimum heat gain or loss and therefore as near ambient temperature as possible. A low-power pump reduces heating by friction and thermal losses from the driving motor, but the circulation time must be short to avoid the introduction of a temperature difference along the pipe due to local variations in ambient temperature. These conflicting requirements were resolved by determining the time constants of temperature change (from the relationship between the heat content and rate of heat loss by radiation and convection), the rate of circulation for various pipe materials and pumps (from frictional losses and pump characteristics) and the heat input to the circuit introduced by the inefficiency of the pump motor. From these values the temperature difference around the intended circuit was obtained. The change in the temperature caused by an ambient temperature change was estimated by assuming a step ambient change over half the circuit.

Table 2. Recommended circulation systems.

Circuit length (m)	Recommended Pipe ID (cm)	Pump rating (W)
0–15	1·3	85
15–30	1·9	85
30–45	3·8	230
45–60	3·8	550
60–75	3·8	950
75–90	5·1	950

The heat gained by a system from the pump motor and the effect of a 5 degree difference in ambient temperature over half the circuit in a wind speed of 0·5 m s^{-1} were considered. Combinations of pump size and of pipe diameter and length giving not more than 0·1 °C difference between the first and last junction under these conditions are shown in table 2. If the system is run in closely controlled ambient conditions where temperature variations are less than 1 °C and draughts are less than 0·25 m s^{-1}, then the maximum circuit length may be extended from 90 m to 200 m. These results have been calculated for copper pipes but can be used for polythene and PVC pipe with insignificant errors.

An experimental system for twelve heat treatment ovens using a circuit 21 m long showed a maximum temperature difference of 0·07 °C and has now been operating satisfactorily for five years in an uncontrolled environment.

6. Fine wire thermocouple extension

To extend widely dispersed rare-metal thermocouples to a central reference point involves the use of base-metal extension wire or actual thermocouple wire. The former gave errors limiting accuracy to about 2 °C; the latter was therefore the only alternative where high accuracy was required.

A practical case was considered for a Creep Laboratory installation where 0·4 mm diameter platinum/platinum—13% rhodium thermocouples are used. The thermocouples have been extended by up to 4 m in 0·1 mm thermocouple wires which had been insulated with Diamel enamel. Calibration of the extension wire showed an agreement within 0·1 °C of the calibration of the 0·4 mm diameter wire at ambient temperatures. Under normal Creep Laboratory conditions a fixed correction could be applied to compensate for this slight deviation.

The installation of a fine-wire system was found to require some care if the accuracy was to be maintained. The connection between the thermocouples and extension wires was made in an isothermal box to prevent any EMF being generated by the intermediate metal introduced at the junction. The 0·1 mm wire is difficult to handle and its installation was therefore made permanent to reduce danger of failure. The installation can be greatly simplified by laying the fine wire in a plastic trunking. At the central reference chamber the fine extension wires are connected to copper leads which are taken to the measuring instrument. A heavier gauge copper wire can be used to simplify installation; however a large number of thermocouples will introduce high thermal load into the reference chamber which must be able to maintain the accuracy of the system. With an ice—water reference the regular maintenance required would mean disturbance of the fragile wires. A controlled chamber is more suitable, particularly one of the thermoelectric cooled types which operate at the ice point.

The fine-wire extension increased the electrical resistance of the thermocouple circuit and consequently caused some reduction of sensitivity of a potentiometer; however if modern high input impedance equipment is used the increase can be ignored. In this installation it was found that the 0·4 mm thermocouples could be made 20 cm shorter since they did not have to enter a reference chamber. This was equivalent in cost to 3·2 m of the fine wire; material costs for the installation were therefore about zero.

Acknowledgments

The paper is published by permission of the Controller, HM Stationery Office. Copyright Controller HMSO, London 1975.

References

Billing B F 1964 *Institute of Engineering Inspection Monograph* 64/1
Caldwell F R 1965 *J. Res. NBS* **69C** 95–101
Sutton G R 1970 *J. Phys. E: Sci. Instrum.* **3** 410–2

High-integrity, small-diameter mineral-insulated thermocouples

A Thomson and A W Fenton
UKAEA Risley Engineering and Materials Laboratory, Risley, near Warrington, WA3 6AT

Abstract. Temperature measurement in some aggressive environments may best be performed using metal-sheathed, mineral-insulated thermocouples. However, when considerations of speed of response or available space require these to be of small diameter, the desired reliability may not be attainable. It is shown how a coaxial design can reduce the risk of failure from several causes at any given cable diameter with some improved speed of response. Some aspects of the development and performance of a coaxial MI thermocouple are discussed.

1. Introduction

Temperature measurements in aggressive environments or critical locations such as the coolant or components of a nuclear reactor are often effected by the use of mineral-insulated (MI) thermocouples. The replacement of faulty units in such applications is costly and possibly hazardous. In consequence, great emphasis is placed upon durability and integrity.

A conventional MI thermocouple assembly consists of a pair of thermoelectric conductors made up into cable form with a continuous metal sheath and powdered mineral insulation (figure 1a). A hot junction is formed at the end of a length of cable by welding the conductors and capping the sheath (figure 1b). In general, the endurance of an MI thermocouple cable may be improved by increasing the cross section of the sheath and conductors. The relative dimensions of these components are dictated by manufacturing considerations, and enhanced reliability is obtained mainly by increasing the cable diameter.

Possible applications range from simple temperature monitoring and process control, to alarm and safety trip systems. Usually, the control and alarm applications require a short thermojunction response time, a condition which is normally achieved by minimizing the cable diameter. It has been shown that response time varies as the square of the cable diameter (Thomson 1965, Bentley and Rowley 1968). In situations requiring temperature measurement in regions of high heat transfer the size of a thermocouple must be limited to minimize heat flux perturbations and the consequent temperature errors. There is thus often a conflict of dimensional requirements when high integrity is required.

This paper reviews the primary causes of failure in conventional MI thermocouples and describes the development of an alternative design. Several features are incorporated which enhance reliability while reducing diameter and response time.

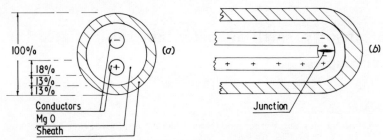

Figure 1. Idealized twin-core cable sections: (a) radial; (b) axial.

2. Common failure mechanisms in conventional MI thermocouples

Discussion will be confined to assemblies with ISA type K (Ni–Cr/Ni–Al) or similar thermoelements and stainless steel or other Ni–Cr–Fe alloy sheaths, operating below 800 °C. It is assumed that the sheath, while intact, completely isolates its interior from the environment and that the insulant purity is satisfactory at the manufacturing stage. The failure in thermocouples may then be attributed primarily to deterioration of the sheath or conductors.

The sheath life is dictated by the rate of corrosion/erosion under operational conditions. Premature failures of sheaths are often caused by small penetrations which are temporarily blocked by oxides or other inclusions so as to pass a simple leak test. Such penetrations may occur in the manufacture of the hot-junction closure weld or as a result of impurities in the original start-tube material or inclusions introduced during the subsequent cable extrusion process.

High-temperature alkali metals and high-pressure water are particularly effective in leaching out such impurities and entering the exposed penetration. The two most likely consequences of such an event are: (i) corrosion of the conductors, leading to open circuit failure; and (ii) reduction of insulation resistance (IR) by contamination, or the formation of conducting bridges. The open-circuit conductor fault is easily identified, but IR faults are not always detectable.

A low IR between conductors and sheath is not necessarily detrimental to thermocouple performance. Between conductors, however, it is difficult to identify and can seriously affect the accuracy. The low-resistance region may constitute a false junction between the conductors and produce a spurious thermoelectric signal. Alternatively the low IR may form an attenuation network in combination with the series resistance of the conductors. In either case, the output EMF can be in gross error.

Open-circuit failure of conductors has been the subject of numerous investigations for over a decade, with conflicting and inconclusive results (Thomas and Meredith 1966, Garrett and Naish 1970). What is clear is that conductors, weakened by grain boundary corrosion, are parted by differential expansion or vibration forces. The positive (Ni–Cr) conductor is not significantly embrittled. The majority of failures occur in the negative (Ni–Al) conductor, close to the fabricated hot junction. Corrosion takes place in the range 400 °C to (perhaps) 700 °C and may be self-limiting in terms of depth of penetration under static stress-free conditions. Under the influence of thermal cycling stresses, the process may proceed to conductor failure.

Figure 2. Radial section photograph, 1·0 mm OD twin-core cable.

Perhaps the most practically significant finding is that corrosion cannot be induced in new, dry, well made cable. Once such cable is cut and exposed to the moist atmosphere for a few minutes, eg, during measuring-junction fabrication, corrosion may occur on heating to 400 °C or above. No drying-out procedure has been found completely effective as a preventative measure.

An inherent feature of the twin-core MI cable is that the thermojunction must be made inside the end of the extruded cable. This is a tedious operation, difficult to control and inspect, particularly in the smaller cable sizes. Expensive closure-weld proving tests must be performed if high sheath integrity is to be expected.

The idealized cross section geometry depicted in figure 1 is seldom realized in practice. A more typical geometry is shown in figure 2. Because of the risk of distortion or displacement of the conductors, it is necessary to have a generous insulant thickness. Hence, the metal thickness of both conductor and sheath is quite commonly a small proportion (12—15%) of the cable diameter. Consequently, where a small-diameter cable is required for fast response or geometry, the predictable life, even assuming no sheath penetrations, may be unacceptably short. For example, to ensure a thermal response time in flowing liquids of less than 100 ms a 1·0 mm diameter thermocouple cable is required. However, in current nuclear reactor designs for example, standard practice requires a 3·0 mm diameter MI thermocouple for direct immersion in liquid coolants. Such thermocouples will have response times approaching 1 second.

3. Alternative design criteria

Consideration of the points discussed above led to the conclusion that to improve both the reliability and speed of response of small diameter thermocouples the following changes must be made:

(i) The measuring junction should be welded before assembly of the cable components. The thermoelements could then be adequately protected during the welding cycle and careful post-weld treatment and inspection should be convenient and effective.
(ii) The conductor thickness should be increased to improve the resistance to fatigue in the event of embrittlement corrosion.
(iii) The sheath thickness should be increased. If this were achieved by the (mechanically convenient) method of drawing one sheath over another, a special advantage would result. Any inclusion or penetration in the outer sheath would require to coincide with

a similar fault in the inner sheath before coolant penetration could occur. This would be a very low probability situation.

(iv) The cross section geometry should be controlled more rigorously in order to reduce the insulation thickness. The insulant thickness (ie, thermal resistance), is significant in determining the response time of a thermocouple.

It was considered that the requirements listed above could be satisfied in principle by adopting a single-core coaxial construction. Since such a system is geometrically very stable during fabrication, lower insulant thickness is possible than in a two-core system. The electrical stresses are normally negligible in a thermocouple cable during operation. Furthermore the concentric geometry would minimize any voltage gradient, eg when the IR is being measured. The thin annular insulant layer results in an improved speed of response compared with conventional cable. Finally, it was clear that in the concentric arrangement the sheath and conductor thicknesses could be larger proportions of the cable diameter, ie the available cross section could be utilized more effectively.

4. The coaxial thermocouple

Typical radial and axial sections of an ideal coaxial thermocouple are depicted in figure 3. The 'hairpin' arrangement (suitably supported) is intended for flowing fluid applications.

The cross section area of the conductor and sheath are about twice the value normally found in the equivalent twin-core cable. The thicker sheath is built up by drawing together two separate tubes at the assembly stage. Cold welding occurs during the extrusion process so that the interface is normally evident only as a zone of fine-grain metal.

The conductor is made of thermoelectric alloy wires, butt-welded to form an integral thermojunction which can be located at any desired axial position. Where the hairpin arrangement is used the junction would normally coincide with the apex of the bend. A second thermojunction may be readily incorporated into the cable, thus providing a differential thermometer function, or allowing the advantage of an integral reference junction.

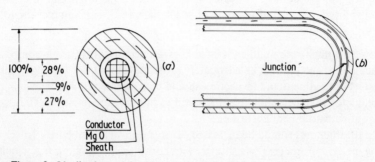

Figure 3. Idealized coaxial cable sections: (*a*) radial; (*b*) axial.

The insulant thickness is about 70% of the minimum value normally found in an equivalent two-core cable. The concentric form contains less insulant yet offers improved thermal and electrical properties. An important feature of the thick-sheathed coaxial system is the high insulant compaction which can be achieved. This results in a disproportionate increase in insulant thermal conductivity and electrical strength.

An evaluation of the insulant requirements for a fast response coaxial thermocouple was undertaken at the outset of the programme with the aid of an analogue simulator. Simulation indicated that sheath thickness does not contribute significantly to the response time in fast-flowing liquids. The principal elements of the thermal response are the thermal mass of the conductor and, to a first order, the radial thermal resistance of the insulant annulus. The former is, of course, proportional to the square of the conductor diameter, while the latter is a direct function of the log-ratio of the diameters of the insulant annulus. For a given sheath diameter and thickness the only variable, assuming constant material properties, is the conductor diameter. In order to minimize the response time the optimum value for this dimension is that which results in an insulant annulus diameter ratio of about 1·7. The amount of insulant in a 1·0 mm cable constructed to the above geometry is more than adequate in terms of electrical insulation strength providing IR tests are performed at less than 100 V.

Both ends of the thermocouple unit would normally be sealed and terminated separately. From an operational point of view, this is not necessarily a disadvantage since a single coaxial termination is the simplest to produce reliably. Individual connectors lend themselves to the concept of plug and socket connectors. With this in mind, a terminal seal has been devised which incorporates a projecting pin of the appropriate thermoelectric material.

5. Manufacture

Given the objective of producing cables of the order of 0·5 mm to 1·5 mm sheath diameter and with radial proportions similar to those in figure 3, it is clear that conventional cable fill/draw techniques are inappropriate. A manufacturing route was needed which took account of the dimensional requirements and permitted testing or inspection at critical stages.

The dimensions of a 'start' assembly and consequently the amount of extrusion required to reach the final size are influenced by several conflicting factors:

(i) As the degree of extrusion is increased, so also are (*a*) the risk of radial geometric distortion and (*b*) the degree of axial 'smearing' of the thermojunction weld region.
(ii) Assembly of the components and the introduction of insulant is both simplest and cheapest with a large-diameter, short-length 'starter'.

The effective axial length of the conductor thermojunction at the butt-welding stage can range between 0·025 mm and 0·1 mm. In order to limit the junction zone length in the finished cable to the desirable figure of 1 or 2 cable diameters, it is necessary to restrict the axial extrusion ratio to about 20:1. The consequent radial dimensions of the start assembly components are such that the insulant may then be introduced reliably only as pre-formed, 'crushable' tubes. Fortunately, such tubes of compacted MgO are available in dimensions which satisfy the 20:1 extrusion criterion.

Figure 4. Radial section photograph, 1·0 mm OD coaxial cable.

Figure 5. Axial section photograph at thermojunction, 1·0 mm OD coaxial cable.

Figure 6. Radial section photograph of sheath, 1·0 mm OD coaxial thermocouple.

A typical 1·0 mm diameter coaxial thermocouple cable section is shown in figure 4. An axial section of a thermojunction region is shown in figure 5 while figure 6 is an enlarged section through a double tube sheath. These results, as with virtually all data quoted here, are from cables assembled using butt-welded conductors of slightly over 1·0 mm diameter 'start' size. The insulant and sheath tubes were appropriately proportioned.

6. Performance

Short-term evaluation tests have been conducted on about 100 laboratory-produced units and these are discussed below. Long-term calibration drift and environmental endurance tests are planned and will involve samples produced under both laboratory and commercial conditions. The latter will be produced by both drawing and swaging techniques.

A batch of 50 units has undergone severe thermal cycling stresses between temperature limits of 400°C and 600°C. The cooling rate was approximately 40 °C s^{-1} for 5 seconds in a cycle period of about 20 seconds. Up to the time of writing, no failures had occurred after 3×10^5 cycles. A further 10 units have been maintained at the mean cycle temperature for the duration of the test to provide metallurgical control information. The criteria for failure are conductor open circuit or insulation resistance deterioration. Similar tests on 3 mm diameter conventional thermocouples showed a mean life of less than 10 000 cycles before failure by open circuit. The inference is that thermal expansion stresses are better resolved in the coaxial form.

The results of insulation resistance measurements compare favourably with twin-core cable provided that the cables are properly dried and sealed. The minimum recorded room-temperature figures are 10^{11} Ω m, falling to $10^7 \Omega$ m at 600 °C.

Radial sections have been taken from each leg of 20 units and the maximum and minimum dimensions recorded for the three components. The total variation in the various dimensions were found to be:

Sheath diameter: ±2%   Sheath thickness: ±6%;   Conductor diameter: ±6%
Insulant thickness: ±20%   Insulant ratio: 1·74 (maximum).

Thermal response times have been determined in a flowing-water system at sufficient velocity (8 m s^{-1}) to ensure a high heat transfer coefficient. A very steep temperature ramp was established in the flowing water and the response of the sample compared with that of a 0·1 mm diameter bare-wire reference thermocouple located close to the sample. Figure 7 is a record of the response of the sample and reference, obtained simultaneously. The reference signal, with a rise time of about 10 ms, represents the combination of the temperature ramp and the thermocouple response time. The sample was located 10 mm downstream of the reference thus introducing a little over 1 ms distance velocity lag. The time constant of the sample recorded in figure 5 is estimated

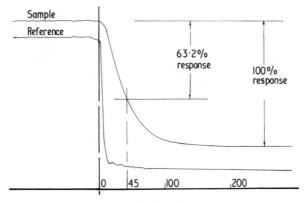

Figure 7. Response time in flowing water, 1·0 mm OD coaxial thermocouple.

at not more than 45 ms. The results obtained by the above method for 10 samples of 1·0 mm diameter coaxial thermocouple ranged from 45 to 50 ms. It is worth noting that when a 1·0 mm butt-welded bare-wire thermocouple was tested in the same facility a time constant of 30 ms was obtained. The following table compares typical relative dimensions and response times of coaxial and conventional 1·0 mm OD/MI thermocouples:

Type	Conductor	Sheath	63·2% Response time
Coaxial	27% OD	27% OD	45–50 ms
Conventional	18% OD	13% OD	50–100 ms

7. Conclusions

A viable mineral-insulated single-core coaxial thermocouple system has been developed which appears to overcome most of the problems associated with conventional twin-core MI thermocouples. The sequence of manufacture is such that no potentially damaging operation need be performed on the completed assembly. The electrical and mechanical integrity and the speed of response are all superior to an equivalent-sized conventional cable and the design eliminates the risk of a 'fail to danger' state occurring. The coaxial form, while suffering the disadvantage of requiring two cable-end seals, is applicable to a variety of thermometry applications when high reliability, small physical size and consistent fast response are required.

References

Bentley P G and Rowley R 1968 *Inst. Meas. & Control J.* 1 100–4
Garrett E L and Naish S T 1970 *UKAEA Report* AERE-R6174
Thomas M W and Meredith K E G 1966 *Br. Corros. J.* 1 199–204
Thomson A 1965 *Soc. Instrum. Tech. Trans.* 17 49

Experimental evidence of erasable EMFs induced by thermal gradients in sheathed chromel–alumel thermocouples during long-term exposure

F Mathieu, R Meier, J Brenez and A Falla
Centre d'Etude de l'Energie Nucléaire, CEN/SCK, Technology Department, B-2400 Mol, Belgium

Abstract. A portion of an annealed chromel–alumel thermocouple cable, which exhibits no noticeable EMFs when tested for thermoelectric heterogeneity by the moving furnace method, is exposed for a long period of time to a high uniform temperature. When subsequently tested again for heterogeneity, residual EMFs are observed at the abscissae where the temperature gradients were located during exposure to temperature. In the following run of the monitoring furnace these EMFs virtually disappeared. For this reason they are called 'erasable EMFs'.

The driving force generating this type of erasable EMF appears to be the temperature gradient dT/dx. Erasable EMFs can be the source of temperature measurement errors. A method for evaluating error magnitudes is proposed.

1. Introduction

The problem raised in this paper falls under the heading of thermocouple drift. When dealing with thermocouple drift investigations one is naturally led to distinguish between permanent (drift) components and non-permanent or erasable (drift) components[†]. Permanent components are those components which remain virtually unchanged after the thermocouple cable has been heated to a high temperature for a short period of time. Non-permanent or erasable components are those components which virtually disappear as a result of heating the thermocouple cable for a short time. These definitions are by no means rigorous. Their only purpose is to facilitate a first contact with the subject.

In recent years powerful methods have been developed for the investigation of permanent components of drift. A comprehensive review can be found in Mathieu *et al* (1974b). The knowledge concerning erasable components on the other hand is very limited at present. For instance it is widely accepted that readings of a thermocouple comply with the standard calibration table only if the sensitive wires have been annealed. This statement implies that any drift which could possibly be observed without annealing is erased by the process of annealing, ie by heating the thermocouple for a short time. At first sight the situation seems to be clear, but as is shown in Mathieu *et al* (1974a), at least in the case of metal-sheathed chromel–alumel thermocouples, experimental contradictions arise immediately. A sheathed thermocouple which has been annealed at a certain temperature is not necessarily annealed for another temperature. For the specific case of annealing there is a lot of room left for discussions on what has to be considered as an erasable EMF, for an unambiguous definition of thermocouple annealing, which resists confirmation by experiment, does not yet exist.

[†] Eventual components due to thermoelectric hysteresis, mechanical stresses, etc, are not considered.

In the previous example erasable EMFs are in the approximate range of 0–100 μV. There are no apparent signs that the temperature gradient dT/dx is a variable with significant influence.

Hereafter we shall focus our attention on a phenomenon which is likely to generate erasable EMFs with higher magnitudes and where the temperature gradient dT/dx appears to be a dominating variable, namely erasable EMFs induced by thermal gradients in sheathed chromel–alumel thermocouples during long-term exposure.

The reported experimental results are limited to the essential minimum. Complementary information will be summarized in Mathieu *et al* (1975).

2. Localization of EMFs in thermoelectric circuits

As has been recognized by several authors (Bloomfield 1974, Fenton *et al* 1967, Mathieu *et al* 1974b, Moffat 1962), it is important to realize that the EMF of a thermocouple is not generated at the hot junction but in the temperature gradient. The thermodynamical demonstration of this fact is given in Mathieu *et al* (1974b).

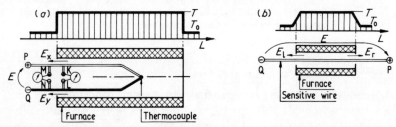

Figure 1. Localization of EMFs in thermoelectric circuits (*a*) and basic experimental set-up (*b*).

Figure 1(*a*) represents schematically a homogeneous thermocouple measuring the temperature T of an ideal furnace with uniform temperature over its length and very steep thermal gradients at the ends. The EMF between terminals K and L situated inside the furnace must equal zero, for otherwise it would be possible to produce mechanical work using a miniature motor without an equivalent energy supply. This of course is inconsistent with the basic principles of thermodynamics. On the other hand the EMF that appears between terminals M and N situated outside the furnace is the same as the EMF between the end terminals P and Q, and it is known from experience that this EMF does not equal zero in the general case. This means that active measurable components of E (E_x and E_y) are generated in the two conductors over the distance separating terminals K and L from terminals M and N — or in other words in the temperature gradient — and only over this distance. No other portion of the thermoelectric circuit is contributing to the signal E. This applies in particular to the hot junction which is merely ensuring galvanic continuity between the two thermoelectric conductors.

As the hot junction does not contribute to the active signal, there are considerable advantages in deleting it definitively from experiments. Accordingly, for one sensitive wire, figure 1(*b*) gives a schematic view of the basic experimental set-up. An EMF E_l is generated in the temperature gradient on the left-hand side of the furnace, whereas an EMF E_r is generated in the gradient on the right-hand side. The following EMF appears between terminals P and Q:

$$E = E_r - E_l$$

which is never zero when $|E_r| \neq |E_l|$. If $E \neq 0$ the wire is said to be thermoelectrically heterogeneous. We shall consider as significant only values of $|E|$ which exceed a certain threshold X. Results will be presented with the sign conventions of figure 1(b), ie polarity of the right-hand terminal P positive.

3. Some details on experimental hardware and operating procedures

3.1. Specification of active sample

The tests are carried out on commercially available, nuclear reactor grade, AISI 304 stainless steel sheathed, MgO insulated, chromel–alumel thermocouple cable with an outer diameter of 1 mm. The cable is annealed, conditioned and finally monitored for heterogeneity. Measurements are done synchronously on both the chromel and the alumel wires.

3.2. Furnaces

For convenience, a distinction is made between a conditioning furnace and a monitoring furnace. A conditioning furnace is used to maintain a portion of the thermocouple cable for a long period at a given uniform and constant temperature. The subsequent analysis of the cable is done with a monitoring furnace which moves slowly – 12 mm min^{-1} – along the cable.

From the experimental point of view it is advantageous to make the monitoring furnace longer than the longest of the conditioning furnaces. Under these circumstances only one gradient at a time is sensing the conditioned portion and no mental exercise is required to separate superposed individual contributions. Unfortunately we did not have the opportunity to realize this condition, but were lucky enough not to run into trouble.

3.3. Working temperature

In order to avoid difficulties with unwanted signals not directly related to the investigated phenomenon, all the tests, annealing included, were executed with nominal furnace temperatures equal to 800 °C.

3.4. Annealing

We shall suppose the thermocouple cable to be annealed for the temperature of 800 °C after the monitoring furnace, at a nominal temperature of 800 °C and moving slowly along the cable, has accomplished a first run.

3.5. Threshold X

From experience it is known that during a second, third, ... run of the monitoring furnace (nominal temperature invariably 800 °C) over the annealed cable, the signals which are observed between the end terminals of the individual wires differ in general from zero but reproduce themselves fairly well.

For the threshold X we shall adopt a value somewhat beyond the maximum of the signal observed on a representative reference length of cable.

3.6. Remarks

It is our intention to discuss only the qualitative aspects here and thus the experimental conditions are not described in detail. For quantitative analysis exhaustive information would be required especially on the temperature gradients of the conditioning and monitoring furnaces (maximum dT/dx: ~500 °C cm^{-1}).

4. Characteristic experimental results

4.1. Sample conditioning

A conditioning furnace (designated con 1) of length l = 500 mm is positioned, as shown in figures 2 and 3, over a test sample of annealed thermocouple cable, both

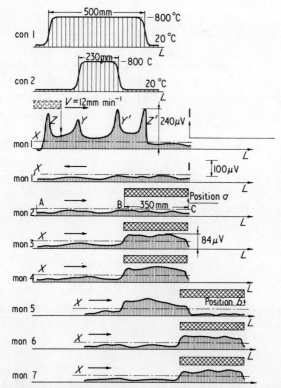

Figure 2. Characteristic experimental results: chromel. The abscissa of the leading gradient of the monitoring furnace at the moment of inversion of the direction of movement is shown on the right-hand side of mon 1.

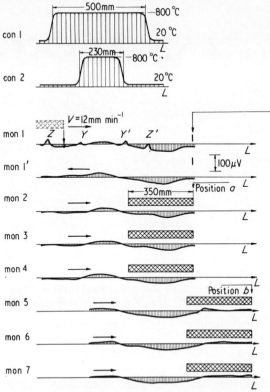

Figure 3. Characteristic experimental results: alumel. The abscissa of the leading gradient of the monitoring furnace at the moment of inversion of the direction of movement is shown on the right-hand side of mon 1.

ends of which are hermetically sealed with silver-alloy plugs in order to prevent air from diffusing inside the sheath and corroding the sensitive wires. The furnace is brought to a nominal temperature of 800 °C which is maintained constant for a period of 28 days.

A second conditioning furnace (con 2) of length $l = 230$ mm is then positioned over the test sample as shown in figures 2 and 3 with the mid-plane at the same abscissa as the first furnace. The temperature of the furnace is raised to 800 °C and maintained constant for a period of 28 days.

4.2. Sample monitoring: chromel

Figure 2 reproduces (cf figure 1b) the chromel signals during successive runs of a monitoring furnace the nominal temperature of which is 800 °C. The starting position of the furnace is situated on the left-hand side of figure 2:

mon 1 corresponds to a run from the left to the right;
mon 1' corresponds to a run from the right to the left without stopping the furnace between mon 1 and mon 1';
mon 2 corresponds to a run from the left to the right. At the end of the run the furnace is stopped in position a and cooled down;

mon 3 is a repetition of mon 2;
mon 4 is another repetition of mon 2;
mon 5 corresponds to a run from the left to the right. At the end of the run the furnace is stopped in position b and cooled down;
mon 6 is a repetition of mon 5;
mon 7 is another repetition of mon 5.

For mon 1, 2, 3, 4, 5, 6 and 7 the signal amplitudes are attributed to the abscissa of the leading end of the moving furnace. For mon 1' they are attributed to the trailing end. No consideration is given to absolute signal amplitudes or to amplitude variations from one run to the next, as long as the threshold X is not exceeded.

Two distinct effects can then be identified in figure 2 in connection with erasable EMFs:

The first is believed to be a 'cooling-law effect'. Before the run mon 3 the test sample comprises a section AB which has been cooled down according to a reference law, ie local cooling of the thermocouple cable by air convection when it leaves the monitoring furnace, and a section BC which is cooled down at the same rate as the furnace in position a. During the run mon 3 the signal amplitude exceeds the threshold X from the moment where the leading gradient of the monitoring furnace engages on section BC. The runs mon 5 and mon 6 confirm that the observed EMFs are erasable. The cooling-law effect is a short-term effect, ie it builds up in short periods of time.

The second is a 'gradient effect' which implies that dT/dx is the driving force. During the run mon 1 peaks are observed when the leading gradient of the monitoring furnace encounters the abscissae where temperature gradients were located during the conditioning phase. Peaks Y and Y' confirm that preliminary 28 day aging at a constant temperature of 800 °C does not have any noticeable inhibiting action. The run mon 1 confirms that the peaks are erased when entering the monitoring furnace, for the trailing gradient of the furnace does not inject negative contributions when sensing the abscissae of the conditioning temperature gradients. Peak amplitudes are presumably a function of the alloy, time and temperature gradients dT/dx of the conditioning and monitoring furnaces and perhaps of other parameters. The maximum amplitude observed for the given set of working conditions is about 240 μV which corresponds to about 6 °C with the sensitivity of the chromel–alumel thermocouple. The gradient effect is a long-term effect, ie it requires long periods of time to build up. At a low nominal temperature of the monitoring furnace – for example 100 °C – the peaks are not erased.

The record mon 1 is a superposition of the cooling-law effect and of the gradient effect.

4.3. Sample monitoring: alumel

The corresponding results for the alumel wire are given in figure 3 for comparison. In this case the signal level hardly exceeds the corresponding threshold X. We therefore limit our comments to two observations: (i) a cooling-law effect, if present at all, cannot be separated from background noise; and (ii) the gradient effect clearly exists but peak amplitudes are low.

5. Measurement of drift attributable to erasable EMFs

The peaks of figure 2 are likely to generate measurement errors in practical applications. A possible approach to the problem of identification of error magnitudes under more or less real conditions with the sensitive wires individually accessible is outlined in figure 4. The thermocouple cable runs through a furnace whose right-hand side has a temperature gradient representative of real service conditions. If necessary, steps are taken to suppress relative movement of the thermocouple cable with respect to this gradient. At the other end of the furnace the cable is protected from undergoing damage by a water-cooled sleeve which pushes back the active gradient into the furnace.

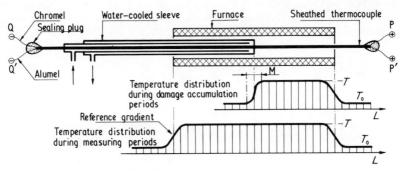

Figure 4. Proposal for the investigation of measurement errors.

The effective error generated in the gradient on the right-hand side is measured periodically with the water removed. The portion M of the cable, which is now at a uniform temperature, does not give any contribution to the signal. Nevertheless care has to be taken with cooling-law effects in the reference gradient.

6. Conclusion

The temperature gradient dT/dx has been identified as being the driving force for the time-dependent generation of erasable EMFs in sheathed chromel–alumel thermocouples. At present it is not known whether erasable EMFs develop in other alloys or whether nature sets an upper limit for the maximum amplitudes. It is also not known whether erasable EMFs are under certain circumstances — ie appropriate alloys, high values of dT/dx, extended periods of time, etc — partially converted to permanent EMFs. It has furthermore been illustrated that different cooling laws can also generate erasable EMFs in the chromel wire. Erasable EMFs are the source of more or less important measurement errors. It is in particular anticipated that they are one of the reasons why small temperature differences resulting from controlled perturbations in a process do not always reproduce themselves identically.

Acknowledgments

The experimental programme has been carried out in the Laboratories of the BR2 Operating Group (GEX)/Technology Department, CEN/SCK, Mol, Belgium. We acknowledge the direct and indirect collaboration of our colleagues as well as the sustained effort by technical staff.

References

Broomfield G H 1974 *Proc. Int. Colloq. on High Temperature In-Pile Thermometry, Petten, Holland, December 1974* to be published
Fenton A W, Dacey R and Evans E J 1967 *UKAEA Risley TRG Report* 1447
Mathieu F, Meier R, Delcon M and Devriendt J P 1975 *Microfilm EUR 5308f* to be published
Mathieu F, Meier R, Soenen M, Delcon M and Nysten C 1974a *NEACRP Specialist Meeting on Reactor Noise, Rome, October 1974* (Oxford: Pergamon)
Mathieu F, Meier R, Vanmassenhove G, Delcon M and De Deyne A 1974b *Proc. Int. Colloq. on High Temperature In-Pile Thermometry, Petten, Holland, December 1974* to be published
Moffat R J 1962 *TMCSI* **3** part 2 33

Inst. Phys. Conf. Ser. No. 26 © 1975: Chapter 4

The automatic calibration of thermocouples in the range 0–1100 °C

T P Jones[†] and T M Egan[‡]

[†] National Measurement Laboratory, CSIRO, University Grounds, City Road, Chippendale, NSW, Australia 2008
[‡] Marconi Elliott Avionics Ltd, Airport Works, Rochester, Kent, England

Abstract. A computer-controlled assembly for the calibration of all types of thermocouples in the range 0 °C to 1100 °C is described. Effects due to inhomogeneities in the thermocouple wires are detected by changing the temperature gradients along the wires during the cooling cycle of the low thermal inertia comparison furnace. The thermocouple EMFs are measured by a digital voltmeter which is corrected for changes in calibration and zero before each scan. These corrected EMFs are further adjusted for temperature drifts during measurement. The EMFs are remeasured if a criterion of relativity between readings is not satisfied. A calibration polynomial and table for each thermocouple are generated using a least-squares method. The uncertainty associated with the calibration procedure and equipment is ± 1 μV up to 11 mV and ± 4 μV to 50 mV, which is small compared with the uncertainty of the thermocouples themselves.

1. Introduction

The calibration of thermocouples is fundamental to precise temperature measurement. However, the time necessary for their calibration is sufficiently large to make the cost of calibration excessive for many purposes. Automatic calibration methods reduce the time involved but to be universally accepted the automatic method should be as accurate as the best manual method. The automatic method of calibration described satisfies these criteria in that, with a minimum involvement of operator's time, it permits calibration in which the quoted uncertainty is limited by the performance of the thermocouples themselves.

2. Calibration methods

Thermocouples are calibrated by measuring their EMF outputs while their measuring junctions are held at known temperatures and then deriving their EMF against temperature relationships. The known temperatures may be derived either by realizing the freezing points of very pure materials and applying the temperatures of the changes of state defined in the International Practical Temperature Scale of 1968 (IPTS-68) or by measuring the temperatures using a calibrated standard thermocouple. Calibration by the first method is known as the fixed-point method and the second the intercomparison method. The uncertainty associated with the fixed-point calibration depends on the method of interpolation between the fixed points. It has been shown (Jones 1972a) that the most accurate means of interpolation is one which fits a curve to the differences between the measured EMF and a smooth reference table (BSI 1973). The lowest uncertainty that can be attained using this method is ±0·3 °C (Jones 1972a) between 0 °C and 630·74 °C and ±0·2 °C (Jones 1968) between 630·74 °C and 1064·43 °C. In the intercomparison method (Roeser and Lonberger 1958) calibration points are generally taken approximately 50 °C apart and the EMF−temperature relationship is determined in a manner similar to the fixed-point method. The uncertainty of this type of calibration is typically ±0·3 °C which is similar to that using the fixed-point method. The main advantage of the intercomparison method is that it can be adapted to the automatic calibration of thermocouples.

3. Equipment

An earlier version (Jones 1972b) of the automatic calibration facility showed the measuring equipment to be satisfactory for the automatic acquisition of data and had indicated the feasibility of extension of the equipment to computer control. These extensions have been incorporated and the final arrangement is shown schematically in figure 1. In the calibration process the EMF output of each test thermocouple is compared with that of the standard thermocouple within a furnace whose temperature is varied in a prescribed manner by a computer which also operates a scanner to select the EMF inputs which are measured on a digital voltmeter (DVM). The computer records and analyses the data and produces the calibration statement.

3.1. Intercomparison furnace

The furnace heater (Kanthal wire, 1·3 mm diameter) is wound on to a Mullite tube, ID 32 mm, OD 39 mm, length 76 cm as shown in figure 2. At each end of the heater a 7·5 cm section is wound with three turns per centimetre to overcome thermal losses at

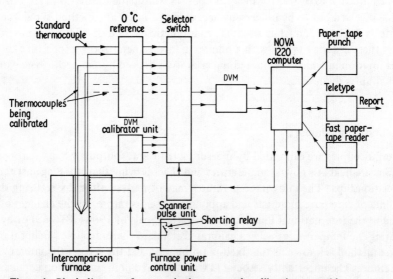

Figure 1. Block diagram of automatic thermocouple calibration assembly.

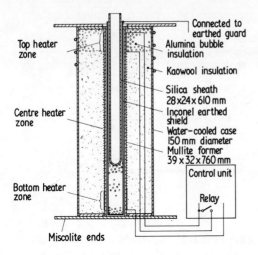

Figure 2. Intercomparison furnace.

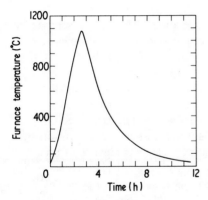

Figure 3. Variation of furnace temperature with time.

the ends whilst the remainder is wound at two turns per centimetre. The furnace has low thermal inertia so that it will heat and cool quickly, permitting a complete calibration to be performed in approximately 12 hours. The variation of furnace temperature with time is shown in figure 3. Inhomogeneities in thermocouples which can affect their calibrations are detected by changing the temperature gradients along the thermocouple wires during heating and cooling by shorting out the top 7·5 cm of the heater during cooling. The temperature differences along the wires when the measuring junctions are at the same temperature for the heating and cooling cycles are shown in figure 4.

The computer controls the power to the furnace by issuing a digital output word which is converted to a proportional analogue DC voltage and supplied to the control winding of a 2 kVA transductor to vary the AC power to the furnace. A separate temperature controller acts as a safety device switching off the power if the furnace temperature exceeds 1100 °C.

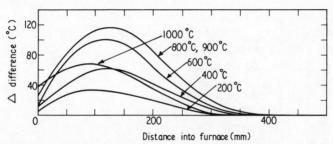

Figure 4. Temperature difference at different positions within furnace for heating and cooling cycle at the measuring junction temperatures shown.

3.2. Measurement circuit

A Solartron LM 1490 DVM has been used for the EMF measurements, but any DVM with a nominal accuracy of 0·01% and a resolution of 0·1 μV could be used. Care has been taken in the EMF measurement circuit to avoid the introduction of spurious EMFs due to extraneous thermal or electrical sources. Copper conductors are used throughout and all connections are made within uniform temperature enclosures. The reference junction at 0 °C is provided by an electronic reference junction (Kaye Instruments, model K 140) which tests have shown to lie within ±0·02 °C of 0 °C. A twelve-position two-pole rotary switch (Leeds and Northrup, type 31) operated by a pulsed rotary solenoid, scans the different thermocouple outputs and reference voltages used to calibrate the DVM. The spurious EMFs developed in the selector switch have been measured to be less than 0·1 μV under all operating conditions. The various voltages used to calibrate the DVM from 0 to 50 mV are produced across calibrated resistors by a mercury battery which is calibrated during the reading cycle by comparison with a standard cell (Muirhead, type k-231 A) temperature controlled at 27·5 °C. Usually seven inputs are used to calibrate the DVM and five to measure the thermocouples.

The measuring junctions of all the thermocouples are welded together into a single bead, and for the calibration of base-metal thermocouples the standard rare-metal thermocouple is peened into a hole bored into the bead. The thermocouples are positioned within a close-ended Inconel tube which is connected to the shield of the multicore cable connecting the ice pot to the scanner. The scanner and DVM calibration unit are shielded and connected to the continuing shield which connects to the DVM. Thus the whole measuring circuit is contained within a single shield which is earthed at one place. Over the whole temperature range, switching off the heaters momentarily does not change the DVM reading of the thermocouple EMFs, indicating that there is no detectable effect due to electric leakage from the heaters.

3.3. Computer and ancillary equipment

The computer (NOVA 1220) has a memory storage of 24 K words. Programs are read into the computer using a fast paper-tape reader (FACIT 4001). The computer can output either to a typewriter (TELETYPE Model 38) or a fast paper-tape punch (FACIT 4070). The computer initiates the reading of the DVM and accepts its digital output. Digital information indicating the scanner position is read into an input register to check the synchronization of the scanner with the program and a digital output word can be produced by the computer for use in both the furnace power control and the interpolation plot for the data.

4. Control program

The program is written in a code which permits the use of system symbols for Assembly language subroutines. The major Assembly language subprogram is a three-word floating-point interpreter (Cassidy 1975) which permits computations with an accuracy of better than 1 in 10^{11}.

Whereas spurious effects on the DVM readings due to extraneous electric signals are rare, each reading is checked to ensure that incorrect readings are eliminated. The individual characters produced by the DVM are checked for validity, ie the only acceptable characters are 0 to 9 and + or −. The calibration correction and zero of the DVM are determined in each scan from the measured calibration voltages. If the measured calibration correction is greater than 1 in 10^3 all the readings of the scan are repeated. The measured EMFs are corrected for zero and calibration correction. Not one character error has been detected in the 2×10^6 characters so far produced in 200 calibration runs and only one scan in approximately 2000 is repeated because of incorrect calibration.

The program commences by measuring the set of thermocouple EMFs twice. The two EMFs for each thermocouple are compared and if any differ by more than 10 μV the two scans are repeated. If the differences are less than 10 μV the second set of corrected EMFs is stored and punched and then a predetermined output word is presented to a digital-to-analogue converter which controls the power to begin heating of the furnace.

The sequence of events in the calibration is determined by the EMF output of the standard thermocouple after conversion to temperature by the application of its calibration relationship. The power of the furnace is increased by the same increment at every 100 °C from 400 °C to 1000 °C. When the maximum temperature 1080 °C is reached, the furnace power is reduced and the top 7·5 cm of the furnace heater is shorted. The power is further reduced for every 100 °C from 1012·5 °C to 612·5 °C at which time it is switched off. Thermocouple calibration readings are taken every 25 °C from 50 °C to 1075 °C during the heating cycle and from 1062·5 °C to 37·5 °C during the cooling cycle. These calibration points are staggered to spread the readings more uniformly over the whole measurement range thus improving the quality of the subsequent curve fitting.

Throughout the calibration the temperature of the furnace is continuously changing. Consequently it is necessary to apply a correction for the change in EMF in the period between measuring the test and standard thermocouples. The rate of change of temperature with time is determined by measuring the time taken for the temperature of the standard thermocouple to change the 5 °C before each calibration point. The temperature difference of each thermocouple from the standard existing at the time of each measurement is then calculated from the measured time of each reading. The product of each temperature difference and the thermoelectric power of the corresponding thermocouple, which is computed for each reading, determines the appropriate EMF correction for temperature drift.

The smoothness of the variation in thermoelectric power with temperature is used to detect large errors in the corrected thermocouple EMFs which may occur due to spurious DVM readings caused by extraneous electrical signals. The thermoelectric power of each thermocouple under test for the latest 25 °C interval is compared with that determined for the 25 °C interval immediately preceding. If the change is greater than 5% a new set of readings is taken. In the calibration of base-metal thermocouples for temperatures below 250 °C the maximum change in thermoelectric power allowed is 10%. The EMFs measured for the standard thermocouple are used to determine the temperature used in computing the thermoelectric powers of each thermocouple, consequently any large spurious errors in the standard or unknown thermocouples will result in the whole set of readings being retaken. At each temperature, during calibration, all relevant information is printed on the teletype for inspection. In all, data for 86 calibration points are stored in the computer for later processing and are punched on paper tape for future use.

The relationship between the measured EMF and temperature for each thermocouple is calculated by a least-squares method. The details of the method are varied depending on the type of thermocouple being calibrated. For the best least-squares fit rare-metal thermocouples are calibrated against standard thermocouples of the same type. The calibrations of the standard thermocouples, either type R or S, are read into the computer from punched paper tapes as two sets of polynomial coefficients. The first set of coefficients allows the EMF of the standard to be converted to temperature during data acquisition and the second set relates the difference in EMF of the standard thermocouple from the EMF of reference tables (BSI 1973) at the same temperature, to the EMF of the standard thermocouple. The difference (ΔE_{tc68}) of the thermocouple being calibrated from the EMF of BSI (1973) (E_{tc68}) corresponding to the temperature of the standard thermocouple is computed. Least-squares orthogonal polynomial techniques are applied to fit ΔE_{tc68} as a function of E_{tc68}. The results for polynomials of powers 2, 3 and 4 are usually insignificantly different and mostly a fourth power relationship is used.

If the calibrations are being performed for formal statements of calibration, tables at 1 °C intervals are calculated and printed. In addition ΔE_{tc68} is tabulated for 100 °C intervals and the coefficients of the polynomials of ΔE_{tc68} as a function of E_{tc68} are given. The deviation of each data point from the fitted relationship is printed by the teletype for inspection and is punched on paper tape for subsequent plotting. The plot shows clearly trends in deviations of the measured EMFs from the fitted values, for example differences between measurements in the heating and cooling cycles.

All base-metal thermocouples are calibrated against rare-metal thermocouple standards. The process is similar to that for rare-metal thermocouples except the least-squares fitting of the data is of the measured EMF as a function of temperature. The degree of polynomials fitted depends on the type of base-metal thermocouple calibrated. For all calibrations 14 additional data points of zero microvolts at 0 °C are introduced before fitting begins to ensure that the fitted EMF at 0 °C is very nearly zero microvolts. The fitted value for 0 °C is stored for subsequent inspection and zero is printed in the calibration table. The maximum values of the EMF at 0 °C before adjustment to zero which have been observed are 1 μV for rare-metal and 10 μV for base-metal thermocouples.

6. Uncertainty of calibration method

The uncertainty of the automatic method of calibration has been assessed in three ways. Firstly the instruments for measuring EMFs were tested by substituting DC voltages from a precision DC potentiometer for the thermocouple EMFs. After application of the zero and calibration corrections for the DVM the indicated voltages agreed with the applied voltages to within 0·5 μV in the range 0 to 11 mV and to within 2 μV in the range 11 to 50 mV. In the second assessment the uncertainty of the method of calibration as well as the instrumentation was investigated by connecting the output of the standard thermocouple to each of the other four available test positions. This thermocouple arrangement was treated as a routine calibration using the usual control program. The differences between the standard and the other test positions were treated in the usual way. For all of the test locations, at all temperatures, the final calibration lay within ±0·5 μV of the standard thermocouple. The measurement of the first thermocouple position occurs approximately 8 seconds before that of the standard, thus the agreement of the same thermocouples in these two positions indicates the adequacy of the furnace drift correction during this period.

For the final check, the calibration of rare-metal thermocouples performed by the most accurate manual methods available were compared with their calibrations by the automatic method. For two type S and two type R thermocouples checked in this way the maximum difference in calibrations between the methods was 1 μV. In the routine automatic calibration of rare-metal thermocouples, a check thermocouple is always included in one of the test positions to monitor the program by comparing its calibration with those obtained earlier. Differences between successive calibrations of this thermocouple have never exceeded 2 μV at any temperature within the calibration range.

All the tests described in this section indicate that the uncertainty which is introduced into rare-metal thermocouple calibrations by the measuring equipment and calibration method is less than ±1 μV in the range 0 to 11 mV. When this is compared with the lowest uncertainty, ±3 μV, which can be applied to rare-metal thermocouples it can be seen that the component uncertainty introduced by the automatic calibration facility and method is not significant in the overall uncertainty of the calibration of rare-metal thermocouples.

The uncertainty in the calibration of base-metal thermocouples contributed by the automatic calibration method is less than 4 μV which is insignificant when compared with the reproducibility of base-metal thermocouples (Burley and Jones 1975).

7. Time taken for calibration

Thermocouple calibrations by manual methods are very time consuming. For the usual fixed-point method, measurements are made at five different immersions in the furnace for each of the eight fixed points in the temperature range 0 °C to 1100 °C.

The time taken for the calibration of a thermocouple depends on the number in each batch. For a single thermocouple it is about 42 man hours but this is reduced if a number of thermocouples are calibrated together. The optimum number in one batch is three, with each thermocouple taking an average time of 24 hours to calibrate. The time taken to produce a calibration table for each thermocouple is 2 hours. In contrast the operator's time to calibrate and produce calibration tables for four thermocouples in the automatic assembly is less than half an hour.

8. Conclusion

The overall uncertainty in the calibration of rare-metal thermocouples by the automatic method described is ±3 μV which is the same as the best that can be achieved by manual methods and in both cases is limited by the performance of the thermocouples themselves. The component uncertainty contributed by the automatic facility and calibration method is less than 1 μV which is small compared to the overall thermocouple uncertainty and indicates that the automatic calibration method is completely satisfactory. The saving in operator's time when compared to the manual calibration methods is enormous. This equipment allows thermocouples to be calibrated with such ease and precision that investigations into the practical performance of thermocouples, previously impracticable because of the large time needed for numerous calibrations, can now be performed (eg Burley and Jones 1975).

References

BSI 1973 *International Thermocouple Reference Tables* BS 4937
Burley N A and Jones T P 1975 this volume
Cassidy G J A 1975 *NML Technical Note* to be published
Jones T P 1968 *Metrologia* 4 80–3
—— 1972a *TMCSI* 4 part 3 1561–7
—— 1972b *Aust. J. Instrum. & Control* 28 91–6
Roeser W F and Lonberger S T 1958 *NBS Circular* 590

Non-contact determination of temperatures by measuring the infrared radiation emitted from the surface of a target

W Heimann and U Mester

Heimann GmbH, 62 Wiesbaden-Dotzheim, Weher Köppel 6, W Germany

Abstract. Fundamentals and basic terms of radiation laws are explained and relevant relations are shown. Different measuring principles and practical procedures for the determination of temperatures with radiation pyrometers are described. Measuring accuracy is primarily determined by the emissivity of the target. Emissivity values of several materials are given and the influence on accurate temperature measurements is studied. Optical materials of lenses and the properties of radiation detectors which determine design parameters for radiation thermometers are discussed. In conclusion, a variety of industrial applications is mentioned.

1. Introduction

It is a well known experience that radiation emitted by objects at temperatures above 100 °C can be felt by the thermal sensors of our skin. At temperatures above 650 °C all objects become incandescent indicating that the radiation covers the visible spectrum of light and is perceived by our eyes. When the temperature is further increased, the colour of the object changes from red to yellow and white. The temperature of the object is directly responsible for the generation of this radiation, which is therefore called 'temperature radiation'. It is emitted by all material objects at temperatures above the absolute zero point.

A characteristic parameter of electromagnetic radiation is the wavelength λ. The described phenomena permit the assumption that temperature radiation is primarily emitted at wavelengths in the invisible infrared region, but at high temperatures also in the visible spectrum. Figure 1 shows the spectrum of electromagnetic radiation

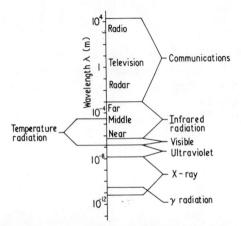

Figure 1. Electromagnetic radiation spectrum.

extending from long radio wavelengths of 10^3 m to the region of x-rays and cosmic radiation with wavelengths of 10^{-12} m.

The temperature radiation covers the region from the middle infrared to the visible light section. The close relationship to the visible spectrum is made evident by the fact that the basic properties of light and temperature radiation are defined by the same terms and obey the same physical laws. Instruments capable of measuring this radiation and providing an output signal calibrated in temperature units are called *radiation thermometers, radiation pyrometers,* or simply *pyrometers.* The scientific study of the determination of temperatures by the non-contact measurement of self-emitted surface radiation is called *radiation pyrometry.*

The method of measuring temperatures with radiation thermometers offers great advantages particularly in the following cases:

(i) The measurement of very high temperatures, when metallurgical reasons forbid the use of thermocouples or resistance thermometers.
(ii) Temperature measurements on materials having a very low thermal conductivity or small thermal capacity.
(iii) Measurements on inaccessible or moving objects.
(iv) Temperature measurements requiring a very short response time.
(v) Measuring very small or very large areas, which can be accomplished by the use of different lenses.

For decades, applications of radiation thermometers had been essentially confined to the measurement of high temperatures as mentioned in section (i). Only in recent years were a number of highly responsive radiation detectors developed. These detectors allow reliable and accurate temperature measurements far below the freezing point of water.

The soaring developments in the electronic field also made possible the processing of small signals in the μA and μV range without any difficulties. The required electronic circuits can be miniaturized to fit into compact, self-contained instruments, which are effectively protected against hostile environmental conditions and fully compensated for changing ambient temperatures.

2. Basic terms and laws of temperature radiation

A radiation thermometer receives the radiant power P, which is emitted from a surface area A into a defined solid angle Ω, and converts it into a signal in terms of either voltage or current. An important parameter is therefore the *radiance N*, which is defined as the radiant power P per unit area A and per unit solid angle Ω:

$$N = \frac{d^2 P}{dA \, d\Omega} \tag{1}$$

in units of W cm^{-2} sr^{-1}. Temperature radiation incident upon a surface is partially absorbed, partially reflected, and partially transmitted. The following relation holds:

$$\alpha + \rho + \tau = 1 \tag{2}$$

in which α is the radiant absorptance, ρ the radiant reflectance, and τ the radiant transmittance of the material object subjected to the radiation. An object which absorbs all incident radiation is called a black body.

Experiments have shown that the radiance N from a real body is always smaller than that from a black body at the same temperature. It depends on the chemical composition of the material, the angle of emission, and the surface condition. Those parameters in turn are dependent on the wavelength of the emitted radiation and the temperature of the object.

The ratio of the radiance N from a real material body to the radiance N_b from a black body at the same temperature is called the emissivity ϵ:

$$\epsilon = N/N_b. \tag{3}$$

In 1860, Kirchhoff had already found that a good absorber was also a good emitter. The law connecting absorptance and emittance is very simple, because both terms are identical

$$\epsilon = \alpha. \tag{4}$$

The construction of a practical black body radiation source is of great importance for testing and calibrating radiation-measuring instruments. A practically perfect black body simulator is a hole in a cavity, provided that the diameter of the hole is small compared to the size of the cavity. Another possibility for the construction of a black body simulator is shown in figure 2. It consists of a metal core kept at a uniform temperature with a cylindrical or conical cavity. The depth of the cavity is much larger than the diameter of the aperture.

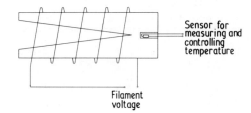

Figure 2. Black body radiation source.

The relation between the temperature T and the spectral radiance $N_{\lambda b}$ (defined as the radiance N_b for a small wavelength increment $d\lambda$) of a black body was first formulated by Max Planck in the year 1900:

$$N_{\lambda b} = \frac{c_1}{\pi \Omega_0 \lambda^5 [\exp(c_2/\lambda T) - 1]} \tag{5}$$

in units of $W\,cm^{-2}\mu m^{-1}\,sr^{-1}$, where $N_{\lambda b}$ is the spectral radiance of a black body, c_1 the first radiation constant; c_2 the second radiation constant and Ω_0 the unit solid angle (1 sr). Figure 3 shows the spectral radiance of black body radiators at different temperatures as a function of the wavelength.

It can be clearly seen that the maxima of the distribution curves are shifted to longer wavelengths for lower temperatures. Furthermore, the area under the curves decreases for lower temperatures. The indicated temperatures are those of well known

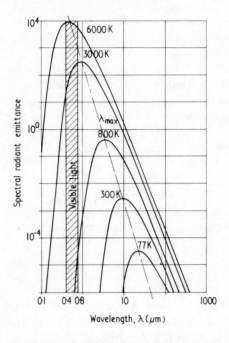

Figure 3. Spectral radiant emittance of several black body radiators.
$W = \epsilon \sigma T^4$ (W cm^{-2}), $W_{\lambda b} = c_1 \lambda^{-5}/[\exp(c_2/\lambda T) - 1]$ (W cm^{-2} μm^{-1}), $\lambda_m = 2900/T$ (μm).

radiation sources: the sun 6000 K; a tungsten-filament lamp 3000 K; a soldering-iron 800 K; the earth 300 K; liquid nitrogen 77 K.

Integration of the spectral radiance N_b over all wavelengths yields the total radiance N_b, which only depends on the temperature of the body (the Stefan–Boltzmann law):

$$N_b = \frac{\sigma}{\Omega_0 \pi} T^4. \tag{6}$$

Application of differentiation gives a simple expression for the temperature T and the wavelength λ_{max} indicating the peak of the radiance curve. This relation is known as the Wien displacement law:

$$\lambda_{max} \approx \frac{2900}{T} \mu m \quad (T \text{ in K}). \tag{7}$$

As previously mentioned, no real body exists with a surface emitting black body radiation. The radiance N is always smaller by a factor ϵ (see equation (3)) than the radiance of a black body at the same temperature. An accurate temperature determination by the measurement of the radiance is therefore only possible if the emissivity of the radiating surface is known. One of the primary objectives in pyrometry is thus the determination of the emissivity of the target surface as a function of wavelength, in particular at the specific wavelength or wavelength band in which the used radiation thermometer responds to radiation. Numerous measured values pertaining to emissivity were published in the *Handbook of Military Infrared Technology* (1965) and some references to this problem are contained in the papers presented at this conference.

3. Classification of radiation pyrometers

Several aspects must be considered to establish classification criteria for radiation pyrometers. Categories can be assigned to characteristic components (optical system, radiation detector), the method of radiation measurement, or the spectral response. Table 1 furnishes a classification scheme on the method of radiation measurement. Brightness pyrometers produce an output signal which is proportional to the radiance from the measured target and to the radiation received by the pyrometer. *Spectral radiation pyrometers* measure radiation only in a narrow wavelength increment (practically at one specific wavelength), *band radiation pyrometers* in a wide wavelength interval, and *total radiation pyrometers* receive more than 90% of the totally emitted radiation from the target which is accomplished by an appropriate design of the optical system and the radiation detector. *Colour pyrometers* work on a different principle.

Table 1. Classification of radiation pyrometers.

Brightness pyrometers	Colour pyrometers
Spectral radiation pyrometers	Colour adjustment pyrometers
Band radiation pyrometers	Two-colour ratio pyrometers
Total radiation pyrometers	

They compare the spectral distribution of the target radiation with the radiation from a black body at the same temperature. Wide applications have been found for the two-colour ratio pyrometers which determine the true target temperature from the ratio of two measured radiances at two different wavelength regions.

4. Measuring principles and practical temperature determination

Figure 4 shows the schematic arrangement of non-contact temperature measurement. A brightness pyrometer measures the radiant power P_m from the target area A. The size and the geometrical configuration of the area A is determined by the optical design of the pyrometer and the working distance between the pyrometer and the target. The radiant power P_m consists of the self-emitted radiant power P_e, the radiant power P_r reflected on the surface of the target, and the radiant power P_{tr} transmitted through

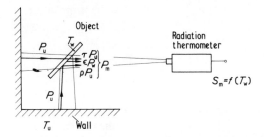

Figure 4. Measuring arrangement.

the target. The latter two portions of radiation are emitted by the surroundings — eg, the walls of the room in which the measurements are made, or the furnace walls when measuring a melt.

$$P_m = P_e + P_r + P_{tr}. \tag{8}$$

The self-emitted radiant power P_e is smaller by a factor ϵ than the radiant power P_w from a black body surface at the same temperature (see equation (3)). The reflected radiant power is smaller than the corresponding radiant powers P_u by a factor ρ and the transmitted radiant power is smaller by a factor τ. These radiant powers are emitted by the foreground and the background respectively:

$$P_m = \epsilon P_w + \rho P_u + \tau P_u. \tag{9}$$

The sum of emissivity ϵ, reflectance ρ and transmittance τ is always 1 (see equation (2)). Thus

$$P_m = \epsilon P_w + (1-\epsilon) P_u. \tag{10}$$

The individual radiant powers can be transformed to the corresponding radiances. N_m is then the apparent radiance from the target, N_w is the radiance from a black body at the true temperature T_w (which is to be determined) and N_u is the radiance from the surroundings.

$$N_m = \epsilon N_w + (1-\epsilon) N_u. \tag{11}$$

The walls of the concurrently measured surroundings are considered to be close approximations of black radiating surfaces. In accordance with Planck's law, the radiances N_w and N_u are clearly related to the temperatures T_w and T_u, provided the relative spectral response $R(\lambda)$, which is determined by the optical system and the radiation detector, is known.

$$N_m = \epsilon \int_{\lambda_1}^{\lambda_2} \frac{c_1 R(\lambda) \, d\lambda}{\Omega_0 \pi \lambda^5 [\exp(c_2/\lambda T_w) - 1]} + (1-\epsilon) \int_{\lambda_1}^{\lambda_2} \frac{c_1 R(\lambda) \, d\lambda}{\Omega_0 \pi \lambda^5 [\exp(c_2/\lambda T_u) - 1]}. \tag{12}$$

The output signal S_m of a brightness pyrometer is proportional to the measured radiance N_m

$$S_m \sim N_m. \tag{13}$$

It is thus permissible to read the measured radiance N_m in terms of a 'measured temperature' T_m and to calibrate the indicating meter for the output signal in temperature units, ie, in degrees Celsius.

$$N_m = \int_{\lambda_1}^{\lambda_2} \frac{c_1 R(\lambda) \, d\lambda}{\Omega_0 \pi \lambda^5 [\exp(c_2/\lambda T_m) - 1]}. \tag{14}$$

The calibration curve $S = f(T_m)$ is given by equation (14) and can be calculated by inserting different temperatures T_m and normalizing the function for the desired temperature range T_1 to T_2

$$S_m = \frac{N_m(T_m) - N_m(T_1)}{N_m(T_2) - N_m(T_1)} \qquad (15)$$

In practical work, a solution of this equation is nearly impossible. Therefore, a characteristic effective centre wavelength λ_c is introduced, at which the weighted average of radiation is emitted in the wavelength interval λ_1 to λ_2 for the temperature range T_1 to T_2. This centre wavelength can be approximated by procedures as given by Ruffino (1973).

Equation (14) simplifies to

$$N_{\lambda m} = \frac{c_1 R(\lambda_c)}{\Omega_0 \pi \lambda_c^5 [\exp(c_2/\lambda_c T_m) - 1]}. \qquad (16)$$

Equation (15) for the temperature scale yields

$$S_m = \frac{1/[\exp(c_2/\lambda_c T_m) - 1] - 1/[\exp(c_2/\lambda_c T_1) - 1]}{1/[\exp(c_2/\lambda_c T_2) - 1] - 1/[\exp(c_2/\lambda_c T_1) - 1]}. \qquad (17)$$

After the temperature T_m has been read from the indicating meter of the radiation thermometer calibrated in accordance with the procedure described above, it is now necessary to calculate the true temperature T_w.

Equation (14) is inserted into equation (12), which is then solved for T_w. The concurrently measured temperature T_u can either be estimated or determined with the radiation thermometer by obtaining a temperature reading from that surface, which is expected to furnish the largest contribution to the reflected radiation. The emissivity ϵ can be taken from a table or it can be determined by the procedure described in §5. The calculation is rather tedious however and is reduced to a reasonable amount upon the introduction of the previously defined characteristic wavelength in the interval λ_1 to λ_2 and for the temperature range T_1 to T_2:

$$\frac{1}{\exp(c_2/\lambda_c T_m) - 1} = \frac{\epsilon}{\exp(c_2/\lambda_c T_w) - 1} + \frac{1 - \epsilon}{\exp(c_2/\lambda_c T_u) - 1}. \qquad (18)$$

Equation (18) must be solved for T_w yielding a function

$$T_w = f(\lambda_c, \epsilon, T_m, T_u). \qquad (19)$$

A well known approximation for a spectral or band radiation pyrometer is obtained when it is assumed that $T_u \ll T_w$ and $c_2 > \lambda_c T_w$. For these cases equation (18) simplifies to

$$T_w = \left(\frac{1}{T_m} + \frac{\lambda_c}{c_2} \ln \epsilon\right)^{-1}. \qquad (20)$$

Equation (20) clearly shows that the indicated temperature T_m is closer to the true temperature T_w the smaller the wavelength λ, at which the measurement is made, and the nearer the emissivity approaches the value 1. The indicated temperature is always lower than the true temperature.

For a total radiation pyrometer, the Stefan–Boltzmann law (6) can be used for the calculations:

$$T_m^4 = \epsilon T_w^4 + (1 - \epsilon) T_u^4 \tag{21}$$

yielding

$$T_w = \frac{T_m}{\epsilon^{1/4}} \left[1 - (1 - \epsilon) \left(\frac{T_u}{T_m} \right)^4 \right]^{1/4}. \tag{22}$$

In practical work it can often be assumed that the concurrently measured temperature of the surroundings is very low compared to the measured temperature T_m. For those cases equation (22) simplifies to

$$T_w = T_m/\epsilon^{1/4}. \tag{23}$$

Equation (22) shows that the indicated temperature T_m of a total radiation pyrometer is closer to the true temperature T_w of the target the higher the emissivity ϵ. In addition, it is obvious that the smaller the difference between the temperatures of the surroundings and the measured target, the less the emissivity affects the measured temperature.

A two-colour radiation pyrometer automatically calculates the ratio of the measured radiances N_{m1} and N_{m2} at the wavelengths λ_1 and λ_2. Thus the output signal of the pyrometer is given by

$$S_m = \frac{N_{m1}}{N_{m2}} = \frac{\epsilon_1 N_{w1} + (1 - \epsilon_1) N_{u1}}{\epsilon_2 N_{w2} + (1 - \epsilon_2) N_{u2}}. \tag{24}$$

In practical work, two-colour pyrometers are primarily used for measurements where $T_w \gg T_u$, thus simplifying the calculation of the true temperature analogous to equations (20) and (23):

$$T_w = \left[\frac{1}{T_m} - \frac{1}{c_2} \frac{\lambda_1 \lambda_2}{\lambda_1 - \lambda_2} \ln \left(\frac{\epsilon_1}{\epsilon_2} \right) \right]^{-1}. \tag{25}$$

It follows that whenever the emissivities ϵ_1 and ϵ_2 at the two wavelengths λ_1 and λ_2 have the same value (as encountered for all so-called grey bodies), the true temperature coincides with the measured temperature. However, when the emissivity ϵ_1 has a larger numerical value than ϵ_2 (as shown later for all metals), the indicated temperature is too high. This is a characteristic feature of a two-colour pyrometer, while brightness radiation pyrometers generally indicate a temperature which is too low. The difference between the colour temperature T_m measured by the pyrometer and the true temperature T_w depends not only on the ratio of the two emissivities, but also on the absolute value and the difference of the two wavelengths at which the measurements are made. Optimum results would be obtained if very short wavelengths and a very large wavelength difference could be selected. However, it must be borne in mind that for larger wavelength differences the reading error of the pyrometer can possibly attain very great values, since the change of the ratio ϵ_1 to ϵ_2 is often more significant than the effect of wavelength difference. Therefore, the reading error primarily depends on the change of emissivity with the wavelength. The special problems encountered in two-colour pyrometry are discussed in detail in one of the following papers presented at this conference.

The theoretically derived equations, giving the relation between the temperature reading of a pyrometer calibrated at a black body radiation source and the true temperature, are rather tedious and cumbersome for practical calculations. A simpler way of determining the true temperature for different emissivities and different temperatures concurrently measured from the surroundings is to use the scale characteristic of the pyrometer. In addition, this procedure permits a convenient estimation of the effect of changing emissivities on the temperature reading.

In analogy to equation (10), the signal S_m can be considered to be composed of two terms:

$$S_m = \epsilon S_w + (1 - \epsilon) S_u. \tag{26}$$

Equation (26) is then solved for S_w

$$S_w = \frac{1}{\epsilon}(S_m - S_u) + S_u. \tag{27}$$

For known emissivities, the true temperature t_w can be obtained by use of the corresponding signal S_w determined with the aid of the scale characteristic. (Lower-case letters denote temperature in °C, capital letters in K.)

The reading S_m is obtained when the radiation pyrometer is directed at the target. Direct measurement of the surface concurrently seen by reflection or transmission gives the signal S_u. This surface can possibly be the wall or ceiling of the room in which the measurement is made, or the walls of a furnace. The difference

$$a = S_m - S_u \tag{28}$$

multiplied by $1/\epsilon$ yields the distance b indicated in figure 5:

$$b = a(1/\epsilon). \tag{29}$$

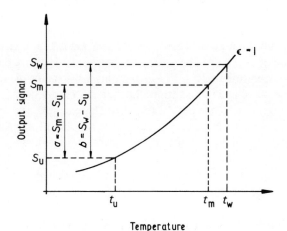

Figure 5. Scale characteristic of a radiation pyrometer. $S_m = \epsilon S_w + (1 - \epsilon) S_u$, $S_w - S_u = \epsilon^{-1} (S_m - S_u)$, $S_w = \epsilon^{-1} (S_m - S_u) + S_u$.

The required value S_w has thus been determined and the true temperature t_w can be read directly from the scale characteristic. Calculations for different emissivities and different background temperatures, and the corresponding measured signals S_u, provide a set of curves with the parameters t_u and ϵ. The less the influence of the temperature of the surroundings, the smaller the signal S_u (obtained by directly measuring the temperature of the surroundings) in comparison to the output signal S_m. This is always the case for target temperatures exceeding by far the temperature of the surroundings.

For pyrometrical measurements near the ambient temperature, thermal radiation from the surroundings considerably affects the reading. Figure 6 shows the scale characteristic for a total radiation pyrometer measuring temperature in the range from $-50\,°C$ to $+50\,°C$.

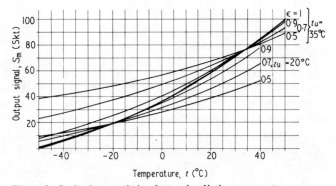

Figure 6. Scale characteristic of a total radiation pyrometer.

The correction curves for two different examples are given in figure 6. In one case, the extraneous radiation is emitted by the surroundings at a temperature of $-20\,°C$, while the other case assumes an ambient temperature of $+35\,°C$. Both temperature values are quite realistic in practical measurements. If a wooden surface is measured out in the open air, an average temperature of $-20\,°C$ can be correlated to the concurrently measured sky radiation (blue sky). The emissivity of wood is approximately 0·9. For a target temperature of $25\,°C$ it can be seen from the diagram in figure 6 that the pyrometer indicates a temperature of $21·4\,°C$, which means a misreading of $-3·6\,°C$. If the same object is then measured in a room with wall temperatures of $+35\,°C$, the pyrometer indicates a temperature which is $1·05\,°C$ too high. This example demonstrates that it is mandatory for exact and accurate temperature measurements near ambient temperatures to give due consideration to the influence of the temperature of the surroundings.

5. Emissivity

An accurate temperature measurement with radiation pyrometers can only be made when the emissivity of the measured target is known. A correction of the indicated temperature value can either be calculated mathematically with the previously derived equations or graphically, as described in the preceding section, with the aid of the scale characteristic. In either case, the emissivity of the measured target must be known.

There are measuring principles which allow temperature determination independently of the emissivity of the radiating target. In particular, two-colour ratio pyrometry offers this advantage. However, it must always be assured that the measured target is nearly a grey body radiator for the two wavelengths in which the measurements are made, ie, that the emissivity is independent of wavelength. Emissivity is generally not only a function of the wavelength and the target temperature, but also of the emission angle and the surface condition. Therefore, a theoretical calculation of emissivity is only feasible in very few idealized model cases. Normally, the emissivity must be determined experimentally.

Materials can be divided into three general groups of different emissivity properties:

(i) *Pure metal surfaces.* Their emissivities decrease with longer wavelengths (see figure 7).

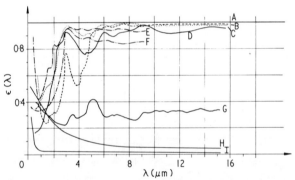

Figure 7. Spectral emissivity of several materials. A, black paint; B, fire-clay; C, wood; D, white paper; E, white paint; F, green leaves; G, aluminium bronze; H, iron, tungsten; I, silver.

(ii) *Oxidized metals and non-metals.* Their emissivities generally increase with longer wavelengths (see figure 7).

(iii) *Transparent materials* such as glass, quartz and plastic foils. These materials exhibit so-called absorption bands in characteristic wavelength regions (see figure 8). In the spectral regions of these absorption bands the emissivity is very high since the reflectivity of plastic foils at wavelengths greater than 3 μm is in general very low.

When pyrometric temperature measurements are limited to the region of such an absorption band, plastic materials and plastic foils can be measured without disturbing background radiation.

Temperature measurements on glass are of particular interest. Figure 9 shows the emissivity, transmittance and reflectance of glass as a function of the wavelength.

It can be recognized that there is a marked region between 4·8 μm and 8·2 μm in which the emissivity of glass is very high. When pyrometric measurements are limited to this spectral region, very accurate surface temperatures of glass can be expected. Unfortunately, there is a strong absorption band of water in this particular spectral

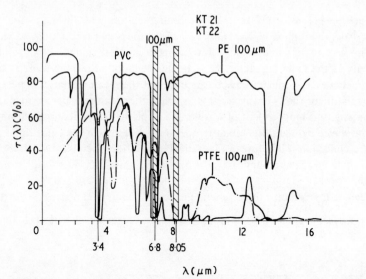

Figure 8. Spectral transmittance of several plastic foils. PVC, polyvinylchloride; PE, polyethylene; PTFE, polytetrafluoroethylene.

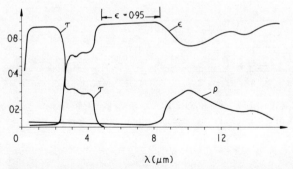

Figure 9. Emissivity ϵ, transmittance τ and reflectance ρ of glass.

region, centred at 6·2 μm. Therefore, in this spectral region, radiation suffers an attenuation caused by the water vapour always contained in the atmosphere. As a result, misreadings are obtained depending on the working distance between the target and the pyrometer. Therefore the spectral response of the pyrometer should be limited to a band from 4·8 μm to 5·6 μm or from 7·8 μm to 8·2 μm. Figure 10 shows the total emissivity of several materials at ambient temperature (300 K).

It appears to be a fortunate coincidence that thermal radiation emitted at the processing temperature of non-metals, plastic materials, rubber, etc (50 °C to 300 °C) is encountered primarily at wavelengths above 5 μm; in this region the emissivity of those substances is very high. By the same token, metals are generally processed and finished at temperatures above 650 °C and thermal radiation is predominantly emitted at wavelengths, where the emissivity of metals, particularly of iron, is also very high. Under

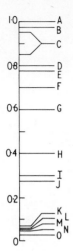

Figure 10. Total emissivity ε of several materials at 300 K. A, black body; B, black velvet paint, wood, masonry, fire-clay, rubber; C, chinaware, ceramics, paper, plaster, oil paints, plastics; D, copper, black oxidized; E, rolled iron; F, rusted iron; G, oxidized iron; H, sand-blasted aluminium; I, silver/aluminium bronze paint, oxidized nickel; J, lead, grey oxidized/iron, dark grey; K, polished tin, polished nickel; L, polished iron; M, polished aluminium; N, polished silver; O, polished gold.

most circumstances the emissivity is even increased by the oxidized surface of the metals.

As mentioned previously, the determination of emissivity is extremely important. In some of the following papers presented at this conference it will be reported how the emissivity is determined for special materials at specific temperatures. Nevertheless, three simple methods for the experimental determination of emissivity are briefly described:

(i) For materials of high thermal conductivity a thermal sensor (thermocouple or resistance thermometer) can be mounted in a hole drilled into the specimen and the true temperature determined. The temperature indicated by the pyrometer in a subsequent radiation measurement then permits the calculation of emissivity with the aid of the appropriate equations previously given or the described graphical procedure.

(ii) For materials of low thermal conductivity a coating of black velvet paint or black soot is applied on the surface of the specimen and the temperature indicated by the pyrometer for the unblackened surface is compared to that for the blackened surface.

(iii) Some radiation pyrometers are equipped with a control permitting the adjustment of emissivity. In those cases the determination of emissivity is very simple, since the control must only be turned until the temperature read-out is the same as the one obtained for the blackened surface. However, due consideration must be given to the temperature of the concurrently measured surroundings for which the control has been laid out. Depending on the design of the instrument, this could be the temperature of the pyrometer housing, a temperature adjustable by a separate control, or a fixed temperature, eg, the starting value of the temperature range.

6. Lens materials and radiation detectors

Lenses and radiation detectors belong to the most important components of radiation pyrometers. Both primarily determine the spectral response and the stability of the instrument. Figure 11 shows the transmittance of materials commonly used for lenses

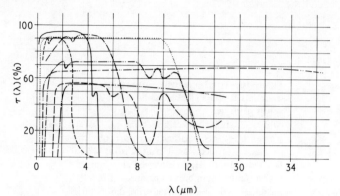

Figure 11. Optical materials. ———, quartz; — — — —, glass; — · — · — · —, LiF; ·········, CaF_2; — ·· — ·· — ·· —, As_2S_3; — — — —, Si; — · — ··· — ··· —, Ge; — — ·· — — ·· — —, KRS-5.

in radiation pyrometers. All listed materials are non-hygroscopic and therefore are well suited for industrial applications. The other component determining the spectral response of a pyrometer is the used radiation detector. Basically, there are two groups of radiation detectors: one exhibits a flat and the other a selective spectral response. Detectors having a flat response furnish a signal which is independent of the wavelength. This condition is approximately met by thermal detectors, such as thermocouples and bolometers. Selective radiation detectors are photoelectric devices, such as photocells, photodiodes, phototransistors and photoresistors.

Figure 12 shows some of the widely used radiation detectors for pyrometers. Silicon or germanium photodiodes and lead sulphide photoresistors are the most commonly used photoelectric detectors. Bolometers and thermocouples exhibit a response independent of wavelength. Although their responsivity is orders of magnitude smaller than that of photoelectric detectors, their performance is superior at longer wavelengths and therefore they are particularly well suited for temperature measurements below 100 °C.

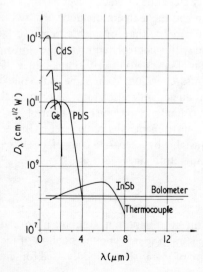

Figure 12. Radiation detectors.

7. Design principles of pyrometers

Figures 13—16 schematically illustrate the design principles of four different commercially available radiation pyrometers. The described pyrometers are representatives for each group which comprises an entire series of similar models. Identical characteristic components are designated by the same number. Figure 13 shows an optical

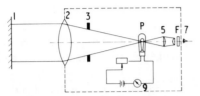

Figure 13. Optical filament pyrometer (spectral radiation pyrometer).

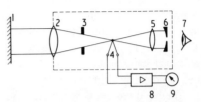

Figure 14. Thermocouple radiation pyrometer (band radiation pyrometer).

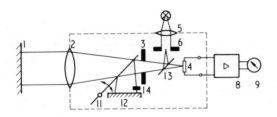

Figure 15. Bolometer pyrometer (total radiation pyrometer).

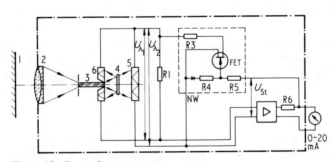

Figure 16. Two-colour pyrometer.

filament pyrometer which belongs to the group of spectral radiation pyrometers; figure 14 describes a thermocouple pyrometer equipped with a glass lens — it is a so-called band radiation pyrometer; figure 15 is the schematic design of a bolometer pyrometer with an optical lens of KRS-5, which works on the principle of optical chopping and is classified as a total radiation pyrometer; and figure 16 presents a two-colour pyrometer.

Spectral radiation pyrometers measure the radiance from the target within a narrow wavelength interval. The radiance measurement is either made objectively in terms of absolute values with a radiation detector or subjectively in terms of relative values obtained by the visual impression of the observer, who compares the brightness of the target with that of a radiation source calibrated at black body temperature. The narrow spectral response is provided by an appropriate filter in the path of the ray. Easily portable and battery operated instruments for the measurement of temperatures above 600 °C usually employ a tungsten filament lamp as a reference source. The image plane of the target coincides with the lamp filament and they can both be simultaneously observed by looking through the eye-piece 5 and the filter F. By adjusting the variable resistor R and the brightness, the temperature of the lamp filament is changed until the filament disappears in the image of the target. In this case both radiances are identical in the spectral region determined by the filter and the spectral response of the eye. The indicating meter, which measures the lamp current, is calibrated together with the reference lamp in such a way that the spectral radiation temperature of the target can be read directly from the scale.

The band radiation pyrometer (figure 14) collects radiation from the target 1 through a lens 2 on to the radiation detector 4. The eye-piece 5 is used for aiming on the target and allows a simultaneous observation of the detector 4.

The bolometer pyrometer in figure 15 employs an oscillating or rotating mirror 11 to reflect the radiation from reference source 12 on to the detector alternately to the radiation received from the target on 4. The output signal is an alternating voltage, which is determined only by the difference of the two radiances. The radiation emitted by the housing of the instrument, which in pyrometers for low target temperatures can assume considerably larger values than the measured target radiation, is not modulated and therefore does not affect the measuring signal. In the pyrometer shown the target is marked by means of an illuminated aperture which is projected on to the target instead of the detector, whenever a mirror 13 is brought into the path of the ray.

Figure 16 demonstrates the measuring principle of a two-colour ratio pyrometer. The target radiation is collected by the objective lens 2 and directed through an aperture on to the optical fibre 3. While passing the optical fibre, the radiation suffers multiple refractions and leaves the fibre evenly distributed without any image contours. The target radiation then falls upon a partially transparent indium phosphide filter, which transmits only radiation above 1 μm, while it reflects all radiation of shorter wavelengths The target radiation is thereby split into two parts and then detected by two silicon photodetectors. Both detectors furnish a certain output voltage whose ratio is determined in the subsequent electronic circuits and indicated on the output meter.

The schematic layouts of the different pyrometers are not conclusive for the various other design considerations, which may have been necessary to meet certain specifications, such as accuracy, stability, high ambient temperatures, power requirements, target size (ratio of target size to working distance), the influence of intermediate transmitting media, easy operation, response time and sales price.

There is no upper limit for measurable target temperatures with total or band radiation pyrometers. As previously shown in the diagrams for the transmittance of window materials and the spectral response of radiation detectors, the lowest, still resolvable temperature depends primarily on the responsivity of the pyrometer at long wavelengths. The lowest temperatures are best measured with pyrometers equipped with reflective mirror systems or refractive lenses of KRS-5 or germanium (temperature measurements down to $-100\,°C$). Band radiation pyrometers with lead sulphide detectors and quartz lenses are capable of measuring temperatures down to approximately 150 °C. Total radiation and band radiation pyrometers are generally used for measuring surfaces exhibiting a high emissivity. At high temperatures this is the case for metals in closed furnaces with walls kept at a uniform temperature, and at low temperatures, all non-metals. Band radiation pyrometers with photoelectric detectors and band radiation pyrometers equipped with glass or quartz lenses are primarily employed in metal processing and finishing industries, while pyrometers with mirror systems or germanium and KRS-5 lenses have found wide application in measuring low temperatures on non-metal targets, such as plastic processing. For measurements on ideally emitting surfaces, ie, when the emissivity is close to 1, the measuring error determined by the accuracy of the instruments amounts to 0·5% to 1·5% of the target temperature. In practical temperature measurements, however, the error is primarily determined by the accurate knowledge of the emissivity of the measured target.

8. Testing and calibration

For testing and calibrating the indicated temperature of a radiation pyrometer, a black body radiation source is required. It must be assured that the target size of the pyrometer for the working distance set up during calibration is smaller than the effective aperture of the black body radiation source. The practical construction of a black body simulator for temperatures up to 1000 °C constitutes no problem. A number of adequate and suitable models, including the necessary temperature controls, are commercially available for that purpose. Problems are encountered for the calibration of pyrometers at temperatures above 1000 °C. Black body radiators have been built for those temperatures but their construction is difficult and expensive. Spectral radiation pyrometers with a peak response at a wavelength of 650 nm can be calibrated by a very simple procedure. So-called ribbon-filament lamps, which have a tungsten ribbon filament, and which are calibrated by the manufacturer or by the national institutes of standards in such a manner that the current passing through the tungsten ribbon is a direct measure for the spectral radiant temperature at 650 nm, are commercially available.

Vacuum lamps are used for measurements up to 1500 °C. For higher temperatures gas-filled lamps are necessary. The measuring error with ribbon-filament lamps calibrated by the national laboratories amounts to approximately ±2 °C for temperatures between 1000 °C and 1200 °C and may reach ±10 °C for higher temperatures. The measuring error for black body simulators primarily depends on the error made in determining the temperature of the simulator with a temperature sensor. Typical attainable values are an accuracy of ±2 °C in the range from 500 °C to 1000 °C. In the range from 0 °C to 500 °C an accuracy of better than 1 °C can be achieved. The effective emissivity of black body simulators, which is the result of optimized cavity shape and surface condition, amounts to 0·99. Of course, far higher accuracies are attained by the national laboratories for specially designed black body radiators and pyrometers. For instance, it is possible to determine the solidification point of gold (1063 °C) with a measuring error of less than 0·1 K.

9. Applications

The non-contact determination of temperatures with radiation pyrometers has a broad application in all branches of industry (see Hudson 1969).

The following is a summary of the most important applications:

(i) Measurements in iron processing and foundry works during casting, rolling and tempering; controlling the temperature of metal melts, etc.
(ii) Measurements on electronic components, such as resistors, diodes, transistors in printed circuits; checking proper mounting of heat-sinks, etc.
(iii) Adjustment of the switching point of bi-metal thermal switches by measuring the temperature of the switching reed in a hot air stream.
(iv) Measuring the surface temperature of transmissions, gears and bearings of all types of motors of all sizes.
(v) Measuring the temperature profile on transformers and other coils.
(vi) Checking the thermal insulation of various materials.
(vii) Monitoring the temperature during thermofixing, drying and laminating processes.
(viii) Temperature measurements on parts and components under high voltage.
(ix) Measuring the temperature on running rubber tyres under load; locating material defects causing hot spots.
(x) Temperature measurements and control during drying and thermal moulding processes in the plastic industry (Mester and Reinhold 1972).
(xi) Temperature measurements on plants, leaves and flowers.
(xii) Temperature measurements in medical work as a diagnostic aid and for therapeutical control.
(xiii) Temperature measurements on flames and hot gases with pyrometers which are responsive at the wavelength of the CO and CO_2 absorption bands.
(xiv) Detection of hot boxes on railroad cars by measuring the journal box covers of a passing train.
(xv) Monitoring the water temperature of rivers and lakes with a pyrometer aboard a helicopter or aircraft to detect excessive or unauthorized thermal pollution.
(xvi) Locating leakages in the pipeline system of district heating plants by measuring the local temperature increase on the ground surface.
(xvii) Finally an application is described (see figure 17) in which a reflector 2 is mounted in such a manner that multiple reflections of the emitted target radiation cause an apparent increase of emissivity (Mester and Seumel 1972).

Since it is impossible to construct a perfectly reflecting surface and even a slight contamination of the surface considerably degrades its reflecting properties, the reflector should be heated close to the target temperature. With this arrangement, the effect of the concurrently measured self-emission of the reflector and variations in emissivity of the target are kept negligibly small.

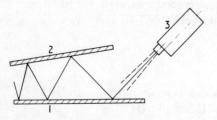

Figure 17. Arrangement for obtaining an apparent increase of emissivity.

References

Handbook of Military Infrared Technology 1965 ed W L Wolfe (Washington, DC: US Govt Printing Office)
Hudson R D Jr 1969 *Infrared System Engineering* (New York: Wiley)
Mester U and Reinhold H D 1972 *Elektrotechnik* 5 10–13
Mester U and Seumel G 1972 *Z. Metall.* 4 342–6
Ruffino G 1973 *VDI Ber. No.* 198

The NPL photon-counting pyrometer

P B Coates
Division of Quantum Metrology, National Physical Laboratory, Teddington, Middlesex

Abstract. Modifications have been made to the NPL photoelectric pyrometer in order to extend its working range to temperatures below 500 °C and to improve the overall accuracy. These include the use of a new photomultiplier with response in the near infrared, and of the photon-counting technique, which is suited to the long measurement times necessary at low temperatures. Tests of the performance of the pyrometer, including measurements of the Au–Cu interval, are given.

1. Introduction

In the region between the triple point of water and the freezing point of gold, 1064·43 °C, the International Practical Temperature Scale, IPTS-68, is deficient in at least two respects. First, the uncertainties in the fixed points are themselves uncertain, and possibly larger than previous estimates. Second, the platinum–10% rhodium–platinum thermocouple is unsatisfactory as an interpolation instrument between 630·74 and 1064·43 °C. These and other deficiencies manifest themselves as discontinuities in the slope of the interpolation functions at 630·74 °C and at the gold point. If the pyrometric scale is continued downwards from the gold point, it diverges from IPTS-68 by an amount increasing to 0·5 °C at 750 °C (Quinn et al 1973).

At NPL an attempt is being made to solve some of these problems by setting up radiation instruments to cover the whole of this range. Between the triple point of water and the zinc point, a total radiation experiment is being assembled (Quinn 1975). This will give absolute temperatures in this range, and also a value for the Stefan–Boltzmann constant. With a known value for the zinc point the scale can then be extended to the gold point by measuring ratios of the spectral density of radiation with a photoelectric pyrometer. Radiation measurements between the zinc and gold points (Hall 1965) and between the antimony and gold points (Heusinkveld 1966, Bonheure 1975), have been recorded, and the main problem is to reduce the uncertainties in both techniques so that a significant improvement upon the existing scale is achieved.

Instead of designing and building a completely new instrument it was decided to modify the existing NPL photoelectric pyrometer (Quinn and Chandler 1972), in order to extend its range. The optical system, shown in figure 1, was retained, although the sectored disc is no longer used. Since the pyrometer is also required for the investigation and calibration of the standard lamps used in the NPL lamp calibration service, possible modifications were limited to variations in the wavelength and bandwidth of the interference filters, and in the photomultiplier cathode, temperature and operating mode, ie, DC or photon counting. Since the optical through-put was unchanged, it was clear from preliminary considerations that the lowest temperature that could usefully

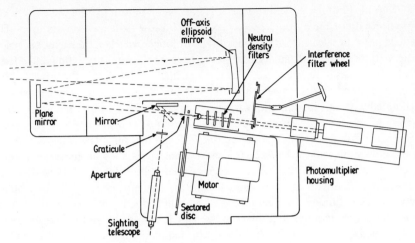

Figure 1. The NPL photoelectric pyrometer.

be measured would be limited by photon statistics, and that this would require experiments of many hours' duration. The modifications aimed therefore at improving, in increasing order of importance, the accuracy, the stability and the sensitivity of the pyrometer.

2. Design considerations

The first and most important design choice, since it determines the useable wavelength range, was that of the photocathode type. Only cathodes which were more sensitive in the near infrared than the standard S-20 were considered. These included the modified S-20, the extended red-sensitive multi-alkali (ERMA or S-25), and the negative electron affinity photocathodes, GaAs and GaInAs. Although sensitivity extending as far as possible into the infrared was desirable, the photomultiplier must also possess certain properties, such as linearity and stability. It appeared, for example, at the time that the choice of detector was made, that photomultipliers containing GaInAs photocathodes did not possess long-term stability, and these were therefore rejected. The GaAs reflection cathode was satisfactory in many respects, but all tubes tested were found to fatigue to an unacceptable extent, even at the low anode currents required. The S-25 cathode was also found to have a very high temperature coefficient in the threshold region, increasing rapidly from $0.1\%/\,°C$ to $5-10\%/\,°C$ at the cut-off wavelength (Cole and Ryer 1972). For this reason it was considered not to be of use above 800 nm.

The photomultiplier eventually chosen was the EMI 9658. This is a plug-in replacement for the standard 9558, and employs the modified S-20 photocathode. To reduce the dark noise, it is necessary either to reduce the effective cathode area by means of the magnetic defocusing technique, or to cool the photomultiplier to about $-25\,°C$. It has been shown (Coates 1975b) that the second method is capable of giving the greater reduction, and has the advantage of not limiting the useable cathode area. In addition, by stabilizing the temperature of the photomultiplier, the main source of drift in non-fatiguing tubes may be considerably reduced.

From the photocathode response, it is possible to determine the wavelength at which the sensitivity of the pyrometer is a maximum for low radiance temperatures. There is little point in selecting interference filters of longer wavelength, partly because the sensitivity then falls off rapidly, and partly because the intensity ratio is then less sensitive to changes in temperature. For the modified S-20 cathode, the wavelength of maximum sensitivity is between 800 and 850 nm. Since standard lamps must still be calibrated with this instrument, it was necessary to include an interference filter with a transmission peak at about 660 nm. Three other filters are at present in place in the pyrometer with peak transmissions at 704, 749 and 812 nm. The filters have narrow half-bandwidths, varying between 0·3 and 1·1 nm, in order to decrease the uncertainty in the effective wavelength from the uncertainties in both the filter transmission and in the spectral sensitivity of the remaining elements of the pyrometer. To interpolate between the antimony and gold points without introducing an error greater than 0·01 °C from this source requires that the effective wavelength is known to about 0·01–0·02 nm. If the scale is extended to the zinc point, this requirement becomes even more stringent. To attain this degree of accuracy, the filters should be calibrated *in situ*; it is advisable also to measure the temperature coefficient and the rate of ageing of the effective wavelength, since these have been found to be high in some cases. Finally, the transmission of the filter should be checked at all wavelengths at which the photomultiplier has significant sensitivity, to avoid errors from out-of-band transmission. It is possible to obtain the transmission characteristics of a 1 nm broad filter to the required accuracy in less than one day. The time for broader filters increases approximately in proportion to their bandwidth, and since several filters may need to be calibrated, it is clear that the time involved for their precise calibration soon becomes prohibitive in practice. The main disadvantage in the use of narrow filters is that the out-of-band transmission must be lower than with broad filters. For those specified for the pyrometer, a transmission of less than 10^{-5} at wavelengths up to $1\,\mu$m was specified.

Although the modified pyrometer has been distinguished by the photon-counting aspect of its design, there is no single outstanding reason for the use of this technique. However, a number of minor advantages together made its adoption worthwhile. It gives a small but significant improvement in sensitivity at low light levels when compared to the DC method (Ingle and Crouch 1972). It also minimizes the effects of leakage currents inside the photomultiplier and around its base, which can be large and erratic at low temperatures because of condensation, and it reduces the dependence of the signal amplitude upon changes in the gain of the electron multiplier. Furthermore, the signal is obtained directly in digital form, which is an advantage for experiments of long duration, whether the data is stored or analysed immediately. The main disadvantage of the photon-counting technique derives from the uncertainty in the dead time of the discriminator and counting circuits, which limits the maximum counting rate which may be used. Since the minimum counting rate is determined by the time needed to reach the required statistical accuracy, the working range may be severely curtailed for measurements of high precision.

3. Instrument performance

The variation in the maximum count rate (ie with no neutral density filters in the beam) is shown for the four interference filters in figure 2 as a function of source

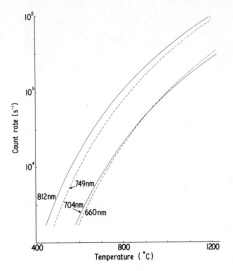

Figure 2. Variation of count rate with temperature.

temperature. At low temperatures the working limit of about 2000 s^{-1} is not determined by the dark count, which is normally about 100 s^{-1}, but by the time required to obtain the statistical accuracy. The count probability distribution was accurately Poissonian for the signal, but the dark-count distribution possessed a noticeable tail towards higher count rates. Although electrical interference may contribute towards this effect, the major source arises from cosmic-ray effects in the photomultiplier window (Coates 1972).

Since the pyrometer establishes the temperature scale by directly measuring intensity ratios, it relies on the linearity of the electron multiplier and the accuracy of the dead time or pile-up correction. The particular 9658 photomultiplier used in the pyrometer was selected for absence of fatigue effects (Coates 1975a) from a small batch of this type. The linearity of its DC response was investigated by the conventional double-aperture technique, with a detecting system consisting of a current amplifier and digital voltmeter. With anode currents of the order of those used in the counting experiments, about 10 nA, no deviation from linearity within the accuracy of the measurements (0·05%) could be detected. At larger currents, a positive deviation increasing to a value of 0·04 ± 0·01% at 1 μA was found. Since in the photon-counting mode the effect of gain variations upon the signal is reduced by a factor between 5 and 10, it was concluded that the photomultiplier linearity was not a serious source of error in these experiments. With the photomultiplier in the photon-counting mode, the double-aperture system may be used to measure the dead time τ of the counting circuits. The weighted mean from these measurements was 40·3 ns, with a standard error of 0·4 ns. This value was checked by a method involving the determination of the apparent attenuation A of a neutral density filter at different count rates. It may be shown that

$$A = A_0 - (A_0 - 1)N_-\tau$$

where A_0 is the true attenuation, or that extrapolated to zero count rate, and N_- the count rate with the filter removed from the optical beam. For filters with A_0 approximately equal to 2, 4 and 64, the dead times obtained were 40·7, 40·4 and 40·0 ns

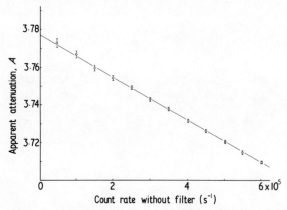

Figure 3. Neutral density filter technique for the measurement of dead time.

respectively. A typical graph is shown in figure 3. Although it is possible to measure dead times with pulse generator systems, it was found that the result depended upon the pulse amplitudes, and a complex calculation was necessary to obtain the correctly weighted dead time. It is therefore preferable to measure this with the pulse height distribution and discriminator setting employed in practice.

The stability of the pyrometer was found to depend almost entirely upon the drift in the temperature of the photocathode. The variations in this have been minimized by a commercial Peltier-cooled housing for the photomultiplier whose internal temperature is stabilized with a thermistor control circuit. In some early experiments in which neither the room temperature nor that of the cooling water to the housing were controlled, it was found from a series of gold melts and freezes that the drift rate was less than 0·01%/hour. This includes statistical errors, which were often of the same order as the differences found.

The photon-counting pyrometer has been in operation for over a year, mainly for the investigation and calibration of standard lamps. As a first step in the programme described in the Introduction, it is being used to measure the temperature intervals between the freezing point of gold and those of copper and silver. The curves obtained with the first copper ingot indicated that some contamination, probably by oxygen, had occurred. The effect diminished after a number of melts, and the values obtained from the horizontal sections of the curves in the last set of 16 runs had a total spread of 0·04 °C. They showed no systematic differences between the melts and freezes, or between the wavelengths defined by the four interference filters. The mean, 1084·75 °C, is significantly lower than the most recent value (Righini et al 1972), but it should be stressed that this was probably due to the presence of impurities, and this temperature is only considered to be a possible lower limit to the correct value for the freezing point of copper. Further work with a second copper ingot and with a silver ingot is in progress.

References

Bonheure J A 1975 *Proc. CCT Meeting, 1974* to be published
Coates P B 1972 *J. Phys. D: Appl. Phys.* **5** 915–30

—— 1975a *J. Phys. E: Sci. Instrum.* **8** 189–93
—— 1975b *J. Phys. E: Sci. Instrum.* **8** 614–7
Cole M and Ryer D 1972 *Electro-Opt. Syst. Des.* **4** 16–9
Hall J A 1965 *Metrologia* **1** 140–58
Heusinkveld W A 1966 *Metrologia* **2** 61–71
Ingle J D and Crouch S R 1972 *Anal. Chem.* **44** 785–94
Quinn T J 1975 *Proc. CCT Meeting, 1974* to be published
Quinn T J and Chandler T R D 1972 *TMCSI* **4** part 1 295–309
Quinn T J, Chandler T R D and Chattle M V 1973 *Metrologia* **9** 44–6
Righini F, Rosso A and Ruffino G 1972 *High Temp.–High Press.* **4** 471–5

Photoelectric direct current standard pyrometers and their calibration at PTB

H Kunz and H J Kaufmann

Physikalisch-Technische Bundesanstalt, Braunschweig, W Germany

Abstract. Two recently developed pyrometer models based on the application of a special photocell with quasi-constant sensitivity are compared with a former model in which a photomultiplier is applied. The comparison shows some advantages of the new types such as maintenance of accuracy as well as simplicity of construction, calibration and on-line operation with a calculator. Furthermore, the description provides a basis for discussion about possible sources of systematic errors in standard pyrometry.

1. Introduction

The state-of-the-art of transferring the International Practical Temperature Scale (IPTS) to tungsten strip lamps with photoelectric standard pyrometers has been demonstrated by the last international intercomparison reported in Lee *et al* (1972). Although the agreement was particularly significant, and in general the differences revealed between the values of the four participating national laboratories were less than the sum of the uncertainties of the calibrations mentioned, small systematic differences could be recognized. This was not so surprising because the pyrometers and methods used were in many respects quite different.

In another intercomparison greater differences could be traced between the average of visual pyrometer calibrations of the six national laboratories (including PTB) and the photoelectric calibration of the PTB in 1962 (Kunz 1964, 1966). In connection with this observation it is remarkable that the pyrometer used by PTB is the same as the one described here under the notation PP60.

Some observations show that additional questions of still greater importance may arise when different types of pyrometer are used for the determination of the fixed points of the IPTS. The measurement conditions are then much more varied than in comparisons of strip lamp calibrations. Furthermore, small systematic errors must also be taken into consideration since they often reveal only the smaller part of their actual magnitude and the basis for a discussion of their causes is a full knowledge of the differences in instruments and methods used in such intercomparisons.

The current paper is therefore an attempt to fill a gap in information concerning the apparatus and methods used at PTB, which have not been described in English scientific literature until now. We consider the instrument used in the above-mentioned intercomparisons and two newly developed instruments, about which only short publications have appeared in the annual report of PTB (Kunz 1973) and which have been used recently in an International Intercomparison of Spectral Irradiance Scales (Ooba *et al* 1964). The main aim of the latter development was the on-line operation with a calculator.

This paper also represents the necessary completion of the theory of our method described in Kunz (1969).

2. The optical systems of the pyrometers

The optical systems to be compared are illustrated in figures 1, 2 and 3. Hereafter they will be distinguished by the notation PP60, PP70 and PP71; the two numerals indicate the year in which the photoelectric pyrometers were put into operation.

In PP60 a photomultiplier EMI 9558A (PMT) is used as radiation detector, but in

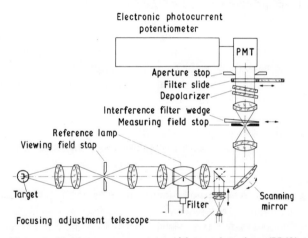

Figure 1. Multipurpose pyrometer with scanning mirror (PP60).

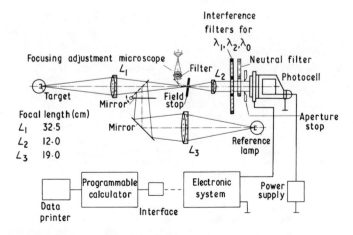

Figure 2. Primitive direct current pyrometer with flap mirror (PP70).

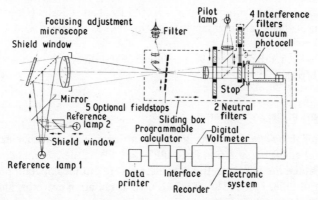

Figure 3. Direct current pyrometer with variable target distance (PP71).

PP70 and PP71 a special vacuum photocell is used. All detectors are to be considered as having a linear response yet only quasi-constant sensitivity. Therefore in all pyrometers at least one built-in reference lamp, ie a tungsten strip lamp, is provided. They are operated with only one constant current value, usually maintaining a radiance equal to the radiance $L_b(\lambda, T_{Au})$ of a black body at the gold point temperature. As the reference lamp is in the main light path, it allows control of the detector sensitivity. If the sensitivity varies slightly, it is readjusted by a potentiometer (see figure 4). The radiance of the target can thus be measured by the detector signal in terms of $L_b(\lambda, T_{Au})$, if the pyrometer is calibrated according to the procedure described in §4. The metrological conditions of this method have already been described in Kunz (1967, 1969). Extensive investigations concerning rotating sectored discs as radiation attenuators have also been discussed. The possibility of using these attenuators was one of the main reasons for retaining the integrating direct current method, the others being high measurement accuracy and simplicity of equipment as a whole.

In all pyrometers the first lens system is a special construction and the first lens itself, made of quartz glass, is not cemented to the second lens. This was done as a matter of precaution in order to reduce a supposed heat influence on the transmittance of the system when measurements were performed in front of a high-temperature furnace.

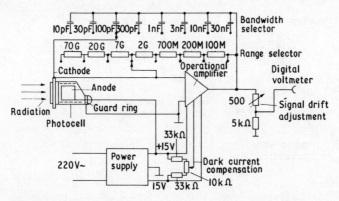

Figure 4. Electronic system of a pyrometer with photocell detector (PP71).

2.1. The special design of PP60 (figure 1)

This pyrometer is a slightly improved version of our preliminary optical arrangement of 1955 (Kunz 1955, 1956). Both of the schemes were originally determined by the idea of producing an objective version of the visually disappearing filament pyrometer.

The viewing field stop, the size of which is larger than the image of the measuring field stop, has been introduced to limit the scattered light flux in the successive light path. The image of the viewing field stop borders on the edge of the filament in the reference lamp, thus avoiding the disturbances of heat exchange between target and filament (Lee 1966). By operating the scanning mirror (Kunz 1956), the images of the viewing field stop and the filament on the measuring field stop can be quickly and precisely interchanged for sensitivity control. The polarization by this mirror was so strong that we had to use two glass plates as depolarizers. Needless to say as in all pyrometers, the measuring field stop − a mirror-like reflecting disc with a slit − and the interference filters are slightly tilted against the optical axis in order to avoid inter-reflections. The removable telescope allows focusing to be controlled with a flap mirror on the measuring field stop as well as in the filament plane and on the viewing field stop.

In PP60, the filter slide and the filter wedge could be used alternatively. Since the scanning mirror is provided with a moving coil such as in a galvanometer, it can be made to oscillate with frequencies up to 100 Hz, thus providing the possibility for operating PP60 in conjunction with the lock-in amplifier technique as an alternating light pyrometer. It is interesting to note that the scanning mirror also provides the possibility of scanning a temperature distribution in a horizontal line of the target, and of applying an objective focusing technique as described in Kunz and Lauterbach (1954, 1955). The latter is of special interest where measurements in the infrared wavelength region, and therefore the lower-temperature region, are concerned.

Nevertheless, in designing PP70 and PP71, these original intentions had to be dropped in favour of more simplicity in the optical and electronic equipment, without sacrificing measurement accuracy for temperatures above the silver point.

2.2. The special design of PP70 (figure 2)

As a consequence of our experiences with PP60 it was decided that a vacuum photocell would fit the conditions mentioned in §1 better than a photomultiplier. It was necessary to initiate a special development that would satisfy the following conditions: sealed-in guard ring electrode, very homogeneous tri-alkali photocathode with elevated red sensitivity, and low saturation voltage. These requirements were obtained by a photocell manufactured by Heimann GmbH†, the local homogeneity of its cathode was found to remain within ±1·5% of the central sensitivity.

To avoid the scattered light corrections − as is necessary with PP60, although there the plane windows of the reference lamps have anti-reflection coatings − the reference lamp was arranged in an optical by-pass.

† The authors wish to acknowledge Professor W Heimann and Dr E L Hoene for assistance in developing this photocell.

The operating mode of the pyrometer is adequately explained by the general system characteristics mentioned above, the following descriptions of the electronic system, and the designation of the separate parts in figure 2.

2.3. The special design of PP71 (figure 3)

The operating mode of this pyrometer is largely similar to that of PP70. The same focal lengths are used as for L_1 and L_2 in figure 2, L_1 usually being arranged to produce an image of unit magnification when using reference lamp 1. Additional optical systems are the built-in pilot lamp for facilitating the adjustment of the whole pyrometer in relation to all kinds of targets and reference lamp 2, which was introduced for operation with variable distances between the target and the first lens. With this special operating mode lamp 2 as well as the sliding box has to be adjusted.

For the discussion in §6, it is important to know that the mirror in front of the first lens is moved horizontally and both of the reference lamp light paths are arranged below the main optical axis.

3. The electronic systems

The block diagrams in figures 1, 2 and 3 show the main differences between the PP60 system and the PP70 and PP71 systems. When PP60 is used, corrections must be made for dark current and scattered light before achieving the photocurrent ratio for which a temperature value can be taken from a calculated table.

The use of a photocell together with a varactor bridge operational amplifier (Model 310 J/K of Analog Devices) for the photocurrent measurement means that the output signal of the amplifier can be fed directly into the chain of digital voltmeter, interface and calculator with built-in data printer, the latter printing the filter wavelength and the temperature when programmed for on-line operation.

Figure 4 shows that in the electronic system dark-current compensation and drift adjustment are provided. The same directions as described in Kunz (1967, 1969) must be followed when selecting the components for the range and bandwidth selector. Since the photocell must be operated only at currents below 3×10^{-9} A in order to avoid drift rates higher than $10^{-4}/10$ min, the cathode resistance must be changed in order to remain in the most favourable operation range of the amplifier, namely 0·1 to 10 V (compare §5). Here an optimum of linearity and signal-to-noise ratio of the order of 10^{-4} is guaranteed. Furthermore it is an advantage with respect to linearity that in the circuit shown in figure 4 the amplifier automatically maintains the cathode potential by current compensation.

4. The primary calibration

The whole calibration procedure can be divided into five steps; these are described in the following subsections.

4.1. The spectral measurements

For calculation of the pyrometer wavelength experimentally determined values of $s_{pr}(\lambda)$ and of $\tau_f(\lambda)$ are needed. $s_{pr}(\lambda)$ is the relative spectral sensitivity of the photoelectric pyrometer without the filter inserted in the light path and $\tau_f(\lambda)$ is the spectral transmittance of the pyrometer filter. It is an essential advantage of the described pyrometer systems that both of the spectral measurements can be done *in situ* since the pyrometer itself can be used as the requisite highly linear radiation detector. Therefore the measurement of the monochromator transmittance can also be avoided as will be shown later.

Since 1970 these spectroscopic measurements have been carried out by using the 3/4 Meter Jarrel–Ash Double Monochromator Model 25-105 as a spectral radiation source. After each exchange of the gratings, the counter readings of the wavelength drive mechanism are calibrated with a direct current neon discharge source by slowly recording the output signal of the pyrometer whereby this is used as a radiation detector of the spectrometer arrangement.

For the determination of $s_{pr}(\lambda)$ values, a pair of 295 grooves/mm gratings is inserted in the monochromator. For the $\tau_f(\lambda)$ measurements, these gratings are exchanged with a pair of 1180 grooves/mm gratings. In both cases, the differences between the spectral line value and the counter readings can be fitted by a straight line.

Within experimental error no systematic differences between the calibration curves taken just before and just after the $\tau_f(\lambda)$ measurements could be proved. The standard deviation for the single counter reading was calculated to be 0·005 nm. The procedure of measuring $s_{pr}(\lambda)$ is as follows.

An achromatic lens system produces an image of the monochromator exit slit on the pyrometer target field, the entrance slit being illuminated by a tungsten halogen lamp. The pyrometer signal caused by this image has to be related to the signal of a neutral radiation detector. A Hilger–Schwarz Thermopile Type FT 12.301 is considered to be sufficiently grey within the pass bands of the pyrometer filters. The comparison is achieved by alternately interposing this thermopile in the optical path at the point of the slit image.

The transmittance $\tau_f(\lambda)$ may be determined by recording the pyrometer signals with the filter inserted and then removed, but for gaining the highest possible accuracy point-by-point measurements were made with 0·2 nm wavelength intervals. Therefore both pyrometer signals at the same counter settings have to be taken and divided. In this procedure, the monochromator slits were set at an opening that gave a wavelength pass band of 0·15 nm. The entrance slit was illuminated by a gas-filled strip lamp.

4.2. The measurement data processing

Only a brief description can be given concerning the processing of the measurement data of $s_{pr}(\lambda)$ and $\tau_f(\lambda)$ by a Siemens Computer 4004. The details will be published elsewhere (Kunz and Steiner 1975). First an outline of the method for finding analytical functions for $s_{pr}(\lambda)$ and $\tau_f(\lambda)$ is described.

4.2.1. The approximations of the pyrometer sensitivity function and the filter transmittance function.
A seventh-order polynomial was fitted to the experimental points of $s_{pr}(\lambda)$ but a more complicated method was applied to the approximation of $\tau_f(\lambda)$, the function which predominantly affects the accuracy of realizing the IPTS. A thirtieth-order Fourier series expansion is considered to be of sufficient accuracy.

At these orders, the differences between the measurement points (40 with $s_{pr}(\lambda)$ and 380 with $\tau_f(\lambda)$) and the approximated function were almost random. The weighted standard deviation $\tau_f(\lambda)$ of the single value was calculated to be 0·1%. In nearly all cases, the observed deviations can mainly be explained by the inaccuracy of the monochromator drive.

4.2.2. The calculation of the mathematical pyrometer wavelength. The notation 'mathematical pyrometer wavelength' is used for emphasizing that here its significance is to be seen preferably in the calculation procedure for the direct current pyrometer. This is done despite the fact that its defining equation is the same as used by Kostkowski and Lee (1962) for their 'mean effective wavelength'. The expression 'effective wavelength' is often used in different but similar relations as in pyrometry, photometry and colorimetry. Therefore, we want to avoid it in a specific mathematical calibration procedure, but it should be applied to the calibration of temperature radiators which have a spectral radiance temperature different from their colour temperature. Further arguments will be given in detail elsewhere (Kunz and Steiner 1975).

Since the mathematical pyrometer wavelength is a function of the reference temperature and the temperature to be measured, we have to find firstly a relation of the wavelength to the pyrometer signal because the knowledge of the wavelength value is necessary for every calculation of an unknown temperature from the pyrometer signal.

The International Practical Temperature Scale of 1968 (IPTS-68) is defined above the freezing point of gold (gold point) by the relation

$$\frac{\partial L(\lambda, T)/\partial \lambda}{\partial L(\lambda, T_{Au})/\partial \lambda} = \frac{\exp[(c_2/\lambda T_{Au}) - 1]}{\exp[(c_2/\lambda T) - 1]} = \frac{L_{\lambda b}(\lambda, T)}{L_{\lambda b}(\lambda, T_{Au})}. \tag{1}$$

Conditions have been described (Kunz 1969) which allow the substitution of relation (1) by the following equations which are valid for photoelectric pyrometers with constant or at least quasi-constant linear detector sensitivity.

$$\frac{L_{\lambda b}(\lambda_{pmn}, T)}{L_{\lambda b}(\lambda_{pmn}, T_{Au})} = \frac{1}{\tau_s} \frac{s_p(\lambda_{pmn}, T)}{s_p(\lambda_{pmn}, T_{Au})} = \Phi_{pn} \tag{2}$$

and

$$\Phi_{pn} = \int_{\lambda_{nb}}^{\lambda_{ne}} s_{pr}(\lambda) \tau_{fn}(\lambda) L_{\lambda b}(\lambda, T) \, d\lambda \bigg/ \int_{\lambda_{nb}}^{\lambda_{ne}} s_{pr}(\lambda) \tau_{fn}(\lambda) L_{\lambda b}(\lambda, T_{Au}) \, d\lambda \tag{3}$$

where τ_s is the transmittance of a neutral attenuator (rotating sectored disc or filter), $s_p(\lambda_{pmn}, T)$ is the pyrometer signal when measuring a black body temperature T with filter n inserted, Φ_{pn} is the calculated pyrometer signal ratio (without attenuator) at a temperature T related to T_{Au}, $\Phi_{pn} = \Phi_p(\lambda_{pmn}, T, T_{Au})$, λ_{pmn} is the mathematical pyrometer wavelength with filter n inserted, λ_{nb} is the wavelength at the beginning of the wavelength region of filter n (usually $\tau(\lambda) \leq 3 \times 10^{-5}$) and λ_{ne} is the wavelength at the end of the region of filter n ($\tau(\lambda) \leq 3 \times 10^{-5}$).

From equations (1), (2) and (3) the following expression for the wavelength λ_{pmn} can be derived:

$$\frac{1}{\lambda_{pmn}} \approx \frac{T}{c_2} \ln\left[1 + \exp\left(\frac{c_2}{\lambda_{pmn} T_{Au}} - \ln \Phi_{pn}\right)\right]. \tag{4}$$

Taking the wavelength value of the centre of the filter transmittance curve as a first approximation to be inserted in the right-hand side of (4), the calibration leads to a second approximated wavelength for a given pair of T and Φ_{pn}. Usually the third approximation is of sufficient accuracy already.

Φ_{pn} is calculated by numerical integration of (3). The calculation of Φ_{pn} and λ_{pmn}, respectively, for the two temperatures at the temperature range limits is usually sufficient for approximating the function by a straight line as calculations have shown. Then the constants of the linear function are determined graphically and are stored in the calculator. From this, the data processing by the computer, a procedure which provides the basis for the programming of the calculator for the on-line operation, is finished. Furthermore only two measurement values are needed: the pyrometer signal at the gold point $s_p(\lambda_{pmn}, T_{Au})$; and the transmittance τ_s of the attenuator.

4.3. The calibration at the gold point

As a gold stabilized black body, two different types of cavities are used, one of them made of ceramic (Marquardtmasse), the other of graphite. The ceramic type has been taken for reasons of convenience during routine work. The cavity length should be at least twice as great as described before (Tingwaldt and Kunz 1958) thus having a theoretical emissivity of about 0·9997. Comparison measurements have shown that an influence of the wall thickness is less critical than the geometry of the cavity. The enlargement of the ceramic cavity has made it necessary to apply braces similar to those used by Lee (1966) and Quinn and Chandler (1972). The total uncertainty caused by deviation of the cavities from ideal conditions is estimated to be less than 0·03 K. The comparison between the cavity radiance at the gold point and the radiance of the reference lamp gives a signal ratio of $s_p(\lambda_{pmn}, T_{Au})/s_p(\lambda_{pmn}, T_r) = s_{pAu}/s_{pr}$. The reference lamp current can be adjusted to give a ratio of nearly 1:1. However, it has proved to be useful that, working in the higher-temperature region with a ratio of 1:10 and in the lower region with 10:1, improved signal-to-noise ratios and therefore a higher degree of accuracy for the radiance ratio measurements can be obtained. When the s_{pAu}/s_{pr} value is determined at the gold point then (for any other measurement) if the pyrometer sensitivity has varied, it can be readjusted by the signal drift adjustment potentiometer (figure 4) in order to provide the same signal s_{pr} as during the calibration at the gold point. Thus an invariant sensitivity can be maintained for all measurements. Needless to say, the lamp characteristic must remain invariable.

4.4. The measurement of the radiation attenuator transmittance

Both types of attenuator, the neutral filter as well as the rotating sectored disc, can be measured *in situ,* the filter by using the built-in reference lamp and the sectored discs by using a lamp outside the pyrometer. A sector is preferred for representing a spectral totally neutral attenuator and for protecting the pyrometer optics against heat influences when measuring in front of high-temperature furnaces with large openings.

4.5. The programming of a calculator for pyrometer on-line operation

The basic formula for programming is derived from equation (4) by eliminating T; that is

$$T = \frac{c_2}{\lambda_{pmn}} \bigg/ \ln\left[1 + \exp\left(\frac{c_2}{\lambda_{pmn}T_{Au}} - \ln \Phi_{pn}\right)\right]. \tag{5}$$

Equations (2), (4) and (5) show that only the values of c_2 and T_{Au}, the two determined constants to the linear wavelength function, and the experimental values of $1/s_{pAu}$ and $1/\tau_s$ have to be stored. Then, according to equation (2), every signal value $s_p(T)$ has to be multiplied by $1/s_{pAu}$ and $1/\tau_s$ respectively to obtain the ratio Φ_{pn}. Secondly, from the programmed wavelength function (4), λ_{pmn} is calculated before the calculation of T according to equation (5). The Wang Model 600 is used as a calculator. The program starts the calculation by using either a single signal value or by using a previously averaged signal series. This means that a variation of the bandwidth, and thus of the signal-to-noise ratio, can also be done by the calculator. At the end of the calculation procedure, values of the temperature and the wavelength λ_{pmn} are printed out. It is a matter of course that the calculator should also be programmed for ratio temperature measurement as well, especially if two significantly spaced wavelength filters are used.

5. Some instrumental parameters of PP71

Only a limited amount of typical data can be specified here (see tables 1–4).

Measurement field diameter	0·73 mm (others <6 mm)
Relative aperture ratio	1:10·5
Target distance	≧750 mm

Table 1.

	Wavelength (nm)		
	540	660	702
Bandwidth (nm)	9·0	9·0	9·0
Cathode current† at the gold point (A)	0·6 × 10⁻¹¹	4 × 10⁻¹¹	4 × 10⁻¹¹
Cathode temperature coefficient (%/K)	−0·066	−0·073	−0·12
Resistor temperature coefficient (%/K)	−0·06 to −0·12		

† After a warm-up time of about one hour, the sensitivity drift generally remains below $10^{-4}/10$ min in the high-temperature region. But since recovery occurs in darkness no remarkable sensitivity change — lower than 0·2% — could be observed after one year of interrupted operation.

Table 2. Signal reproducibility after various position changes†.

At change of position of:	Signal reproducibility
Reference lamp	± 0.0005%
Fieldstop	± 0.001%
Filter	± 0.001%

† Size of source correction for target of 1.6 mm width and 50 mm length = ± 0.12%.

Table 3. Range parameters above the gold point at a wavelength of 660 nm and a time constant $R_c C_c$ of 1 second ($R_c = 3 \times 10^9$ Ω).

	Temperature (K)		
	1337	1880	
Attenuators τ_s	–	–	0.1
Average cathode current (A)	4×10^{-11}	3×10^{-9}	3×10^{-10}
Relative total noise	6×10^{-5}	2×10^{-5}	2×10^{-5}
Corresponding temperature resolution (mK)	5	3	3
Estimated uncertainty of IPTS realization (K)	0.04	0.1	

	Temperature (K)			
	2300		2856	3800
Attenuators (τ_s)	0.1	0.002†	0.002	0.002
Average cathode current (A)	4×10^{-9}	8×10^{-11}	5×10^{-10}	5×10^{-9}
Relative total noise	2×10^{-5}	4×10^{-5}	3×10^{-5}	2×10^{-5}
Corresponding temperature resolution (mK)	5	10	100	120
Estimated uncertainty‡ of IPTS realization (K)	0.25		0.4	0.7

† Composed of 0.1 and 0.02.
‡ This uncertainty has until now been caused (mainly) by the uncertainty of the wavelength determination of 0.05 nm, which results from the radiance inhomogeneities in the monochromator exit pupil and the local spectral transmittance inhomogeneities within the effective filter diameter. The influence of both will be essentially reduced by new high-quality filters.

Table 4. Temperature resolution below the gold point. λ = 660 nm, measurement field diameter 0.73 mm, time constant 3 s†.

	Temperature (K)		
	1235	1100	1000
I_c (A)	1×10^{-11}	1×10^{-12}	1×10^{-13}
R_c (Ω)	3×10^{10}	10^{11}	10^{11}
Relative total noise	7×10^{-5}	5×10^{-4}	1×10^{-2}
Corresponding temperature resolution (K)	0.004	0.05	0.5

† At λ = 702 nm, measurement field diameter 2 mm, $R_c = 10^{11}$ Ω, and a time constant of 10 s, a temperature resolution of 0.01 K at T = 1000 K may still be expected.

6. Conclusions

The description of these pyrometers has been presented as a report on experiences which may be of some interest in the future to designers of pyrometers by other institutes. The simplicity of achieving a primary calibration for on-line operation, the simplicity of the construction of PP70, and the flexibility of PP71 without making any remarkable sacrifice in measurement accuracy should be considered as special advantages.

Furthermore the comparison of the three pyrometer types gives rise to the following statements:

(i) Each of these and many other types of pyrometer can be used with the substitution method, as described by Quinn and Ford (1969), when the pyrometers or radiators are mounted so that they can be moved.
(ii) A built-in reference lamp has the advantages of a fast sensitivity control and a comparison that needs no adjustment operation and thus provides a higher degree of reproducibility by use of high precision slides.
(iii) The unfavourable experiences of Lee (1966) with built-in reference lamps provide no proof that the disturbing drift cannot be avoided. Our experiences with PP60 were much better than those of Lee. The reasons for this are to be seen in the fact that the reference lamp is not moved and the mirror which is to be moved is fully illuminated by the comparison light and by the target light. Furthermore, all the critical surfaces are arranged vertically.
(iv) An arrangement such as that of PP70, with vertical optical surfaces only, should be preferred to that of PP71, in which the bulb of lamp 1 is a dust precipitator. The horizontal shield window was arranged to get a surface which could be cleaned more easily through a covered slit of the housing if dust settled there. It is supposed that dust-settling is increased by convection in the vertical light path. Therefore the other shield window was arranged so that the mirror housing was dust-tight.
(v) A dust-tight housing and a mounting of the critical parts of a pyrometer in a dust-free atmosphere is recommended. Precautions for radiation screening must be taken in order to avoid heating of the housing and therefore the mirror. A variation of the mirror reflection with a slow response was observed, the temperature coefficient being $+0.03\%/K$.
(vi) When the sector is inserted in the light path just in front of the first lamp it must have an opening which is great enough to avoid small diffraction influences depending more or less upon the size and radiance structure of the target.

Acknowledgments

The authors would like to thank Mrs D Helwig and Mr R Todtenhaupt who assisted in the measurement and the evaluation, Mr H Breite of the Main Workshop who constructed most of two pyrometers and Mr O Steiner for his continued assistance in programming. The authors also wish to acknowledge the invaluable advice and encouragement given by Professor U Schley and Professor W Thomas.

References

Kostkowski H J and Lee R D 1962 *TMCSI* **3** part 1 449
Kunz H 1955 *Wiss. Abh. Phys.-Tech. Bundesanst.* **7** 28
—— 1956 *Wiss. Abh. Phys.-Tech. Bundesanst.* **8** 29
—— 1964 *Proc. CCT 7th Sess.* 79
—— 1966 *Verein Deutscher Ingenieure Ber.* No. 112
—— 1967 *Thesis* Technische Hochschule, Hanover
—— 1969 *Metrologia* **5** 88
—— 1973 *Jahresber. Phys.-Tech. Bundesanst.* 152
Kunz H and Lauterbach 1954 *Deutsches Bundespatent* 961767
—— 1955 *Diplomarbeit (Lauterbach)* Technische Hochschule, Braunschweig
Kunz H and Steiner 1975 to be published
Lee R D 1966 *Metrologia* **2** 150
Lee R D *et al* 1972 *TMCSI* **4** part 1 377
Ooba N 1974 *Interim Report on the International Intercomparison of Spectral Irradiance Scale, Tokyo* CIE-TC 1.2
Quinn T J and Chandler 1972 *TMCSI* **4** part 1 295
Quinn T J and Ford M C 1969 *Proc. R. Soc.* A **312** 31
Tingwaldt C and Kunz H 1958 *Optik* **15** 333

Standard lamps and their calibration

J W Andrews and P B Coates
Division of Quantum Metrology, National Physical Laboratory, Teddington, Middlesex, England

Abstract. The suitability of the various forms of the calibration curves fitted to the measured data are considered. The residuals obtained in the fitting process indicate measurement errors, while the coefficients obtained for lamps of a given type provide an indication of the reliability of the calibration and the behaviour of the lamp.

1. Introduction

The prime function of a lamp calibration service is to provide for the user accurate values of the radiance temperature of a lamp when it is observed under given conditions. This information may be presented in tabular form or as a polynomial, in terms of an electrical variable which can be easily realized and accurately measured. In practice the current through the lamp is normally given.

At NPL the lamp is calibrated against a standard lamp of a similar type. The procedure is to set the current through the standard to give an appropriate radiance temperature, and the current through the uncalibrated lamp is then adjusted until its intensity, and hence its radiance temperature, is the same as that of the standard. This current is recorded and the procedure repeated at a number of points throughout the operating range of the lamp. The calibration in its final form is then obtained by fitting a chosen analytical expression through these points. This also serves to smooth out random experimental errors introduced during the calibration.

This paper discusses the factors governing the selection of the analytic expression, and the additional information that may be obtained during the curve-fitting process.

2. The energy balance model

The suitability of analytical expressions based upon physical models of lamp behaviour were first considered. These have the potential advantage that variations in the performance of lamps of a given type might then be related to the conditions of manufacture or operation, and this in the long run could lead to more stable and reliable lamps.

Russell and Schofield (1960) have discussed in considerable detail the factors affecting the characteristics of strip filament lamps, although they deal mainly with the temperature distribution along the filament. This paper also covers the earlier work of Ribaud and Nikitine (1927). It was found, however, that simple expressions derived from this work did not follow the experimental curves to within the accuracy required for the calibration service. If extra parameters were introduced in order to reduce these systematic differences, then not only was the physical significance of the parameters lost but the computing time required for the least-squares fit to the data became excessive.

A simple and rather more successful model was also developed. It will be described in this section applied to the vacuum black body lamp designed by Quinn and Barber (1967). The lamp (figure 1) was taken to consist of a tube at a constant temperature,

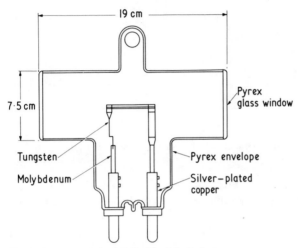

Figure 1. Construction of the black body lamp.

T, equal to the radiance temperature observed along the axis of the tube. This was connected by two low-resistance supports to a massive base, which is at a controlled temperature T_0. The energy entering the lamp is measured from the potential V across the lamp and the current I through it. Most of this energy is lost by radiation from the tube. It may be noted that, although the glass envelope is not transparent at all wavelengths, its temperature remains low and little energy is returned to the tube, which has a small area for adsorption, by re-radiation. The remaining energy is conducted through the supports to the cooling water flowing through the base. Since this is small at radiance temperatures above the gold point, it was assumed that this could be approximated by a linear term $C(T - T_0)$. This 'energy balance' model therefore led to the equation

$$VI = \sigma A T^4 \epsilon(T) + C(T - T_0)$$

where σ is Stefan's constant, A the effective radiating area of the tube and $\epsilon(T)$ the hemispherical total emissivity.

This equation was fitted with an unweighted least-squares program to a set of 12 measurements of V and I taken at radiance temperatures between the gold and platinum points, using a cubic polynomial to describe $\epsilon(T)$. The mean residual was equivalent to about 0·2 °C, and the value for C (4·2 × 10^{-3} W K^{-1}) agreed well with direct measurements of the heat transferred to the cooling water. The emissivity polynomial

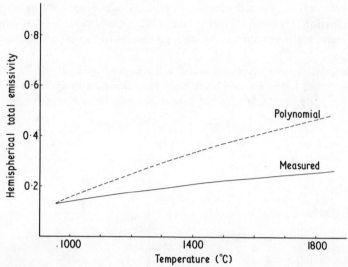

Figure 2. Comparison of emissivity polynomial with measured values.

is shown in figure 2, compared to direct measurements (Touloukian and De Witt 1970). Although the agreement is initially very good, the curves diverge as the temperature increases. This was mainly due to the fact that the black body tube supports, which are made of thin sheet, become very hot at high radiance temperatures, thereby increasing the effective area A. In addition, it should be noted that the radiance temperature T is only approximately equal to the real temperature of the tube, and that the black body radiation from the open ends of the tube was neglected.

3. Chebyshev curve fitting

The expression given above is unsuitable for practical applications, since it involves the measurement of two electrical variables V and I, adding to the difficulty of realization of the calibration for the user. Also, because the resistances of all types of lamps are small, typically $0 \cdot 1\ \Omega$, the effects of lead and contact resistances make voltage measurements much less reliable than those of current. The search for a physically significant expression has therefore been abandoned for the moment, and the choice made on the basis of practical criteria. Ideally, a function was required which not only provided a very accurate least-squares fit to the data, but also possessed a minimum of adjustable parameters, especially if it was nonlinear in form. The problem of computer time is not in general serious, since relatively few data points are fitted for each lamp and the number of lamps calibrated at any one time is not large.

Following the accepted strategy in such a case, an unweighted least-squares fit with a low-order polynomial was first tried. To avoid a possible loss of numerical accuracy, the current I through the lamp was not expressed in terms of the radiance temperature T, but of a reduced variable x, defined by

$$x = (2T - T_{\max} - T_{\min})/(T_{\max} - T_{\min})$$

where T_{min}, T_{max} are the minimum and maximum temperatures at which a lamp of a particular type might be calibrated. Although any polynomial in this variable would give the same fit, a convenient program† was fortunately already available in NPL. With this, the lamp current was written as a sum of Chebyshev polynomials

$$I = 0 \cdot 5 a_0 T_0(x) + a_1 T_1(x) + a_2 T_2(x) + \ldots$$

where

$$T_0(x) = 1, \qquad T_1(x) = x, \qquad T_2(x) = 2x^2 - 1,$$

$$T_3(x) = 4x^3 - 3x, \qquad T_4(x) = 8x^4 - 8x^2 + 1.$$

With a simple driver program, least-squares fits to the data could be made with degree i, which increased from zero up to a specified maximum k. For each degree, the $(i + 1)$ Chebyshev coefficients and the root mean square residual s_i were printed. (The residual is the difference between the measured current and that calculated from the fitted polynomial.) For the maximum degree k, a table of the residuals at each data point was provided. The program could also be instructed to construct a calibration table of currents at a given temperature interval over the operating range of the lamp.

Most of the lamps calibrated at NPL are of the black body, high-stability and cylindrical-envelope types. These are available either evacuated, or gas-filled, in order to increase the working temperatures above 1700 °C. The results accumulated over several years have been analysed for each type of lamp.

The maximum degree k was determined from the behaviour of the root mean square residual s_i obtained with typical sets of calibration data. It was found that, as i increased, s_i fell rapidly at first, and then at degree 3 for vacuum lamps or 4 for gas-filled lamps reached a nearly constant value. This was taken to indicate the closest polynomial approximation justified by the accuracy of the data. For the sake of uniformity, all lamp calibration data has been fitted with k equal to 4. Mean values of the Chebyshev coefficients a_j for each type of lamp are shown in table 1. It will be seen that the coefficients decrease rapidly in size; since Chebyshev polynomials are restricted to a range of values between -1 and $+1$, this means that the contribution of the higher terms to the current is small over the whole range from T_{min} to T_{max}. An additional advantage is that it is not necessary to retain large numbers of significant figures in the coefficients. Each may be reduced to contain the same number of decimal places, equal to the required accuracy in the sum with an extra digit to guard against rounding errors.

All information produced by the curve-fitting process has been found to be of value. First, a large residual at a single datum point indicates in general that an operator error has occurred, for example an instrument may have been misread, or a number incorrectly copied into the data file. If the source of the error is not found, that point in the calibration is repeated. It was also found that the root mean square residual for a given lamp type was roughly constant. This is not a measure of the accuracy of the

† M G Cox and J G Hayes *Curve fitting: A guide and suite of programs for the non-specialist user* NPL Report NAC 26.

Table 1. Chebyshev coefficients and standard deviations (GF = gas-filled).

Lamp type		a_0	a_1	a_2	a_3	a_4	Lamps tested	RMS residual (°C)
Black body	Vac	42.15	9.25	0.50	−0.05	0.003	29	0.18
		4.4 (−)	4.8 (1.0)	4.5 (1.5)	7.4 (4.3)	150 (146)		
	GF	87.11	17.02	0.73	−0.03	−0.02	25	0.52
		4.4 (−)	4.8 (1.3)	8.9 (6.9)	62 (61)	120 (118)		
High stability	Vac	16.37	3.91	0.24	−0.02	0.005	11	0.11
		7.3 (−)	7.7 (0.5)	6.0 (2.6)	10 (11)	28 (28)		
	GF	28.84	4.34	0.20	0.001	−0.003	4	0.41
		2.8 (−)	2.8 (0.5)	2.9 (2.8)	400 (113)	90 (90)		
Cylindrical envelope	Vac	11.65	3.64	0.58	−0.07	−0.01	22	0.66
		5.3 (−)	3.8 (1.9)	7.7 (3.3)	13 (12)	67 (50)		
	GF	28.97	6.60	0.56	−0.03	0.008	20	0.36
		3.7 (−)	4.3 (1.1)	8.5 (6.0)	41 (41)	104 (103)		

curve fitting, but represents the random errors of calibration. Mean values for these are given for each type of lamp in the last column of table 1, converted into an equivalent uncertainty in temperature at the midpoint of the recommended operating range of the lamp. It will be seen that the uncertainty for vacuum lamps is much less than that for the gas-filled types. This is partly due to the higher operating temperatures of the latter, and partly to the presence of varying convection currents in the heated gas. (The high value recorded for the vacuum cylindrical envelope type is artificial, and arose from the fact that the calibration range of these lamps extends down to 700 °C. Until recently, the scale below 1064 °C was based on measurements made with a disappearing filament pyrometer, and was therefore much less precise than the scale above this temperature.) The larger residuals from the black body lamps, in comparison with the corresponding high-stability types, are thought to be real and due to the large angular variation of intensity from these lamps. A root mean square residual from a particular lamp calibration significantly higher than the mean for that type is taken to indicate that either the comparator or the lamp was unstable at the time of the calibration. If the residual is particularly high the calibration may be partly or wholly repeated to investigate the origin of the instability.

In addition, the Chebyshev coefficients a_j provide a useful source of information. For a given lamp type, it was found that the coefficients were very reproducible from lamp to lamp. The figures to the left below the coefficients in the table are the standard deviations, expressed as a percentage of the mean coefficient of the distribution of values obtained. The narrow spread of values is remarkable, especially when it is considered that the third term, for example, represents at the most a few per cent of the total lamp current. The variance is largest in the last two coefficients, since these are often not much larger than the random errors in the measurements. Moreover, on further analysis it became clear that the coefficients were not independent. If they were divided by a_0, the spread of the values of the ratio a_j/a_0 was considerably less than that of the coefficients alone. The standard deviations of the ratios are given in parentheses to the right beneath the coefficients. Other manipulations of the coefficients may be performed to further reduce the spread. However, none of these have much effect upon the variances for the last two coefficients, since they are mainly determined by experimental errors.

The significance of the reproducibility of the coefficients for a given lamp type may be illustrated by the following points. Without making any measurements at all upon a vacuum black body lamp, for example, a calibration accurate to 200 °C with 90% certainty can be given for its complete recommended operating range. A single calibration point reduces the error to 10 °C, while two, assuming that they are made with the usual accuracy achieved with the comparator system and that the standard lamp is itself accurately calibrated, produces a calibration accurate to within 1 °C. For other lamp types the errors are similar in magnitude but larger. Unfortunately this does not mean that the lamp calibration procedure may now be performed more rapidly. Much time is taken by the ageing and stabilization of new lamps, and in aligning them on the comparator. Reducing the number of data points from 12 to 2 saves only a small fraction of the overall time.

The similarity of the Chebyshev coefficients is useful, however, as a further check upon the accuracy of the calibration. In particular, there appears to be a correlation between significant deviations from the mean coefficients or ratios and the presence of unsatisfactory characteristics in the lamp — instability and sensitivity to mechanical shocks etc. This correlation is not yet well established since few faulty lamps are found, but it is being further investigated by checking all lamps in which significant deviations are noted.

4. Errors in lamp calibrations

In conclusion, the source of some of the errors in the calibration and use of these lamps will be briefly described. Since all lamps are calibrated against standard lamps at equal intensities, the error introduced in the comparison is likely to be small. However, indirect errors, which may be overlooked, arise in the calibration of the standard lamps or in subsequent use of the calibrated lamps, because these may involve instruments with characteristics different from those of the comparator. The transfer standard lamps are calibrated on the NPL photoelectric pyrometer by comparison with a gold-point black body. Since it is difficult to eliminate entirely differences in effective aperture and angular sensitivity between the comparator and the pyrometer it is essential to use a lamp whose output does not vary rapidly with angle. For this reason high-stability strip lamps are more suitable in this application than black body lamps. In addition a correction for the effective wavelength of the pyrometer must be made. On the whole, the calibration of a high-stability or black body lamp is more stable than is often thought. While the drift rate at the maximum operating temperature of the lamp is normally quoted, at lower temperatures this drift is often insignificant. A standard black body lamp, V6, has been in use with the photoelectric pyrometer for about six years. It is often left running at the gold point for months at a time, but its calibration has altered over this period by less than 1 °C at that temperature.

One source of error which becomes significant at temperatures below 1000 °C is the dependence upon the base temperature. While the coefficient for this may be neglected for vacuum high-stability and black body lamps above 1200 °C it increases rapidly below this, and may reach a level of 0·5 °C/ °C at 700 °C.

All pyrometric lamps must be operated using a DC current and, as reversal of this current has the effect of displacing the temperature distribution in the filament, the operating polarity must always be specified. In the long term, polarity reversal can change the calibration of a strip lamp by its effect upon the surface structure of the tungsten (Quinn 1965). In addition, it is important to focus upon the correct area of the filament from the direction specified in the calibration report (Barber 1946). With black body lamps, the calibration is strictly only valid at the optical aperture (f/11) given.

When a lamp is moved or transported some strain may be introduced into the filament. This is particularly true of black body lamps, but it has been found that any errors introduced could be eliminated by running the lamp at its maximum operating temperature for 1 hour before use. After this initial period, for temperatures above 1000 °C, the lamp should attain an equilibrium temperature within 15 minutes. Below 1000 °C, 30 minutes or longer may be required depending on the lamp type.

References

Barber C R 1946 *J. Sci. Instrum.* **23** 238
Quinn T J 1965 *Br. J. Appl. Phys.* **16** 973-80
Quinn T J and Barber C R 1967 *Metrologia* **3** 19
Ribaud G and Nikitine S 1927 *Ann. Phys. Paris* **7** 5-34
Russell D C and Schofield F H 1960 *Phil. Trans. R. Soc.* A **252** 463-98
Touloukian Y S and De Witt D P 1970 *Thermophysical Properties of Matter* (New York: Plenum) **7** 776-7

Increasing precision in two-colour pyrometry

G Ruffino
Leeds and Northrup, Torino, Italy

Abstract. Two-colour pyrometry has been affected by two main drawbacks. First, lack of resolution, as compared with monochromatic pyrometry, which is particularly important when the two wavelengths are close to each other. Second, lack of reproducibility due to the influence of ambient temperature on the effective wavelengths.

A design of an improved two-colour pyrometer is presented. It is based on a rigorous definition of the effective wavelengths, which clarifies their dependence from source temperature. This definition allows a precise calibration of the instrument and leads to a fitting equation for the calibration points.

Two main advantages of two-colour pyrometry are presented. First, the possibility of measuring the temperature of targets approaching grey-body conditions or with emissivity which is a well defined function of target temperature. Second, reduced dependence, as compared with the monochromatic pyrometer from target temperature non-uniformity and foreign matter obstructing the optical path between target and instrument.

1. Introduction

Two-colour pyrometry is theoretically based on the measurement of the spectral radiance ratio at two wavelengths, according to the expression based on Planck's law:

$$Q = \frac{\lambda_1^{-5} \exp[(C_2/\lambda_1 T) - 1]}{\lambda_2^{-5} \exp[(C_2/\lambda_2 T) - 1]} \quad (1)$$

which may be simplified, according to Wien's approximation, to the form:

$$Q = \left(\frac{\lambda_2}{\lambda_1}\right)^5 \exp\left(\frac{C_2}{\Lambda T}\right) \quad (2)$$

where Λ, called the equivalent wavelength, is given by the expression:

$$\frac{1}{\Lambda} = \frac{1}{\lambda_1} - \frac{1}{\lambda_2}. \quad (3)$$

From equations (2) and (3) we deduce the temperature resolution and the influence of wavelength imprecision on temperature imprecision of this kind of instrument.

1.1. Temperature resolution

If we derive T with respect to Q in equation (2) and substitute differentials with small increments we get the temperature resolution:

$$|\Delta T| = \frac{T^2}{C_2 \Lambda} \left|\frac{\Delta Q}{Q}\right| = A \left|\frac{\Delta Q}{Q}\right|. \quad (4)$$

The parameter A is inversely proportional to Λ. From equation (3) it is apparent that $\Lambda > \lambda_1$ and $\Lambda > \lambda_2$ and, for practical separation of the two wavelengths, it is much bigger than both of them. It follows that the temperature resolution of two-colour pyrometry is much smaller than in monochromatic pyrometry operating in the same band or, in other words, a given temperature precision exacts much higher precision in radiation flux ratio measurements.

1.2. Wavelength precision required by two-colour pyrometer

The derivative of T with respect to λ_1 and λ_2 in equations (2) and (3) is:

$$\frac{\partial T}{\partial \lambda_1} = -\frac{T(C_2/\lambda_1 - 5T)}{C_2(1 - \lambda_1/\lambda_2)} = D$$

$$\frac{\partial T}{\partial \lambda_2} = \frac{T(C_2/\lambda_2 - 5T)}{C_2(\lambda_2/\lambda_1 - 1)} = E.$$

(5)

Substituting differentials with small increments, we get the imprecisions ΔT_1 and ΔT_2 caused by wavelength imprecisions $\Delta\lambda_1$ and $\Delta\lambda_2$. If wavelength errors are accidental, then $\Delta\lambda_1 = \Delta\lambda_2 = \Delta\lambda$, with random sign. Therefore the temperature error is:

$$\Delta T = (D^2 + E^2)^{1/2} \Delta\lambda = C\Delta\lambda. \tag{6}$$

For λ_1 and λ_2 included in the visible and near-infrared region of the spectrum, with a spacing of 150 nm (which is usual in two-colour pyrometry), the parameter C ranges from 4700 K μm^{-1} to 7700 K μm^{-1} in the temperature range from 600 °C to 1600 °C. It is apparent that stability of temperature reading of 1 K requires wavelength stability of the order of 0·1 nm.

A third problem arises from the need of using finite bandwidths instead of single wavelengths in order to attain usable signal-to-noise ratios. This means that the simple equation (2) is of no practical use. As a consequence, all commercial pyrometers, up to the present date, were simply calibrated in terms of instrument signal versus temperature. The above analysis gives an account of the tremendous variation in readings of such instruments.

2. Effective wavelength

Two-colour pyrometers use photoelectric detectors which yield, for each band, a signal of the form:

$$S = K \int_0^\infty \lambda^{-5} \exp(C_2/\lambda T)\, \tau(\lambda)\, \sigma(\lambda)\, d\lambda \tag{7}$$

where: T is the source temperature; $\tau(\lambda)$ is the transmittance of the optical system, including filter; $\sigma(\lambda)$ is the detector spectral responsivity; and K is a parameter which includes the first Planck's constant and a geometrical factor. Within the limits of detector linearity K is a constant.

Under this latter hypothesis, the radiation flux ratio is equal to the signal ratio, if both fluxes are detected by the same sensor. Then the ratio is:

$$Q = \frac{\int_2 \lambda^{-5} \exp(-C_2/\lambda T) \tau(\lambda) \sigma(\lambda) \, d\lambda}{\int_1 \lambda^{-5} \exp(-C_2/\lambda T) \tau(\lambda) \sigma(\lambda) \, d\lambda}. \tag{8}$$

The indexes 1 and 2 refer to integration limits which include the two bands.

Now we define (Righini *et al* 1972) the effective wavelength λ_e as the one which satisfies the equation:

$$\int_0^\infty \lambda^{-5} \exp(-C_2/\lambda T) \tau(\lambda) \sigma(\lambda) \, d\lambda = \tau(\lambda_e) \sigma(\lambda_e) \lambda_e^{-5} \exp(-C_2/\lambda_e T). \tag{9}$$

This procedure is similar to the one which leads to the definition of effective wavelength in monochromatic pyrometry (Kostkowski and Lee 1962).

Equation (9) has a solution provided we refrain from imposing any further condition on λ_e:

$$\lambda_e = \frac{\int_0^\infty \lambda^{-5} \exp(-C_2/\lambda T) \tau\sigma \, d\lambda}{\int_0^\infty [\lambda^{-5} \exp(-C_2/\lambda T) \tau\sigma/\lambda] \, d\lambda}. \tag{10}$$

In practice, the integration limits are the ends of the range within which $\tau(\lambda) \neq 0$. This wavelength range is the band of the pyrometer filter.

The effective wavelength is a function of source temperature. Its evaluation can be performed by means of an automatic data acquisition system (Pasta *et al* 1975).

A good measurement procedure consists of the following steps.

(i) A strip lamp at high temperature T_0, which does not need to be known, but must be constant (about 2000 °C), illuminates the entrance slit of a monochromator. The pyrometer is focused on its exit slit. We sort out a pyrometer signal corresponding to one of its filters. The terminals of this signal are connected to an automatic data acquisition system. A computer drives the wavelength drum of the monochromator and stores the signal which is permanently recorded on a punched strip. The pyrometer signal is

$$S = \epsilon(\lambda, T_0) \mathcal{R}(\lambda, T_0) \tau_M(\lambda) \tau_P(\lambda) \sigma(\lambda) \tag{11}$$

in which ϵ is the source emissivity, $\mathcal{R}$ is its spectral radiance and τ_M and τ_P are the transmittances of the monochromator and of the pyrometer.

(ii) The same band is scanned using a thermal detector (such as a thermopile), instead of the pyrometer, with the source at the same temperature T_0. The thermopile signal is:

$$S' = k\epsilon(\lambda, T_0) \mathcal{R}(\lambda, T_0) \tau_M(\lambda) \tag{12}$$

where k, constant with wavelength, is the responsivity. These data are fitted in a series of polynomials, which are stored in the computer.

(iii) The computer calculates a function Φ which is the ratio S/S':

$$\Phi(\lambda) = K\tau_P(\lambda)\sigma(\lambda). \tag{13}$$

This quantity is a function of wavelength only and is peculiar to any individual pyrometer.
(iv) The effective wavelength at any temperature can be calculated through the formula:

$$\lambda_e = \frac{\int_0^\infty \Phi(\lambda)\mathcal{R}(\lambda,T)\,d\lambda}{\int_0^\infty [\Phi(\lambda)\mathcal{R}(\lambda,T)/\lambda]\,d\lambda} \tag{14}$$

by means of numerical integration.

The advantage of this method consists in the fact that the function $\Phi(\lambda)$ is calculated at high source temperature with a convenient signal-to-noise ratio. Besides, temperature plays no role in the determination of $\Phi(\lambda)$. In this way, the computation of λ_e can be extended to low temperatures where the direct experimental measurement is impractical. Another advantage of high source temperature is that it allows detection and measurement of lateral transmittance peaks of interferometric filters.

Generally $1/\lambda_e$ fits with very small deviations in a linear function of $1/T$ (T being the Kelvin temperature):

$$\frac{1}{\lambda_e} = a + \frac{b}{T}. \tag{15}$$

The two-colour pyrometer measures a signal ratio which is:

$$Q = \frac{\Phi(\lambda_{e2})}{\Phi(\lambda_{e1})}\left(\frac{\lambda_{e1}}{\lambda_{e2}}\right)^5 \exp\left[\frac{C_2}{T}\left(\frac{1}{\lambda_{e1}} - \frac{1}{\lambda_{e2}}\right)\right]. \tag{16}$$

In principle this equation allows absolute measurements of temperature with a two-colour pyrometer. A practical difficulty arises from the high precision required in wavelength measurements: according to equations (5) and (6), we need an imprecision $\Delta\lambda = 0.14$ nm to achieve $\Delta T = 1$ K at 1100 °C.

3. Pyrometer calibration

The aim of a two-colour pyrometer calibration is to find an interpolating equation which relates the signal ratio Q to target temperature T (Pasta *et al* 1975).

We remark that, as a consequence of equation (15), the effective equivalent wavelength, given by the equation

$$\frac{1}{\Lambda_e} = \frac{1}{\lambda_{e1}} - \frac{1}{\lambda_{e2}} \tag{17}$$

may also be written:

$$\frac{1}{\Lambda_e} = e + \frac{d}{T}. \tag{18}$$

We have also:

$$\left(\frac{\lambda_{e1}}{\lambda_{e2}}\right)^5 = \left(\frac{a + b/T}{a' + b'/T}\right)^5 \simeq \left(\frac{a}{a'}\right)^5 \left(1 + \frac{5(a'b - b'a)}{aa'T}\right). \tag{19}$$

The approximation is allowable since the second term within brackets is of the order of 10^{-3} and therefore we may neglect higher powers of it.

If the pyrometer filters have a rectangular transmittance curve, a fact which is quite common, the quantity $\Phi(\lambda_{e2})/\Phi(\lambda_{e1})$ is fairly constant. Then we have:

$$\frac{\Phi(\lambda_{e2})}{\Phi(\lambda_{e1})} \left(\frac{\lambda_{e1}}{\lambda_{e2}}\right)^5 = f + \frac{g}{T} \tag{20}$$

with f and g constant.

Putting equations (18) and (20) in equation (16) and solving with respect to T, we get:

$$T = \frac{C_2}{(e + d/T) \ln Q - \ln(f + g/T)}. \tag{21}$$

Again neglecting high powers of small quantities, equation (21) may be written:

$$T = \frac{B}{\ln Q - A} + C \tag{22}$$

where

$$A = \ln f \qquad B = eC_2 + \frac{g}{f} \qquad C = \frac{d}{e}.$$

The calibration consists of reading a number of ratios $Q_1, Q_2 \ldots Q_n$ corresponding to temperatures $T_1, T_2 \ldots T_n$ and finding the coefficients A, B, C through a least squares method. The standard deviation is very small provided the linear relationship (7) is valid. The coefficients a and b are of the order of 1 and 10. Under this circumstance equation (22), with a pyrometer having nominal wavelengths of 0·85 μm and 1 μm, fits temperatures with maximum deviation of 8 mK in the temperature range of 600–1600 °C. An actual calibration gave deviations of 1·5 K or even more for certain units of the same series. A deeper insight shows that this inconvenience arises from departure from linearity of $1/\lambda_e$ versus $1/T$, a departure which is caused by secondary transmittance peaks. Therefore the interference filters to be used in precision two-colour pyrometry must be specified for secondary peaks not to exceed a certain fraction (typically 0·1%) of peak value.

4. Influence of sensor temperature

The response curve of a quantum sensor, like the silicon photodiode, has a roughly triangular shape with the steeper side towards the longer wavelengths in the cut-off region (figure 1). A temperature rise causes a shift to the left of the steep side thus decreasing the cut-off wavelength. At the same time the other side is raised, thus increasing sensitivity. The opposite happens when temperature decreases.

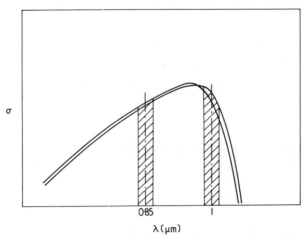

Figure 1. Sensor responsivity curves at different temperatures.

The same figure represents, with dashed strips, the transmittance bands of the filters. They are centred on 0·85 μm and 1 μm. Unfortunately, wavelength spacing and resolution requirements make it impossible to avoid the region above the peak, where the most substantial responsivity variations with temperature occur.

It is qualitatively apparent that sensor temperature variation causes the effective wavelengths to shift in opposite directions, so that their influence on temperature is added. Due to the high value of parameter C in equation (6), it may be expected that a change in sensor temperature has a big influence on pyrometer reading.

A thorough analysis of the error is complicated and, on the other hand, it is meaningless, owing to the spread of sensor characteristics. Therefore, an experimental evaluation has been preferred. The pyrometer has been focused on a target at constant temperature $T = 986\ °C$. The sensor temperature T_s has been made to change from 25 to 47 °C and, for each temperature, a pyrometer reading T_1 has been taken. The results are reported in figure 2. For $T_s = 40\ °C$, we find that $\Delta T_1/\Delta T_s = 5$. We conclude that the sensor must be thermostated within 0·2 K in order that the spread of the instrument reading may not exceed 1 K.

Sensor temperature control is particularly critical in two-colour pyrometry.

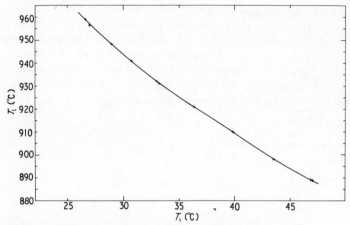

Figure 2. Influence of sensor temperature on pyrometer reading.

5. Influence of target conditions

We analyse in this section two more obvious cases.

5.1. Emissivity inequality for the two pyrometer bands

In the general case the radiance ratio in a two-colour pyrometer is

$$Q = \frac{\epsilon_1}{\epsilon_2} K_e \exp\left(\frac{C_2}{\Lambda_e T}\right) \tag{23}$$

where ϵ_1 and ϵ_2 are the emittances within the two bands and

$$K_e = \frac{\Phi(\lambda_{e2})}{\Phi(\lambda_{e1})} \left(\frac{\lambda_{1e}}{\lambda_{2e}}\right)^5.$$

It is an obvious advantage of two-colour pyrometers that their reading is independent of emissivity if $\epsilon_1 = \epsilon_2$, namely for grey bodies. If this is not the case, neglecting the emissivity difference is equivalent to introducing a fractional error of flux ratio:

$$\frac{\Delta Q}{Q} = \frac{\epsilon_1 - \epsilon_2}{\epsilon_2}. \tag{24}$$

This causes an error which may be evaluated through equation (4). The De Vos (1954) data on tungsten at 1600 K are: $\epsilon_1(0\cdot85\,\mu\text{m}) = 0\cdot426$ and $\epsilon_2(1\,\mu\text{m}) = 0\cdot390$. It is therefore $(\epsilon_1 - \epsilon_2)/\epsilon_2 = 0\cdot09$, which causes a temperature difference $\Delta T = 63$ K. If emissivities are known, this correction can be brought to the reading. In this case the resulting error derives from emissivity imprecision. If the latter, for instance, is 2% for each value, then the error is reduced to 2·5 K.

5.2. Non-isothermal targets and obstruction of the optical path

The first case happens with an iron ingot or a beam covered with scales and the second one is the case of dust particles located between target and pyrometer.

In order to analyse the first case (Ruffino 1974), we consider a target model consisting of two portions: one, at temperature T_1 with fraction α of the total area, and the other at temperature T_2 with fraction β. We have $T_1 > T_2$ where T_1 is the desired value of target temperature and T_2 is a disturbing quantity. Then the pyrometer reads the flux ratio:

$$Q = \frac{\alpha\lambda_2^{-5}\exp(-C_2/\lambda_2 T_1) + \beta\lambda_2^{-5}\exp(-C_2/\lambda_2 T_2)}{\alpha\lambda_1^{-5}\exp(-C_2/\lambda_1 T_1) + \beta\lambda_1^{-5}\exp(-C_2/\lambda_1 T_2)} \tag{25}$$

which corresponds to temperature T'. Here $T_1 > T' > T_2$. In addition, $T' = T_1$ when $\beta = 0$ (temperature uniformity) or $T_2 = 0$. This means that the temperature error $\Delta T = T_1 - T'$ is limited.

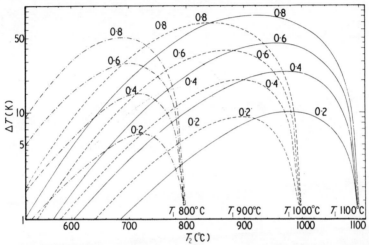

Figure 3. Pyrometer error caused by target 'cold spots'. T_1 is temperature of hot substrate of target; T_2 is temperature of cold spots. Curve parameter is fractional area, β, of cold spots.

Analysis of equation (25) yields ΔT as a function of the lower temperature of the target for different values of T_1 and β. Results are represented in figure 3. From the curves we draw the following conclusions:

(i) for T_2 decreasing below T_1, ΔT rapidly reaches a maximum (roughly 100 °C below T_1);
(ii) the maximum value of ΔT increases with increasing β (fraction of 'cold spots');
(iii) if the 'cold spots' do not emit visible radiation ($T_2 < 500$ °C), for $T_1 = 800$ °C, the error is below 1%, a value which rapidly fades out for higher values of T_1.

The same analysis applies when the optical path between target and pyrometer is obstructed by solid materials which cover a fraction β of the cross section. They do not adversely affect the reading provided β is not too large and their temperature is substantially lower than the one of the target.

The final conclusion is that the two-colour pyrometer discriminates within a non-uniform radiation beam, emphasizing the target zones at higher temperature. The other zones only cause disturbances, which are negligible if temperature and percentage of 'cold fluxes' are moderately low.

References

De Vos J C 1954 *Physica* **20** 690
Kostkowski H J and Lee R D 1962 *TMCSI* **3** part 1 449
Pasta M, Ruffino G and Soardo P 1975 *High Temp. High Press.* to be published
Pasta M, Ruffino G, Soardo P and Toselli G 1973 *High Temp. High Press.* **5** 99
Righini F, Rosso A and Ruffino G 1972 *TMCSI* **4** part 1 413
Ruffino G 1974 *High Temp. High Press.* **6** 223
—— 1975 *Rév. Internat. des Hautes Temp. et de Refract.* to be published

On the state of ratio pyrometry with laser absorption measurements

H Kunz

Physikalisch-Technische Bundesanstalt, Braunschweig, W Germany

Abstract. Variations of the optical arrangement for surface temperature determination by laser absorptance measurements are discussed. One of these is provided for by using commercially available instruments. Measurements show that the method can also be applied to very different sorts of material and different sizes of the sample up to high temperatures.

1. Introduction

For the determination of the true temperature T of a non-black body by a ratio pyrometer the knowledge of the ratio of the spectral surface emittances at the two effective pyrometer wavelengths λ_1 and λ_2, ie, $r_e = \epsilon(\lambda_1)/\epsilon(\lambda_2)$, must be available. Then T can be calculated according to the formula

$$\frac{1}{T} = \frac{1}{T_r} + \frac{\ln[\epsilon(\lambda_1)/\epsilon(\lambda_2)]}{c_2(1/\lambda_1 - 1/\lambda_2)}$$

where T_r is the ratio temperature measured by the pyrometer and c_2 is the second radiation constant of Planck's radiation law.

A previous paper (De Witt and Kunz 1972) has shown how the emittance ratio, equal to the absorptance ratio according to Kirchhoff's relation, can be indirectly determined by the observation of the temperature variations caused by absorption in the sample surface if it is irradiated by laser light of wavelengths λ_1 and λ_2. This method should be applied whenever the sample surface is not specularly reflecting or whenever black body conditions cannot be provided, since the method is largely independent of the surface conditions.

The principal purpose of this paper is to develop a better understanding of the present capabilities of 'laser absorption pyrometry' by describing simpler optical arrangements than given in the previous paper, by discussing some theoretical aspects of measurement uncertainties, and by presenting new experimental results concerning laser power requirements for different kinds of material and different sample dimensions. Due to a shortage of time in the 5th Temperature Symposium we could only report on preliminary measurements.

2. Optical arrangements of laser absorption pyrometry

For a better understanding of the further discussion of the method, the relative simple optical arrangement of figure 1 should be considered first. Contrary to the pre-

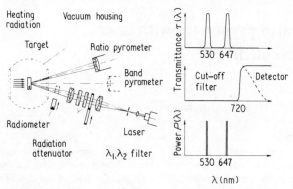

Figure 1. A simple arrangement for laser absorption pyrometry.

vious arrangement in De Witt and Kunz (1972), equipment is used here that might also be commercially available. The deviations from the ideal conditions of the previous arrangement can be assumed to be not critical on the whole, since a symmetric angular radiation characteristic of the target can often be taken for granted. As well as this, in the pyrometer and in the laser light path, only ratio measurements are made, which have an intrinsic quality of equalizing error. The target, eg a material sample at which thermophysical properties are to be investigated, may be arranged in vacuum or, especially at higher temperatures, in a gas atmosphere. In the latter case radiance fluctuations caused by convective disturbances must sometimes be taken into consideration. Preference should be given to the kind of heating, eg electric current, electron bombardment or infrared laser radiation, that may cause the smallest radiance fluctuations on the target and the smallest amount of stray light in the λ_1 and λ_2 wavelength regions.

The ratio pyrometer has to be provided with narrow band interference filters having a peak wavelength as near as possible to λ_1 and λ_2. They should be adjustably mounted for precise adjustment of the effective pyrometer wavelength until they are equal to λ_1 and λ_2. This is done by tilting the filters against the optical axis. Generally the laser wavelength cannot be arbitrarily selected yet high-performance CW dye lasers developed recently make it possible as far as the requirements of accuracy allow their application.

The band pyrometer measures the relatively small (eg 20%) radiance rise in the λ_0 region. It has to be provided with a cut-off filter as described in De Witt and Kunz (1972). Furthermore it should also have a radiation sensitivity of at least five times that of the ratio pyrometer depending on the accuracy requirements and on the available laser power. This radiance rise is produced by absorption when the target is irradiated with the laser light at λ_1 or λ_2. The cut-off filter has to totally suppress the laser light reflected at the target during irradiation.

For the measurements reported below the ratio pyrometer and the band pyrometer were combined into one instrument, namely, the direct current pyrometer PP70 (see Kunz and Kaufmann 1975). This paper also stated that the pyrometer type with linear detector sensitivity could be easily calibrated and programmed for on-line operation. A linear pyrometer characteristic should at least be provided for all ratio pyrometers with a temperature range above 2000 K in order to save the high expenses for high-temperature black bodies which otherwise would be necessary for calibration purposes.

With respect to the cut-off filter a special observation is of considerable interest for instrument design. It was found that directly reflected laser light of 647 nm wavelength may cause luminescence in the Schott filter glass RG 715. Therefore the 'reflection effect' described in De Witt and Kunz had only an indirect relation to the reflection of the laser light. The magnitude of the luminescence effect seems to vary with the charge of the filter glass. It could be reduced by a factor of 30 when inserting the λ_0 filter into the space between the first pyrometer lens and the measuring field stop. Further investigations seem necessary, although this correction term could be made negligible by a very small rotation of the tungsten strip lamp which has been used as a target for these experiments.

In the laser light path only linearly polarized light is used, in contrast to the optical arrangement of De Witt and Kunz. This is permissible if in the ratio pyrometer polarizing filters with the same orientation are also provided. Application of polarized laser light gives the advantage of using the lasers most effectively. Besides this the whole equipment is simpler. If the laser amplitude stability control is sufficiently reproducible then it may also be sufficient to apply two polarizers as an attenuator whereby only one of them has to be calibrated. The other is inserted to maintain the polarizing angle. One polarizer is adjusted to balance the laser powers at λ_1 and λ_2 so that they cause the same radiance rise in the λ_0 region. For achieving the highest degree of measurement accuracy an additional radiometer with high-resolution capability should be used to measure the ratio of the laser powers in the balanced condition. This ratio is equal to the required $\epsilon(\lambda_1)/\epsilon(\lambda_2)$. It may be a photoelectric type of good reproducibility. Its relative spectral characteristic has to be calibrated *in situ* with a high-precision radiometer as described in Möstl and Bischoff (1975).

On the right-hand side of figure 1 the spectral characteristics of the ratio pyrometer, the band pyrometer, and the laser beam are represented for the case when a krypton laser is applied. Only a brief description can be given of the many other optical arrangements which can be produced by different combinations of the main parameters, that is: one laser, one pyrometer; two lasers, one pyrometer; two lasers, two pyrometers. We have discussed 12 variations, thereby taking into consideration the application of beam splitters, dichroic mirrors, optoelectronic shutters, optomechanical shutters and rotating interference filter disks. For future work those arrangements which allow a nearly simultaneous measurement of the ratio temperature and the absorptance ratio will be of special interest. This variation of the method may be achieved if chopper operation is used in both light paths. By applying phase-sensitive circuits and the corresponding optical shutter technique a phase shift of 180° between the opening times of the light paths is provided. Then the signal of λ_1 is taken in that part of the synchronized periods in which the target is irradiated with light of wavelength λ_2. No disturbance by reflection may be encountered and the measuring time can be considerably reduced. A more detailed discussion will be published elsewhere.

3. Measurement uncertainty in different laser wavelength combinations

The basic formulae for uncertainty calculations are discussed in De Witt and Kunz (1972). Figure 2 represents the calculated temperature uncertainty δT as a function of T for some combinations of significant laser wavelengths. Curves A–D are related to a

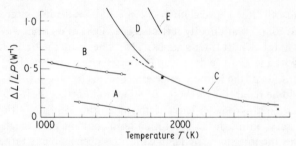

Figure 2. Uncertainty δT of a ratio pyrometer at different wavelength combinations. Curve A, $\lambda = 658$ nm = effective wavelength of a spectral pyrometer. For the ratio pyrometer: curve B, $\lambda_1 = 514\cdot5$ nm, $\lambda_2 = 1060$ nm; curve C, $\lambda_1 = 514\cdot5$ nm, $\lambda_2 = 647\cdot1$ nm; curve D, $\lambda_1 = 500$ nm, $\lambda_2 = 580$ nm; curve E, $\lambda_1 = 514\cdot5$ nm, $\lambda_2 = 647\cdot1$ nm. Full curves, $\delta\Phi/\Phi = 0\cdot002$; broken curve, $\delta\Phi/\Phi = 0\cdot005$; and $\delta\lambda = 0\cdot025$ nm.

radiance ratio uncertainty $\delta\Phi/\Phi = 0\cdot002$ and curve E to $\delta\Phi/\Phi = 0\cdot005$. Since the absorptance ratio uncertainty $\delta r_e/r_e$ has the same weight in the total uncertainty the curves can also be used for considering the influence of δr_e on the sum of both uncertainties. Only the combination of a YAG laser and an argon ion laser (curve B) provides an uncertainty nearly comparable with that of a spectral pyrometer. Curve B can also be taken for a YAG laser alone but with a frequency doubler. The curves C and E are representative of the combination of an argon and a krypton laser.

A YAG laser, with high-power output especially, would be of interest when realizing the method discussed in the previous section and when using a silicon diode as a radiation detector. But until now, of the commercially available lasers, only ion lasers have sufficient amplitude stability.

For high-precision measurements much attention should be paid to using only interference filters that have a high degree of homogeneity.

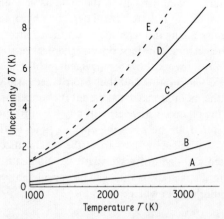

Figure 3. Variation of relative radiance decrease per watt with temperature, $\Delta L/LP$. Curve A, end surface of a platinum rod in the open air, 3·5 mm diameter, 100 mm length; curve B, Globar rod in the open air, 20 mm diameter, 300 mm length; curve C, tungsten strip in argon (0·7 bar), 0·02 mm thickness, 2 mm width, 35 mm length; curve D, tungsten strip in vacuum, 0·02 mm thickness, 2 mm width, 35 mm length; curve E, tungsten strip in argon (0·7 bar), 0·02 mm thickness, 1 mm width, 35 mm length.

4. Measurements concerning energy requirements

In the preliminary measurements of the previous paper only the energy requirements for tungsten ribbon samples of different sizes could be investigated. Some measurements which were considered as being suspect at the very outset had apparently indicated a nonlinear variation of the radiance rise with the irradiance level. Therefore these measurements were repeated. It was found that the nonlinearity does not exist, at least within the investigated power limits of 25 mW to 145 mW. The apparent nonlinearity of the former measurements was a result of the nonlinearity of a commercial radiometer. Besides this the repeatability of the relative radiance rise per watt, $\Delta L/LP$, has proved to be within the measurement uncertainty limited by the relative low-power output of the krypton laser used here. The crosses of curve C (figure 3) are correct values of the former measurements.

The other curves of figure 3 represent samples of different materials and different dimensions, thus positively answering the open question of whether laser absorption pyrometry could be applied to bulky samples of different kinds of material.

5. Conclusion

The current paper, as a supplement to the previous one, provides further fundamental aspects for the design of laser absorption pyrometers thus showing that, with ratio pyrometers as well as with lasers, an instrument performance is achieved which is high enough to be applied in laser absorption pyrometry. At least in the temperature region below 2000 K the expense of the laser should no longer be a severe hindrance to application.

Acknowledgments

I am very much indebted to Mr M Rasper who built the pyrometer used. I also wish to acknowledge the invaluable cooperation of D P De Witt. Without his courage in undertaking the previous investigations, first theoretical derivations of the method might have remained in my desk up to this very day.

References

De Witt D P and Kunz H 1972 *TMCSI* 4 part 1 599
Kunz H and Kaufmann H J 1975 this volume
Möstl K and Bischoff K 1975 to be published

Determination of the difference between the thermodynamic fixed-point temperatures of gold and silver by radiation thermometry

H J Jung

Physikalisch-Technische Bundesanstalt (PTB), Institut Berlin, 1 Berlin 10, Germany

Abstract. The interval between the thermodynamic temperatures of the gold and the silver point has been found to be 0·13 K less than between the fixed-point values assigned by the IPTS-68. The measurements consisted of alternating comparisons of the spectral radiance of two black bodies at wavelengths of about 650 nm and 546 nm. One black body had been submerged in a gold bath, the other in a silver bath. Melts and freezes were run through simultaneously.

1. Introduction

Radiation measurements concerning the thermodynamic fixed-point temperatures of gold and silver published in the past ten years are reviewed in table 1. All authors except Blevin and Brown (1971) compared spectral concentrations of the radiances† of black bodies starting from certain reference temperatures (Zn point, Sb point, Au point).

Table 1. Results obtained by other authors. $\Delta T = [T_{Au} - T_{Ag}] - [(T_{68})_{Au} - (T_{68})_{Ag}]$.

Author	$T_{Au} - (T_{68})_{Au}$	$T_{Ag} - (T_{68})_{Ag}$	ΔT
Heusinkveld (1966)	0·30 ± 0·3 K	0·52 ± 0·3 K	−0·22 K
Hall (1965)	−0·43 ± 0·6	−0·43 ± 0·5	0·0
Blevin and Brown (1971)	−0·31 ± 0·4†		
Bonhoure (1973)	−0·05 ± 0·16	+0·08 ± 0·13	−0·13 ± 0·2
Bonhoure (private communication)	direct comparison		−0·19 ± 0·13
Quinn et al (1973)	$(T_{68})_{Au} = T_{ref}$	0·12 ± 0·1	−0·12

† Obtained from a measurement of the total radiation flux emitted by a black body at T_{Au} based on the best theoretical value of the Stefan–Boltzmann constant.

For the radiation constant c_2 the value adopted in the IPTS-68 was used. In order to obtain comparable values some of the results cited have been corrected for the refractive index of air. The reference temperature 903·82 K (Sb point) used by Heusinkveld (1966) was also substituted by the IPTS-68 value.

† In the following text referred to briefly as spectral radiances.

All these results indicate that measurements with reduced uncertainties are necessary to determine possible errors of the values assigned to the gold point and silver point by the IPTS-68. One method can be a direct photoelectric comparison of the spectral radiances of two black bodies, one of them submerged in a gold melt and the other in a silver melt. From the measured ratio of spectral radiances the difference between the thermodynamic temperatures of both fixed points will be computed with $c_2 = 0.014\,388$ mK as adopted in the IPTS-68.

2. General description of the apparatus

The arrangement is shown schematically in figure 1. Two horizontally radiating black bodies B_1 and B_2 made from graphite are arranged in opposite positions. One of them is submerged in a gold bath and the other in a silver bath. Each refractory contains 25.5 cm^3 of metal of 99.999% purity. Both refractories are heated by a resistance

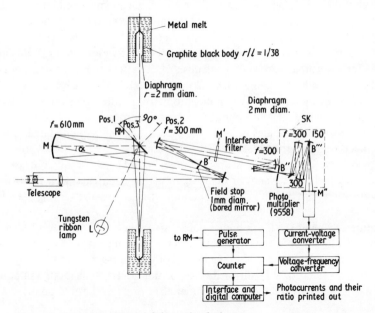

Figure 1. Schematic sketch of the entire device.

furnace to a temperature about 2 or 3 K above or below the freezing point of the metal. After reaching a stationary state simultaneous melting or freezing equilibria of between 20 and 40 minutes' duration were run through by augmenting or reducing the heating power of the furnaces by about 18 and 9% respectively.

The radiation emitted by the black bodies is reflected by the rotatable plane mirror RM to the focusing spherical mirror M (see figure 1). The mirror RM turns periodically between its two well defined positions 1 and 2 and rests for about 12 seconds in each one. This way it is possible to measure the black body radiation emitted at the gold point and the silver point temperature periodically in turn. After four full periods RM stops once in an intermediate 45° position for a dark current measurement. In each position after the mirror has stopped completely the measurement of the photocurrent is carried out by integrating over 10 seconds. The ratio of the spectral radiances is given by the ratio of the photocurrents. All mirror rotations are controlled by a simple mechanical pulse generator which also triggers the electronic components.

Along the path of rays in B' an image of the black body hole appears with its natural size of 2 mm diameter which is stopped to 1 mm diameter by a plane mirror with a sharp-edged circular bore. The distance from $B_{1,2}$ to M is equal to 1150 mm and the diameter of the spherical mirror M is 40 mm. This means that the maximum aperture approximately corresponds to the ratio 1:29, so there are no essential astigmatic distortions in B' caused by the angle of reflection $\alpha = 5 \cdot 7°$ at M.

Hereafter the beam is made parallel and passes a temperature-stabilized interference filter which essentially defines the effective wavelength. The filter is located in the first image M' of M and can be shifted out. The beam is then made to converge and enters another arrangement of mirrors SK (the broken line in figure 1) which broadens the beam to about 11 mm diameter in the plane M'' where an image of M is located. Exactly in this plane we placed a diaphragm of 9 mm diameter directly followed by the cathode of the photomultiplier (EMI 9558). In every case this diaphragm serves to illuminate the very central region of the photocathode of which the spectral response had already been measured. This way no changes of the effective wavelength caused by partial irradiation of cathode regions with unknown spectral response can occur. The sub-unit SK can be easily replaced by a monochromator in order to measure the spectral transmission of the interference filter under working conditions.

Adjustment of the black bodies is facilitated by a telescope accommodated to their image B'. Thus centering of the black bodies with respect to the 1 mm field stop is controlled by observation with a telescope. In order to obtain sufficient contrast between the black body cavity and the diaphragm there is an intense, well focused auxiliary light source (not drawn in figure 1) illuminating the cavities during the adjustment procedure.

Temperature setting of the black bodies before an equilibrium run is controlled by radiance comparison, the reference light source being the tungsten strip lamp L. It is located at the same distance from M as the black body diaphragm. L is 'seen' by the device after turning the rotatable mirror RM to position 3. L was calibrated in position 3 as a pyrometric standard at a brightness temperature equal to the gold point.

3. Black bodies and furnaces

The central part of a furnace containing the metal-filled black body refractory is shown to scale in figure 2. The length of the resistance-heated commercially available furnaces is 600 mm. They are mounted on movable desks and are adjustable in a vertical direction and in a horizontal plane. The black body refractories consist of the radiator centred to the crucible by a screw, all components being turned from high-purity graphite (EK 506 from Ringsdorff-Werke, less than 10 ppm solid impurities according to the manufacturer). The cavity has a cone-shaped bottom; in order to obtain diffuse reflection (Lambert's law) the interior has been oxidized by injecting air into the hot cavity.

The residual reflectance ρ_c of the cavity was calculated after Quinn (1967) and Bauer and Bischoff (1971) for the case of a flat bottom. The emittance of graphite

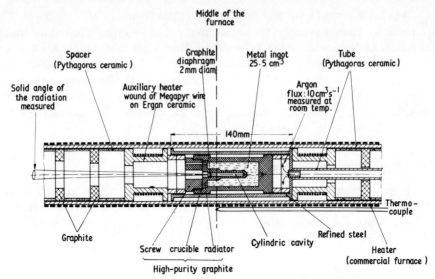

Figure 2. The central part of a furnace containing the black body (sketch according to a scale). The cylindrical cavity has dimensions of 38 mm length, 4 mm diameter and 2 mm wall thickness.

was assumed to be 0·85 at the silver point T_{Ag} and 0·84 at the gold point T_{Au}. With the cavity dimensions, length $l = 38$ mm, inner radius $R = 2$ mm and diaphragm radius $r = 1$ mm, one obtains (in sufficient agreement between the formulae given in Quinn (1967) and Bauer and Bischoff (1971))

$$\Delta \rho_c = \rho_c(T_{Au}) - \rho_c(T_{Ag}) = 0{\cdot}000114 - 0{\cdot}000106 = 0{\cdot}000007.$$

Thus the correction δ to be applied to the measured $T_{Au} - T_{Ag}$ is +0·5 mK at 546 nm and +0·6 mK at 650 nm if Lambert's law holds.

The temperature drop $\Delta_b(T)$ in the cavity bottom is another systematic error to be discussed. It is produced by the total radiation of the bottom through the diaphragm. By equating the conducted with the radiated heat it can be shown that $\Delta_b(T)$ is approximated by

$$\Delta_b(T) \approx \epsilon_{tot} \sigma T^4 (d/\lambda_b)(r/l)^2. \tag{1}$$

Using the data $d = 2$ mm (thickness of the bottom), $\lambda_b \approx 36$ W m^{-1} K^{-1} (thermal conductivity of EK 506 at 1000 °C) and $\epsilon_{tot} \approx 0{\cdot}9$ (total emissivity of EK 506 at 1000 °C) one obtains

$$\Delta_b(T_{Au}) = 6{\cdot}3 \text{ mK} \qquad \Delta_b(T_{Ag}) = 4{\cdot}6 \text{ mK}.$$

The correction δ_b to be applied to the measured $T_{Au} - T_{Ag}$ is given by

$$\delta_b = \Delta_b(T_{Au}) - (T_{Au}/T_{Ag})^2 \Delta_b(T_{Ag}) = +0{\cdot}9 \text{ mK}. \tag{2}$$

Another source of error may be the temperature difference between the ends of the refractory. To overcome this the refractory is put into a cylinder of refined steel and auxiliary heaters are located at its two ends. These heaters have been powered in such a way that temperature inhomogeneities measured in the axis of an empty axially bored refractory did not exceed 0·3 K in the stationary state. During the freezings and meltings different powers were applied to the auxiliary heaters which had been chosen so that temperature inhomogeneities less than 0·8 K occurred either during heating or cooling the empty refractories.

4. Filters and spectral response

Four different interference filters were used. They are referred to in the form wavelength (nm)/halfwidth (nm), namely, 650/40 or 546/10·5. The transmission D of three of them is shown in figure 3 together with the relative spectral response s of the

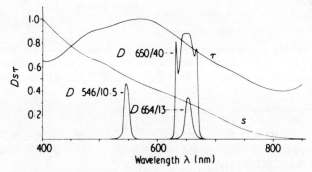

Figure 3. Spectral response of filters, multiplier and mirror set.

photomultiplier and the 'overall' relative transmission τ of the mirror set shown in figure 1. All filters and mirrors have been measured in working position; the minimum values of D taken into account were in the 10^{-6} region.

One advantage of the broadband dielectric multilayer filters 650/40 and 650/33 is their sharp-edged transmission curve. For this reason alone the effective wavelength is about three times less dependent on uncertainties of τ or s than it is for a gaussian filter with the same halfwidth (Jung and Verch 1973). Another advantage is the improved transmission.

The measurement of s was done by comparing the multiplier response to that of a radiation thermocouple (Hilger and Watts FT 17.1) as black reference.

To measure the mirror set transmission τ, the black body B_1 was replaced by a tungsten strip lamp L_1. Then the image located in B''' (see figure 1) was radiance-compared by means of another rotating mirror like RM and a monochromator to another strip lamp L_2. From the ratio of the signals τ was calculated because L_1 and L_2 are calibrated standards of spectral radiance.

The estimated uncertainty of the measured relative slope of $s\tau$ is $5 \times 10^{-4}\,\mathrm{nm}^{-1}$. Therefore the uncertainty of the effective wavelength is less than ±0·07 nm for the broadest filter (see Jung and Verch 1973). The corresponding uncertainty in temperature is ±12 mK.

5. Photomultiplier and electronic components

The observed sensitivity drift of the photomultiplier (EMI 9558) did not exceed $2 \times 10^{-4}\,h^{-1}$. The multiplier was temperature-stabilized at 25 °C. Its nonlinearity was measured by the light addition method (Jung 1971) down to 10^{-9} A.

Another source of nonlinearities could be the voltage to frequency converter (HP 2212 A-M 3). Its nonlinearity was determined with a precision voltage source. The current to voltage converter contains an electrometric operational amplifier (AD 310 K) with a high-stability feedback resistor of 10 MΩ. No significant nonlinearity could be detected. All corrections applied due to the components of the apparatus are shown in table 2.

Table 2. Corrections for systematic errors.

Wavelength (nm)	Corrections		
	Non-linearities	Black body errors ($\delta_B + \delta$)	Total
650	−7 mK	+2 mK	−5 mK
546	−4	+1	−3

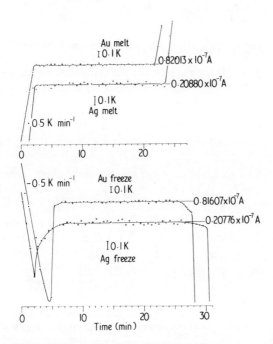

Figure 4. Two pairs of simultaneously taken melts and freezes of photomultiplier current against time, filter: 650/40. Initial heating/cooling rate: about $0.5\,K\,min^{-1}$.

6. Results

One pair of simultaneously measured melts and another pair of freezes are shown in figure 4. The photomultiplier currents $i(\text{Au})$ and $i(\text{Ag})$ in the figure are averages taken from the horizontal parts of the curves. Distortions due to supercooling, recalescence and other distortions at the end of the freezes have been omitted. Least-square calculations showed the melting ranges were found, in most cases[†], to be less than 13 mK for gold and less than 16 mK for silver which is nearly equal to the statistical uncertainty. The estimated temperature uncertainty due to impurities and time-dependent temperature gradients is approximated by the melting ranges observed. The chemical analysis showed a total contamination of 27 atomic ppm (Au 72)[‡] and 12 atomic ppm (Ag 72). The liquidus depressions estimated from this are 11 mK (Au 72) and 3 mK (Ag 72); the corresponding estimates for the melting ranges (see McLaren 1962) are 40 mK (<13 mK observed) and 14 mK (<16 mK observed). The discrepancy for Au 72 may be due either to impurity segregation into the analysed part of the ingot or to the uncertainty of the liquidus and solidus data taken from alloy diagrams (Hansen 1958).

In order to evaluate temperatures from the measured photocurrent ratios

$$Q = i(\text{Au})/i(\text{Ag}) \tag{3}$$

an approximation described in Jung and Verch (1973) was used which transforms the exact equation

$$Q \int_{\lambda=0}^{\infty} L_{s,\lambda}(T_{\text{Ag}}) D s_T \, d\lambda = \int_{\lambda=0}^{\infty} L_{s,\lambda}(T_{\text{Au}}) D s_T \, d\lambda, \tag{4}$$

where $L_{s,\lambda}(T)$ is the spectral radiance of a black body at temperature T, to

$$\log Q + \frac{b}{T_{\text{Ag}} - c} = \frac{b}{T_{\text{Au}} - c}. \tag{5}$$

The constants b and c characterize the filter for a given spectral response s_T and have been calculated so that the error produced by equation (5) becomes zero at T_{Au} and T_{Ag}.

In this way T_{Au} was computed by equation (5) from any measured photocurrent ratio with the reference temperature $(T_{68})_{\text{Ag}} = 1235 \cdot 08$ K and $c_2 = 0 \cdot 014388$ mK as assigned in IPTS-68, which is implicitly contained in b and c. Table 3 shows the difference ΔT between the thermodynamic and the IPTS-68 interval between the gold and the silver point. Approximately the same number of melts and freezes have been used because there was no significant difference between their results. Column 3 contains the uncorrected values as measured in air; ΔT_{air} represents the mean taken from all

---

[†] Higher 'melting ranges' have been identified to be due to poor optical adjustment or sensitivity drifts before reaching a constant temperature of the photomultiplier. The corresponding runs have not been taken into account.

[‡] Name of the sample; the number refers to the year when the measurements were performed.

Table 3. Results. $\Delta T = [T_{Au} - T_{Ag}] - [(T_{68})_{Au} - (T_{68})_{Ag}]$.

Filter	Number of runs	ΔT_{air} (K)	Correction for n_{air} = 1·00028	ΔT (K)	Standard deviation, single run (K)	Standard deviation, mean (K)
Samples Au 72/Ag 72						
650/40	5	−0·170		−0·139	0·004	0·002
650/33	5	−0·177	0·031 K	−0·146	0·008	0·004
654/13	23	−0·166		−0·135	0·014	0·003
546/10·5	15	−0·170		−0·139	0·025	0·006
		Mean: −0·171[†]		Mean: −0·140		
				Weighted mean: −0·139 (standard deviation 0·002)		
Samples Au 74/Ag 74						
650/40	6	−0·161		−0·130	0·007	0·003
650/33	6	−0·158	0·031 K	−0·127	0·009	0·004
654/13	7	−0·151		−0·120	0·020	0·008
546/10·5	7	−0·159		−0·128	0·028	0·011
		Mean: −0·157		Mean: −0·126		
				Weighted mean: −0·128 (standard deviation 0·003)		

[†] First result as published in Jung (1973).

equilibrium runs with one filter. Column 4 gives the correction to be applied for the refractive index of air (see Blevin 1972, Quinn 1974) and column 5 the corrected ΔT values. The standard deviation in column 6 is that of one equilibrium run to the mean of all runs taken with the same filter. Column 7 contains the corresponding standard deviations of the mean with respect to the number of runs.

The observed standard deviations clearly demonstrate the improvement of the signal-to-noise ratio when higher transmitting and broader filters are used. The insignificant difference between the weighted mean and the weighted mean of ΔT indicates that the measurements with the narrow filters agree well with those made with broad filters which are strongly preferred in the weighted mean.

The systematic uncertainties are given in table 4.

Table 4. Estimated systematic uncertainties.

Number	Uncertainty due to	Estimate X_i (mK)
1	residual nonlinearities (= ±0·5 × 10^{-4})	4
2	contaminations, time-dependent temperature gradients (observed as melting range)	12 (Au) 16 (Ag)
3	effective wavelength (= ±0·07 nm)	12
4	diffraction; stray light from imperfect optical surfaces	4
5	radiation constant $\Delta c_2/c_2$ = ±31 ppm (see *Codata* 1973)	4
6	different right−left response of the mirror set (see figure 1); different transmission of the furnaces as possibly generated by hot gas schlieren	30

The total systematic uncertainty is given by $\left(\sum_{i=1}^{6} X_i^2 \right)^{1/2} = 39$ mK.

The final results as averaged from the weighted means in table 3 is

$$[T_{Au} - T_{Ag}] - [(T_{68})_{Au} - (T_{68})_{Ag}] = -0.13 \text{ K}$$

with a systematic uncertainty of ±0.04 K. The statistical uncertainty given by three standard deviations of the weighted mean (see table 3) is ±0.01 K.

7. Conclusion

The present result agrees well with those cited above[†]. Because of its reduced uncertainty it is more evident that the gold—silver interval should be decreased in a future version of the International Practical Temperature Scale.

References

Bauer G and Bischoff K 1971 *Appl. Opt.* **10** 2639
Bonhoure J 1973 *Proc. Verb. CIPM* **41** 52
Blevin W R 1972 *Metrologia* **8** 146
Blevin W R and Brown W J 1971 *Metrologia* **7** 15
Codata 1973 **11** 6
Hall J A 1965 *Metrologia* **1** 140
Hansen M 1958 *Constitution of Binary Alloys* 2nd edn (New York: McGraw-Hill)
Heusinkveld W A 1966 *Metologia* **2** 61
Jung H J 1971 *Z. Angew. Phys.* **31** 170
—— 1973 *PTB Jahresber.* 204
Jung H J and Verch J 1973 *Optik* **38** 95
McLaren E H 1962 *TMCSI* **3** part 1 185
Quinn T J 1967 *Br. J. Appl. Phys.* **18** 1105
—— 1974 *Metrologia* **10** 115
Quinn T J, Chandler T R D and Chattle M V 1973 *Metrologia* **9** 44

[†] It also agrees within the given uncertainties to $\Delta T = -(0.17 \pm 0.059)$ K, recently obtained by A Coslovi, A Rosso and G Ruffino (private communication).

Radiance temperature of molybdenum at its melting point

A Cezairliyan
National Bureau of Standards, Washington, DC, USA
L Coslovi, F Righini and A Rosso
Istituto di Metrologia 'G Colonnetti', Torino, Italy

Abstract. Radiance temperature (at two wavelengths, 653 and 995 nm) of molybdenum at its melting point was measured using a subsecond-duration-pulse heating technique. Specimens in the form of strips with initially different surface roughnesses were used. The results do not indicate any dependence of radiance temperature (at the melting point) on initial surface or system operational conditions. The average radiance temperature at the melting point of molybdenum is 2531 K at 653 nm and 2331 K at 995 nm, with a standard deviation of about 0·6 K and a maximum absolute deviation of 1·2 K in both cases. The total inaccuracy in radiance temperature is estimated to be not more than ±8 K.

1. Introduction

In earlier studies (Cezairliyan 1973, Cezairliyan and Righini 1975) it was shown that the radiance temperature[†] (at 0·65 μm) of niobium and zirconium at their respective melting points was independent of the specimen's initial surface conditions and was reproducible and was constant for a given metal during its initial melting period.

The objective of the work described in this paper is to extend the measurements to molybdenum. A novel feature of this work is that the measurements were performed at two wavelengths (653 nm and 995 nm). The measurements at 653 nm were performed at NBS, and those at 995 nm were made at IMGC.

The method is based on rapid resistive self-heating of the specimen from room temperature to its melting point in less than one second by the passage of an electrical current pulse through it; and on measuring specimen radiance temperature with a high-speed pyrometer. At NBS, temperature is measured at the rate of 1200 measurements per second with a photoelectric pyrometer (Foley 1970) operating at an effective wavelength of 653 nm. The bandwidth of the interference filter is 10 nm and the circular area viewed by the pyrometer is 0·2 mm in diameter. At IMGC, temperature is measured at the rate of up to 13 000 measurements per second with an infrared pyrometer operating at 995 nm. The bandwidth of the interference filter is 50 nm, and the circular area viewed by the pyrometer is 0·2 mm in diameter. Details regarding the construction and operation of the two measurement systems are given in earlier publications: the NBS system in Cezairliyan (1971) and Cezairliyan et al (1970b), and the IMGC

[†] Radiance temperature (sometimes referred to as brightness temperature) is the apparent temperature of the specimen surface corresponding to the effective wavelength of the measuring pyrometer.

system in Righini et al (1972b). It may be noted that some improvements (optical and electronic) have been made in both systems since their general description in the earlier publications.

2. Measurements

The measurements of the radiance temperature of molybdenum (99·95% pure) at its melting point were performed on 28 specimens (14 specimens for each wavelength) in the form of strips. The manufacturer's typical analysis indicated the presence of the following impurities (in ppm by weight): W, 150; Cr, Fe, Ni, Si, Sn, 50 each; Al, Ca, Cu, Pb, Mg, Mn, 10 each. The nominal dimensions of the strips were: length, 51 mm; width, 6·3 mm; and thickness, 0·25 mm. Before the experiments, the surface of most of the specimens was treated using abrasive; three different grades of abrasive were used yielding three different surface roughnesses (ranging from approximately 0·2 to 0·5 μm) for different specimens. In some experiments, specimens with 'as received' surface conditions (approximately 0·1 μm in roughness) were also used.

All the experiments were performed with the specimen in an argon environment at atmospheric pressure. Because of the high vapour pressure of molybdenum at its melting point, it was not possible to conduct meaningful experiments with the specimen in a vacuum environment. The heating rate for different specimens was in the range 1000 to 5200 K s^{-1}, corresponding to specimen heating periods (from room temperature to its melting point) in the range 0·3 to 1 s.

Variation of the radiance temperature as a function of time near and at the melting point of molybdenum for typical experiments corresponding to wavelengths 653 and 995 nm is shown in figures 1 and 2, respectively. The magnitude of the spikes before the melting plateau is probably related to the degree of initial surface roughness of the specimen. However, regardless of the differences in the initial conditions, radiance temperature at the melting plateau is approximately the same for all the specimens at a given wavelength. The peak temperatures of the spikes were higher than the plateau tempera-

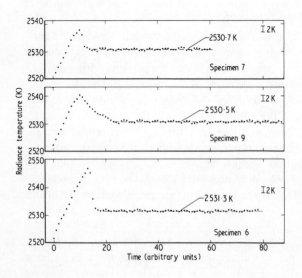

Figure 1. Variation of radiance temperature (at 653 nm) as a function of time near and at the melting point of molybdenum for three typical experiments. 1 time unit = 0·833 ms.

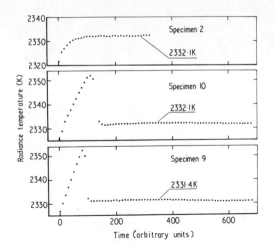

Figure 2. Variation of radiance temperature (at 995 nm) as a function of time near and at the melting point of molybdenum for three typical experiments. One temperature out of ten is plotted. 1 time unit = 0·151 ms.

tures in various experiments by about 4 to 16 K in the case of measurements at 653 nm and by about 10 to 52 K in the case of measurements at 995 nm.

3. Results

All temperatures reported in this paper are based on the IPTS-68 (1969). The radiance temperature of molybdenum at its melting point for the 28 specimens and other pertinent results corresponding to wavelengths 653 and 995 nm are presented in tables 1 and 2, respectively.

A single value for the radiance temperature at the plateau for each specimen was obtained by averaging the temperatures at the plateau. The number of temperatures at the plateau ranged from 25 to 99 for the measurements at 653 nm and from 122 to 596 for the measurements at 995 nm. The variation in the number of temperatures in measurements at a given wavelength depended both on the melting rate and on the behaviour of the specimen during melting. The standard deviation of an individual temperature from the average was in the range 0·3 to 0·4 K for the measurements at 653 nm and was in the range 0·1 to 0·3 K for the measurements at 995 nm. Similar values (for standard deviations) were obtained when data in the pre-melting period were fitted to a polynomial function in time. This indicates that during melting no undesirable effects took place, such as vibration of the specimen, development of hot spots in the specimen and random changes in the specimen surface conditions.

To determine the trend of measured temperatures at the plateau, temperatures for each experiment were fitted to a linear function in time using the least squares method. The slopes of the linear function do not show any significant bias with respect to sign. The detailed results are reported in tables 1 and 2. The temperature difference between the beginning and the end of the plateau (corresponding to the slope in the plateau) is in the range 0–0·5 K for the experiments at 653 nm and is in the range 0–1·1 K for the experiments at 995 nm.

The final results corresponding to measurements at the two wavelengths are summarized below.

Table 1. Summary of measurements of radiance temperature (at 653 nm) of molybdenum during melting.

Specimen number[a]	Surface roughness[b]	Premelting period		Melting period				
		Heating rate[c] ($K\ s^{-1}$)	Standard deviation[d] (K)	Number of temperatures[e]	Slope at plateau[f] ($K\ s^{-1}$)	Plateau temp. difference[g] (K)	Radiance temperature[h] (K)	Standard deviation[i] (K)
1	B	3500	0.3	39	8.9	0.3	2530.6	0.3
2	B	3400	0.2	32	−8.2	−0.2	2529.3	0.3
3	C	3300	0.3	49	−3.7	−0.2	2530.5	0.3
4	A	3300	0.3	25	0	0	2530.2	0.2
5	B	2500	0.3	67	6.6	0.4	2530.7	0.3
6	C	2500	0.4	58	0	0	2531.3	0.3
7	A	2400	0.4	47	5.6	0.2	2530.7	0.3
8	A	2400	0.3	75	0	0	2530.4	0.3
9	B	2500	0.3	78	0	0	2530.5	0.3
10	C	2500	0.3	48	9.3	0.4	2529.2	0.4
11	A	2400	0.4	45	7.7	0.3	2530.7	0.3
12	B	1200	0.4	99	0	0	2530.4	0.4
13	A	2600	0.4	28	0	0	2530.7	0.4
14	A	2500	0.4	51	11	0.5	2530.9	0.4

[a] Also represents the experiments in chronological order.
[b] The notations correspond to the following typical roughnesses in μm: A, 0.2; B, 0.4; C, 0.5.
[c] Heating rate evaluated at a temperature approximately 10 K below the melting point.
[d] Represents standard deviation of an individual temperature as computed from the difference between the measured value and that from the smooth temperature versus time function (quadratic) obtained by the least squares method. Data extend approximately 100 K below the melting point.
[e] Number of temperatures used in averaging the results at the plateau to obtain an average value for the radiance temperature at the melting point of the specimen.
[f] Derivative of the temperature versus time function obtained by fitting the temperature data at the plateau to a linear function in time using the least squares method.
[g] Maximum radiance temperature difference between the beginning and the end of the plateau based on the linear temperature versus time function.
[h] The average (for a specimen) of measured radiance temperatures at the plateau.
[i] Standard deviation of an individual temperature as computed from the difference between the measured value and that from the average plateau radiance temperature.

Table 2. Summary of measurements of radiance temperature (at 995 nm) of molybdenum during melting.

Specimen number[a]	Surface roughness[b]	Premelting period		Melting period				
		Heating rate[c] ($K\,s^{-1}$)	Standard deviation[d] (K)	Number of temperatures[e]	Slope at plateau[f] ($K\,s^{-1}$)	Plateau temp. difference[g] (K)	Radiance temperature[h] (K)	Standard deviation[i] (K)
1	A	1700	0·3	138	−11·4	−0·4	2330·5	0·1
2	D	3300	0·1	175	0	0	2332·1	0·1
3	D	3100	0·1	298	−21·4	−1·0	2331·0	0·3
4	B	3300	0·1	172	12·5	0·3	2331·9	0·1
5	C	1200	0·1	303	−15·1	−0·7	2330·9	0·2
6	A	2400	0·1	370	6·5	0·4	2332·4	0·1
7	C	2300	0·1	596	−1·2	−0·1	2330·7	0·1
8	B	2400	0·1	280	−11·3	−0·5	2330·1	0·2
9	A	2300	0·1	316	2·0	0·1	2331·4	0·1
10	C	1700	0·1	380	−1·7	−0·1	2332·1	0·1
11	B	2100	0·1	181	−8·6	−0·2	2331·0	0·1
12	C	2000	0·1	466	16·3	1·1	2331·6	0·3
13	B	5200	0·1	122	21·9	0·4	2332·3	0·1
14	C	1000	0·1	296	−1·4	−0·1	2330·5	0·2

[a] Also represents the experiments in chronological order.
[b] The notations correspond to the following typical roughnesses in μm: A, 0·2; B, 0·4; C, 0·5; D, 0·1 (as received).
[c] Heating rate evaluated at a temperature approximately 10 K below the melting point.
[d] Represents standard deviation of an individual temperature as computed from the difference between the measured value and that from the smooth temperature versus time function (linear) obtained by the least squares method. Data extend approximately 100 K below the melting point.
[e] Number of temperatures used in averaging the results at the plateau to obtain an average value for the radiance temperature at the melting point of the specimen.
[f] Derivative of the temperature versus time function obtained by fitting the temperature at the plateau to a linear function in time using the least squares method.
[g] Maximum radiance temperature difference between the beginning and the end of the plateau based on the linear temperature versus time function.
[h] The average (for a specimen) of measured radiance temperatures at the plateau.
[i] Standard deviation of an individual temperature as computed from the difference between the measured value and that from the average plateau radiance temperature.

3.1. At 653 nm

The average radiance temperature at the melting point for the 14 molybdenum specimens is 2530·4 K with a standard deviation of 0·6 K and a maximum absolute deviation of 1·2 K. The results are presented in figure 3. The two experiments (corresponding to specimen 2 and specimen 10) with relatively large negative deviations (about −1·1 K and −1·2 K) have caused a downward shift of the average by about 0·2 K. In order to reduce the contribution of these two experiments, the average temperature is rounded-off to the higher value. It may be concluded that the radiance temperature (at 653 nm) of molybdenum at its melting point is 2531 K.

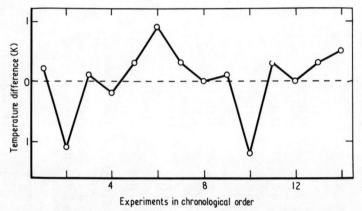

Figure 3. Difference of radiance temperature (at the melting point of molybdenum, at 653 nm) for individual experiments from their average value of 2530·4 K (represented by the 'zero' line).

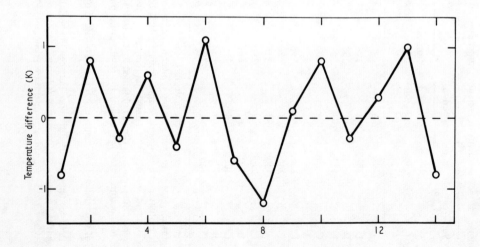

Figure 4. Difference of radiance temperature (at the melting point of molybdenum, at 995 nm) for individual experiments from their average value of 2331·3 K (represented by the 'zero' line).

3.2. At 995 nm

The average radiance temperature for the 14 molybdenum specimens is 2331·3 K with a standard deviation of 0·7 K and a maximum absolute deviation of 1·2 K. The results are presented in figure 4. It may concluded that the radiance temperature (at 995 nm) of molybdenum at its melting point is 2331 K.

Considering 2894 K for the melting point of molybdenum (Cezairliyan *et al* 1970a) and the radiance temperatures measured in this work, values for the normal spectral emittance of molybdenum at its melting point are obtained; 0·335 at 653 nm and 0·298 at 995 nm.

4. Estimate of errors

The details of sources and estimates of errors in temperature measurements in high-speed experiments using the present systems are given in earlier publications: the NBS system in Cezairliyan *et al* (1970b), the IMGC system in Righini *et al* (1972a, b). Summaries of estimated inaccuracies (combined random and systematic errors) in radiance temperature measurements pertinent to the work described in this paper are given in tables 3 and 4 for the measurements at 653 nm and 995 nm, respectively. Specific items in the error analysis were recomputed whenever the present operational conditions or experimental set-ups differed from those in the earlier studies. The estimated inaccuracy in radiance temperature measurements is about 5 K for the system at NBS and about 6 K for the system at IMGC. In an earlier study (Cezairliyan *et al* 1970a), the effect of impurities (about 0·1%) on the melting point of molybdenum was estimated to be not more than 2 K. It may be concluded that the inaccuracy in the reported radiance temperature of molybdenum at its melting point is not more than ±8 K.

Table 3. Estimated inaccuracy in radiance temperature measurements (at 2500 K and at 653 nm) with the high-speed photoelectric pyrometer at NBS.

Source	Inaccuracy (K)
Standard lamp calibration	3
Drift in standard lamp calibration	1
Radiation source alignment	2
Neutral density filter calibration	1
Window calibration	1
Pyrometer calibration stability	2
Effective wavelength calibration	2
Total inaccuracy (root sum square of above items)	5

Table 4. Estimated inaccuracy in radiance temperature measurements (at 2400 K and at 995 nm) with the high-speed infrared pyrometer at IMGC.

Source	Inaccuracy (K)
Gold-point calibration	1·4
Neutral density filter calibration	1·5
Window calibration	1
Radiation flux ratio calibration	4·5
Pyrometer calibration stability	2
Effective wavelength calibration	2
Total inaccuracy (root sum square of above items)	6

5. Discussion

The present results have shown the constancy and reproducibility of the radiance temperature of molybdenum at its melting point for a number of specimens with different initial surface conditions in experiments performed under different operational conditions. This, in addition to similar earlier results on niobium (Cezairliyan 1973) and zirconium (Cezairliyan and Righini 1975) suggests the evaluation of the possibility of using the radiance temperature of selected metals for *in situ* calibration or checking of optical temperature-measuring equipment. This scheme has practical advantages compared to elaborate black body cavities (for obtaining true melting points) at temperatures above 2000 K, especially for high-speed measurement systems. However, in the former case, a knowledge of the dependence of radiance temperature on wavelength is required to assess the magnitude of errors that may result in calibrating equipment whose effective wavelength departs, by an unknown amount, from that used in the original determination of the radiance temperature.

An estimate for the wavelength dependence of the radiance temperature of molybdenum at its melting point is obtained in the following paragraph.

Using Planck's law, the relation for radiance temperature, $T_r(\lambda)$, as a function of wavelength, λ, at the melting point may be expressed as

$$T_r(\lambda) = c_2 \left[\lambda \ln \left(1 + \frac{\exp(c_2/\lambda T_m) - 1}{\epsilon(\lambda)} \right) \right]^{-1} \tag{1}$$

where c_2 is the second radiation constant (0·014388 mK), T_m is the melting point and $\epsilon(\lambda)$ is the normal spectral emittance at the melting point. The quantity $\epsilon(\lambda)$ is evaluated using a linear relation between the values of the present work corresponding to wavelengths 653 nm and 995 nm. Considering the value 2894 K for the melting point of molybdenum (Cezairliyan *et al* 1970a), one obtains the wavelength dependence of the radiance temperature of molybdenum at its melting point. The results are presented graphically in figure 5 for two wavelength regions (near those used in the present work).

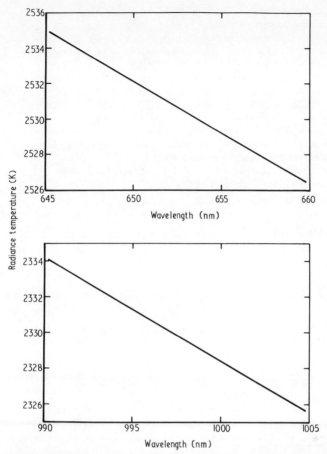

Figure 5. Wavelength dependence of the radiance temperature of molybdenum at its melting point near 0·65 and 1 μm (computed using equation (1)).

The results indicate that the radiance temperature dependence on wavelength is nearly linear over the ranges considered, and that radiance temperature decreases with increasing wavelength by approximately 0·6 K nm^{-1}. This simplified computation provides an estimate of the error in radiance temperature measurement for an uncertainty in the effective wavelength of the measuring equipment. It also allows the comparison of the literature results obtained from measurements at different wavelengths near each other.

Only two papers reporting measurements of the radiance temperature of molybdenum at its melting point near 0·65 μm were located in the literature. Their results are summarized in table 5. No measurements near 1 μm were found in the literature. It may be seen that the value of the present work is in reasonably good agreement (within 6 K) with that of Berezin *et al* (1971). However, the value of the present work is 26 K higher than that of Bonnell *et al* (1972).

In conclusion, the results of the present work in addition to those of the earlier works on niobium (Cezairliyan 1973) and zirconium (Cezairliyan and Righini 1975) suggests the possibility of using the radiance temperature at the melting point of selected metals

Table 5. Radiance temperature, T_r, of molybdenum at its melting point near 0·65 μm reported in the literature.

Investigator	Purity %	As reported[†]		Converted[‡]	
		λ (nm)	T_r (K)	λ (nm)	T_r (K)
Berezin et al (1971)		656	2523·7	653	2525·5
Bonnell et al (1972)	99·9	645	2510	653	2505
Present work	99·95	653	2531	653	2531

[†] Radiance temperature as reported by the investigators corresponding to the effective wavelength of the pyrometer used.
[‡] Radiance temperature corresponding to 653 nm. Conversion is made using the factor 0·6 K nm^{-1} obtained in this work.

for secondary calibration and checking of optical temperature measuring equipment in high temperature systems. However, the final assessment will require additional accurate work on the same and other metals.

Acknowledgment

The work performed at NBS was supported in part by the US Air Force Office of Scientific Research.

References

Berezin B Ya, Chekhovskoi V Ya and Sheindlin A E 1971 *High Temp. High Press.* **3** 287–97
Bonnell D W, Treverton J A, Valerga A J and Margrave J L 1972 *TMCSI* **4** part 1 483–7
Cezairliyan A 1971 *J. Res. NBS* **75C** 7–18
—— 1973 *J. Res. NBS* **77A** 333–9
Cezairliyan A, Morse M S and Beckett C W 1970a *Rev. Int. Hautes Temp. et Réfract.* **7** 382–8
Cezairliyan A, Morse M S, Berman H A and Beckett C W 1970b *J. Res. NBS* **74A** 65–92
Cezairliyan A and Righini F 1975 *Rev. Int. Hautes Temp. et Réfract.* in press
Foley G M 1970 *Rev. Sci. Instrum.* **41** 827–34
IPTS-68 1969 *Metrologia* **5** 35–44
Righini F, Rosso A and Ruffino G 1972a *TMCSI* **4** part 1 413–21
—— 1972b *High Temp. High Press.* **4** 597–603

Temperature measurement of filaments above 2500 K applying two-wavelength pyrometry

W Lechner
Philips Forschungslaboratorium Aachen GmbH, Aachen, W Germany
O Schob
NV Philips Gloeilampenfabrieken, Light Division, Eindhoven, The Netherlands

Abstract. A simple combined brightness and two-wavelength pyrometer has been developed for temperature measurements on filaments of incandescent lamps. Methods and instrumentation are described, as well as a calibration aid for the temperature region above 2800 K using the radiance of tungsten at its melting point. Both theoretical considerations and experimental results show that in the temperature range from 2500 to 3660 K — at least for industrial purposes — reproducibility and accuracy, especially for the determination of temperature differences, are quite satisfactory. Under certain conditions measuring errors arising from uncertainty regarding emittance and transmittance can be detected and eventually eliminated by employing brightness and two-wavelength pyrometry alternately.

1. Introduction

The development of incandescent lamps nowadays necessitates accurate determination of the true temperature of the filaments, while small, sometimes in fact extremely small, local temperature differences have to be determined with quite considerable accuracy, even when the level of the true temperature is not known with great precision. Reference may be made in this connection to some research on burning-out mechanisms in incandescent lamps in which temperature differences along the tungsten wire or between adjacent turns of a coil play an important role (Hörster et al 1971, 1972, Schob et al 1973).

Brightness pyrometers are commonly used to determine the true temperature of a radiator by means of a single wavelength, but they are devices, which require, however, the spectral emissivity of the radiating source to be known with great accuracy. The lack of reliable values for this physical quantity has in the past led many investigators to develop methods which are independent of it. Thus a considerable number of publications deal with the possibility of overcoming this disadvantage by employing two-wavelength (or two-colour) pyrometry, or even multi-colour pyrometry (Ackermann 1962, Elder 1962a, Hecht 1962, Hill 1962, Hornbeck 1962, DeWitt and Kunz 1972).

Our aim here was to use both brightness pyrometry and two-colour pyrometry to improve measuring results wherever possible. This, however, is not done by simply determining the colour temperature, assuming grey body conditions verified with tungsten radiators. Instead, the true temperature is determined by deliberately considering known values of the spectral emissivity of tungsten at the respective measuring wavelengths.

2. Description of the pyrometer

As an example of a simple and inexpensive pyrometer, which fulfils all the requirements for incandescent lamp research and development, a combined brightness and two-colour pyrometer is shown in figure 1. The aim with this laboratory instrument was to achieve the greatest possible simplification and omit all complicated optical and electronic equipment without any loss of sensitivity or reliability. A simple camera lens (with a built-in iris diaphragm) is used to project an image of the radiator onto the entrance slit of an integrating sphere. After two effective measuring wavelengths (in our case at 450 and 630 nm) have been filtered out the radiation is collected by the

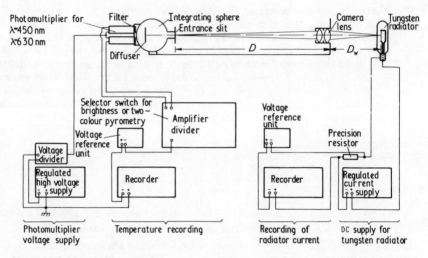

Figure 1. Schematic diagram of the optics and electronics of the combined brightness and two-colour pyrometer. If, for example, a 1:4/35 camera lens (focal length 35 mm) is used, with $D = 600$ mm and $D_W \approx 35$ mm, the magnification is 20×, which is suitable for most temperature measurements on incandescent lamps.

photocathodes of two separate photomultipliers (Valvo XP 1016, with spectral characteristic S 20). After each photomultiplier signal has passed through a double-stage amplifier, the blue-red ratio signal is formed precisely by using the Analog Devices Divider type 427 K, while the resulting signal is recorded on a standard recorder. By switching to either the red or the blue measuring wavelength on its own the signal can be recorded with sufficient sensitivity at temperatures above 2000 K without any further amplification. The other electronic units shown in figure 1 for controlling and measuring voltage and current are commercially available laboratory equipment. Briefly, the device combines simplicity and ease of operation with high sensitivity and accuracy from the viewpoint of reproducibility. Indeed, the technical problems are negligible compared with the problems arising from the chief uncertainties of applied pyrometry, namely the effects of emittance and transmittance on the determination of true temperatures.

3. Definitions and theory

Using the Wien approximation to the Planck radiation law, the following relations can be derived for the response of the instrument:

Brightness pyrometry:

$$s = K\tau\epsilon I(\lambda_{1,2}, T) = K\tau\epsilon c_1 \lambda_{1,2}^{-5} \exp(-c_2/\lambda_{1,2} T)$$

or taking logarithms:

$$\ln s = -\frac{c_2}{\lambda_{1,2}} \frac{1}{T} + \ln [K\tau c_1 \lambda_{1,2}^{-5}] + \ln \epsilon. \tag{1}$$

Two-colour pyrometry:

$$r = K_r \bar{\tau} \bar{\epsilon} I(\lambda_1, T)/I(\lambda_2, T)$$

$$= K_r \bar{\tau} \bar{\epsilon} (\lambda_1/\lambda_2)^{-5} \exp\left[\frac{c_2}{T}\left(\frac{1}{\lambda_2} - \frac{1}{\lambda_1}\right)\right]$$

or taking logarithms:

$$\ln r = -c_2\left(\frac{1}{\lambda_1} - \frac{1}{\lambda_2}\right)\frac{1}{T} + \ln [K_r \bar{\tau} (\lambda_1/\lambda_2)^{-5}] + \ln \bar{\epsilon}. \tag{2}$$

In the above equations $I(\lambda, T) = c_1\lambda^{-5} \exp(-c_2/\lambda T)$ is the spectral radiance, T the true temperature, s, r, the response of the pyrometer, K, K_r the instrument parameters, λ_1, λ_2 the effective wavelengths of the instrument, c_1, c_2 the first and second constants of Planck's law, $\tau = \tau(\lambda_{1,2})$ the transparency of envelope, $\epsilon = (\lambda_{1,2}, T)$ the spectral emissivity and $\bar{\tau} = \tau(\lambda_1)/\tau(\lambda_2)$ and $\bar{\epsilon} = \epsilon(\lambda_1, T)$ are abbreviations.

The second terms in equations (1) and (2) are constants, ie they are independent of temperature. Furthermore, it can be shown by careful analysis of the available data (de Vos 1954) that the temperature dependence of the spectral emissivity of tungsten is so small that the terms $\ln \epsilon$ in equation (1), and $\ln \bar{\epsilon}$ in equation (2) can also, for practical purposes, be considered constants. Thus, equation (1) and (2) may be written:

$$\ln s = -\frac{c_2}{\lambda_{1,2}} \frac{1}{T} + \text{constant}$$

$$\ln r = -c_2\left(\frac{1}{\lambda_1} - \frac{1}{\lambda_2}\right)\frac{1}{T} + \text{constant}$$

and thus will, in a $\ln s$ (or $\ln r$) against $1/T$ diagram, be respectively represented by straight lines. This is actually shown in figure 2, where, in the dark region, straight lines are indeed obtained with our instrument for characteristic measuring conditions of incandescent lamps.

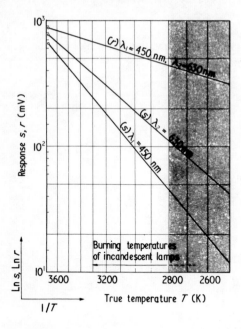

Figure 2. Plot of a characteristic response (s, r) against the true temperature (T) of a tungsten radiator, respectively the $\ln s$ (or $\ln r$) against $1/T$ diagram, obtained with the pyrometer for different measuring wavelengths: (s) brightness pyrometry, (r) two-colour pyrometry at the specified measuring wavelengths. The temperature range for incandescent lamps is indicated. The region of calibration temperatures accessible with tungsten strip lamps is restricted to the shadowed area. The response lines end at the melting point of tungsten indicated by an open circle.

4. Calibration

The only calibration standards available today are tungsten strip lamps with which temperatures exceeding 2800 K cannot be calibrated without damaging the standard. This means that the temperature region in which incandescent lamps are operated, namely up to about 3400 K as indicated in figure 2, cannot be directly calibrated. To allow measurements in this region, the $\ln s$ (or $\ln r$) against $1/T$ lines had to be extrapolated. Both these lines are obtained for the dark region of figure 2 by recording the radiance signals from a strip lamp standard. To convert the radiance temperatures of the standard into true temperatures of the filaments, the emissivity data of de Vos (1954) and the relevant values for the spectral transparency of the envelopes are used. The $\ln s$ (or $\ln r$) against $1/T$ lines found in this way are indeed straight lines which can be extrapolated for both brightness and two-colour pyrometry because of the conditions already discussed above.

To improve this extrapolation method, we employ a method of correction which, though useful, is restricted to our case, namely measurement of the temperatures of incandescent lamps. By melting the tungsten filament inside a halogen lamp locally and recording the signal, we can obtain a valuable fixed temperature point. If the hottest turn of a coil is identified and the lamp current slowly increased, the tungsten wire can be melted locally without destroying the filament. A halogen lamp is used because it prevents the condensation of evaporated tungsten onto the inner surface of the bulb. On the SEM picture (figure 3) of the coil of a halogen lamp treated in this way, the melted and subsequently solidified part of a single turn can be clearly distinguished. The differences between the signals found in this way can be expressed in Kelvin and, as a series of tests has shown, they do not exceed 10 K.

It must be borne in mind, of course, that this method is proposed not as a means of determining the temperature at which tungsten melts, but merely in order to generate

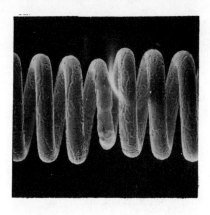

Figure 3. Coil of a halogen lamp with the melted and subsequently re-solidified part of a turn. The lamp was used to record the response of the pyrometer to the spectral radiance of the melting tungsten.

a signal at a reproducible fixed point on the temperature scale. Under approximately the same conditions prevailing in our investigations into incandescent lamps, we can use this signal point (s, r in mV for the accepted melting temperature of tungsten, ie, 3660 K) to determine with quite considerable accuracy the difference between any other recorded temperature signal (in mV) and the melting point, without knowing the exact value of the latter.

5. Accuracy and range

The sensitivity of the instrument is indicated by the slopes of the relevant $\ln s$ (or $\ln r$) against $1/T$ lines. Examination of equations (1) and (2) shows that the slopes can be expressed by the term $-c_2/\lambda_{1,2}$ for brightness pyrometry and $-c_2(1/\lambda_1 - 1/\lambda_2)$ for two-colour pyrometry, neither of which contains the instrument parameters K and K_r. A variation of these parameters by, for instance, varying the iris diaphragm and the size of the entrance slit, entail a parallel shift of the plotted lines in figure 2 without influencing the slope and the sensitivity, at least for temperatures not lower than 2000 K. Apart from c_2, the slope terms contain only the effective wavelengths. Thus the sensitivity to temperature changes is determined solely by the choice of the measuring wavelengths (cf figure 2). With the wavelengths chosen in our instrument, the sensitivity of two-colour pyrometry is about two to three times smaller than that of brightness pyrometry, as can also be seen from figure 2. As the sensitivity cannot be influenced by the parameters of the instrument, ie they can only shift the level of the signals, it is therefore convenient to specify the accuracy obtained with this instrument in terms of the reproducibility of the signals, expressed in Kelvin. In the temperature range from about 2200 to 3660 K and with a filament surface area as small as 2×10^{-4} mm^2 the signals reproduce with an accuracy corresponding to ±1 K for brightness pyrometry and ±2 K for two-colour pyrometry. This compares very well with the accuracy claimed for a number of other laboratory electronic pyrometers with more sophisticated technical set-ups.

Advantage can be taken of this fact to obtain, for instance, very accurate temperature profiles by scanning filaments, even if the relevant spectral emissivity of the filaments is not known exactly. Provided that other means, eg scrutinizing the filament with a microscope, can be taken to ensure that the surface conditions are the same everywhere, temperature differences can be determined with the accuracy mentioned, without exact knowledge of the level of the true temperature.

6. Application

Measurements made on industrially manufactured incandescent lamps are unavoidably performed under conditions quite different from those applying, for example, elaborate calibration standards. For instance, plane tungsten surfaces are seldom present, and wires and coils cannot always be sighted perpendicular to the surface. Furthermore, the following conditions may be encountered:

(i) Multiple reflections within grooves in the filament's surface, between turns of the coil and/or between the glass envelope and the filament;
(ii) Angular dependence of radiance;
(iii) Variation of the measuring area;
(iv) Disturbance of the optical path due to blackening of the envelope or to glass defects.

When two-colour pyrometry is employed the above shortcomings, as might be expected, can be overcome when forming the signal ratio if these effects on the original spectral radiance of the filament are of the same order and sign at the two measuring wavelengths. With this consideration in mind some investigations were carried out, with the following results:

(i) Multiple reflections can cause serious errors, especially when they occur within a coil. If brightness pyrometry is employed, an apparent rise of the temperature is found which originates from the well known increase of radiance when a black body cavity is approached. When the blue-red ratio is formed in two-colour pyrometry an apparent temperature drop of about the same magnitude as the rise with brightness pyrometry is observed. As for tungsten $\epsilon_{\lambda_1}(\lambda_1 = 450\,\text{nm}) > \epsilon_{\lambda_2}(\lambda_2 = 630\,\text{nm})$, (cf de Vos 1954) approaching black body conditions means that the emissivity at 630 nm will show a larger increase than at 450 nm. This explains the apparent drop in temperature (see also figure 4). A similar, though less-pronounced effect can be found with filaments possessing a rough surface.

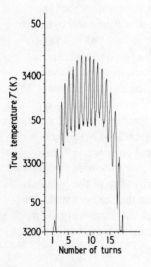

Figure 4. Temperature profile of a halogen lamp (12 V, 55 W) obtained with two-colour pyrometry. The maxima represent the true temperature of the single turns of the coil (15 turns). The minima, which in fact originate from the more intense radiance from within the coil (black body effect), are obtained by forming the blue-red ratio.

A different situation occurs if the radiation is reflected from the envelope back on to the tungsten radiator. In this case two-colour pyrometry practically completely eliminates an otherwise serious measuring error.

To sum up, errors caused by multiple reflections between tungsten and tungsten cannot be eliminated. In the case of multiple reflections between the envelope and tungsten, two-colour pyrometry offers distinct advantages. By comparing the results of the different measuring methods, the nature of the errors can be detected and allowance made for them.

(ii) With regard to the angular dependence of tungsten radiance, two-colour pyrometry might be expected to eliminate the increase in radiance that occurs when the viewing direction deviates from the perpendicular (deviation from Lambert's cosine law). As experiments showed (see figure 5), the 450 and 630 nm measuring wavelengths used

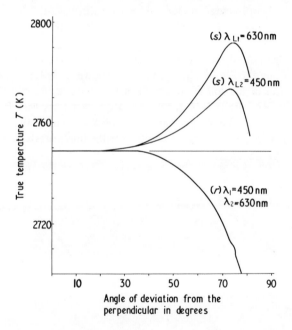

Figure 5. Angular dependence of the radiance of tungsten at a true temperature of about 2700 K. For comparison of the different measuring errors which might result from the application of brightness (s) or two-colour (r) pyrometry, the signals obtained have been converted to the corresponding number of 'true temperature' Kelvin.

in our instrument are still separated too far apart to compensate for this increased radiance at viewing angles exceeding 40° from the perpendicular. As figure 5 shows, the temperatures obtained from the blue-red ratio signal in this range seem to drop at about the same rate at which they rise when brightness pyrometry is employed. This, therefore, was an error which could not be eliminated in our measurement conditions (temperature and wavelengths chosen).

(iii) With respect to variations of the measuring area, two-colour pyrometry shows distinct advantages. Such variations can be caused by a change of wire geometry during an investigation, with the consequence that the entrance slit is not completely covered by the source area. When the measuring area at the same temperature varied up to about 50% in size, measurements with our two-colour pyrometer did not show any perceptible deviation. Brightness pyrometry, however, gave variations of the signal proportional to the measuring area.

(iv) With regard to disturbances of the optical path, we confined our activities to investigating the influence on transparency of blackening of the envelopes by evaporated tungsten, and of glass or quartz defects. In the matter of the former problem the use of two-colour pyrometry yielded little improvement. The condensed tungsten apparently influences the emerging radiation by selective absorption and dispersion in a manner which is too complicated to permit any appreciable systematic improvement of the measurement.

The situation, however, is somewhat better with respect to glass defects, as long as they have small dimensions, eg tiny bubbles, schlieren. These can affect brightness pyrometry measurements in two different ways: the temperature may appear either higher or lower, due to a converging, diverging or scattering of the radiance by the glass defects. Measurements with two-colour pyrometers under the same conditions are much less influenced. Consequently the nature of these errors can be identified by using both methods in turn.

Finally, as an example of the application of temperature measurements a record of the temperature changes which occur during the life of a halogen lamp is shown in figure 6. The temperature of the individual turns of the coil at different stages of the life of the lamp burning at an average temperature of about 3400 K are obtained by scanning the coil as indicated in figure 4. As will be seen, the temperature differences occurring can be recorded with sufficient accuracy to give an informative picture of the burn-out of the halogen lamp.

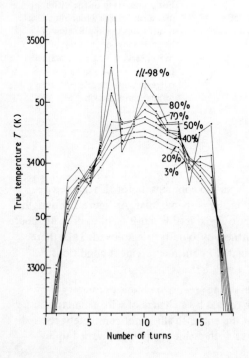

Figure 6. Temperature distribution along the coil of a halogen lamp (12 V, 55 W; cf figure 4) at different stages of its life, the time t for which the lamp has burnt being expressed as a percentage of the life l. The various temperature profiles were obtained by repeated scanning of the coil, employing two-colour pyrometry as shown in figure 4.

References

Ackermann S 1962 *TMCSI* **3** part 2 839, 849
DeWitt D P and Kunz H 1972 *TMCSI* **4** part 1 599
Elder S A 1962a *TMCSI* **3** part 2 859
—— 1962b *TMCSI* **3** part 2 873
Hecht G J 1962 *TMCSI* **3** part 2 407
Hill W E 1962 *TMCSI* **3** part 2 419
Hornbeck G A 1962 *TMCSI* **3** part 2 425
Hörster H, Kauer E and Lechner W 1971 *Philips Tech. Rev.* **32** 155
—— 1972 *J. Illum. Eng. Soc.* **1** 309
Schob O, de Bie J R and Klomp W G A 1973 *Light. Res. Technol.* **5** 29
de Vos J C 1954 *Physica* **20** 669, 690

Temperature measurement using a Plumbicon[†] camera tube

J R de Bie and W G Klomp

Lighting Division, NV Philips' Gloeilampenfabrieken, Eindhoven, The Netherlands

Abstract. The Plumbicon camera tube possesses excellent photoelectric properties for the measurement of temperature.

Equipment has been developed upon the basis of the following principles. The modulation along every line of the scan indicates the radiant intensity of the video signal on the line. The radiant intensity of an object is a known function of its temperature and thus, with the aid of an appropriate circuit, the temperature distribution over the object for each line of scan can be observed. In the temperature range 1700–1950 K the signal is a linear function of temperature. Above 1950 K appropriate filters have to be used. The circuitry used also permits the selection of a small part of the total information, eg the temperature of a preselected spot. Another advantage of the method is the fact that the objective measurement of temperature can be combined with the visual observation of the radiating object.

The accuracy is determined by the inhomogeneity of the photosensitive layer and is better than 0·5% above 1700 K. A standard light source is used as a reference.

1. Introduction

Mainly because their temperatures are high, their dimensions small and their structure complicated the only really suitable method of determining the temperature of light sources is to measure their radiation. A photoelectric detector is an obvious means of doing this.

The use of special TV camera tubes for this purpose, especially in a low-temperature range, is common practice. It will be shown that a normal Plumbicon camera tube can also be used for temperatures above 1000 K. For a measuring instrument it is necessary for the relation between the true temperature of the incandescent source and the signal voltage to be known.

2. Basic considerations

The basic principle of radiation measurement is shown in figure 1.

By definition (Rutgers and de Vos 1954, de Vos 1953 and the 5th Temperature Symposium), the radiance temperature is the temperature of a black body radiating at a certain wavelength (λ) which has the same radiant intensity as the incandescent source considered. The relation between the radiance temperature (T_{black}) and true temperature (T) as shown in figure 1 is therefore given by:

$$L(\lambda, T_{\text{black}}) = \tau_1(\lambda)\, \epsilon(\lambda, T)\, L(\lambda, T) \tag{1}$$

[†] Registered trade mark.

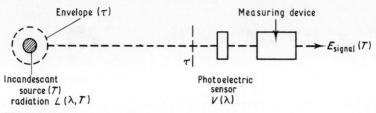

Figure 1. $\tau_{i=1,\ldots n}$ = transmittance, $V(\lambda)$ = spectral response, $L(\lambda, T)$ = spectral radiance and $\epsilon(\lambda, T)$ = emissivity.

where $L(\lambda, T)$ is the spectral radiance according to Planck's law and is equal to $c_2 \lambda^{-5} [\exp(+c_2/\lambda T) - 1]^{-1}$, T is the true temperature of the source, T_{black} the radiance temperature, $\tau_1(\lambda)$ the transmittance of the envelope and $\epsilon(\lambda, T)$ the emissivity of the source.

Under measuring conditions the photoelectric device has a spectral sensitivity over a certain range $V(\lambda)$. The transmittance of the atmosphere and the optical system have also to be considered. This means that the relation between the radiance temperature and the true temperature is now given by:

$$\int_{\lambda_1}^{\lambda_2} V(\lambda) \left(\prod_{i=2-n} \tau_i(\lambda) \right) L(\lambda, T_{\text{black}}) \, d\lambda = \int_{\lambda_1}^{\lambda_2} V(\lambda) \left(\prod_{i=1-n} \tau(\lambda) \right) \epsilon(\lambda, T) L(\lambda, T) \, d\lambda$$

$$\sim E_{\text{signal}}. \qquad (2)$$

$\lambda_1 - \lambda_2$ is the range of spectral sensitivity of $V(\lambda)$.

The requirements for measurements can now be more precisely formulated as follows:

(i) $E_{\text{signal}} = f(T)$, ie the electrical signal, must only be a function of the temperature.
(ii) For simple operation of the equipment the sensitivity in the temperature range to be considered (dE_{signal}/dT) must have a constant value. In the case of an approximated linear function for T, $E_{\text{signal}} = c_0 T$, c_0 must be a maximum for optimum sensitivity.
(iii) With regard to accuracy, the minimum value $\Delta T/T$ that can be detected has to be specified. For example, $\Delta T/T < 1\%$.

3. Use of camera tubes

Three different camera tubes and their properties will be compared with reference to the foregoing specifications (de Bie et al 1972, Dollekamp et al 1970, van Doorn 1966, van der Drift et al 1963, 1964, Heyne 1960, Levitt 1970).

From the given relation (2) we can conclude that the spectral sensitivity $V(\lambda)$ and the electrical signal–light sensitivity are the most important characteristics of the camera tubes.

Figure 2 shows the essential features of the orthicon, the vidicon and the Plumbicon camera tubes. The spectral responses $V(\lambda)$ are given in figure 2(a) while figure 2(b) shows the signal–light ($E_{\text{signal}} - E_f$) curves under identical measuring conditions. The signal–light relation can be given by $E_{\text{signal}} = E_f^\gamma$, where E_f is the luminous flux and γ the exponent.

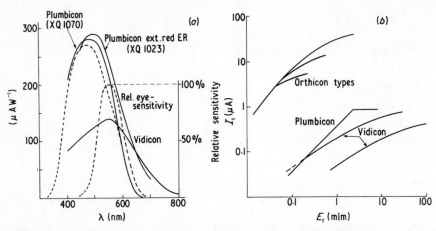

Figure 2. (*a*) Spectral response of some types of camera tube. (*b*) Signal–light characteristics of different types of camera tube.

From these figures it can be seen that the Plumbicon is much more sensitive than the vidicon tube and that it also has a more linear signal–light characteristic ($\gamma \approx 1$).

Table 1 also shows that it has a low dark current and a better response time in the decay mode.

These properties make the Plumbicon camera tube very suitable for radiation measurements. The more complicated construction and operation of the orthicon tube make it somewhat difficult to use in a simple measuring device.

Table 1. Properties of camera tubes.

Type of camera tube	Photo-layer	Sensitivity ($\mu A\, lm^{-1}$)	γ	Dark current (nA)	Maximum face-plate illumination (lx)	Decay residual signal > 200 ms
Plumbicon (XQ 1070)	PbO	300–400	~1	<3	500	<0.6%
Vidicon (XQ 1010)	Sb_2S_3	~20	~0.7	20	5000	8%

Using the spectral response curve in figure 2(*a*) we can now calculate the properties:

$$\int_{\lambda_1}^{\lambda_2} V(\lambda)\, L(\lambda, T)\, d\lambda \quad \text{and} \quad \int V_{\text{eye}}(\lambda)\, L(\lambda, T)\, d\lambda$$

as a function of the temperature. $L(\lambda, T)$ is the radiation energy of tungsten ribbon (emissivity values taken from de Vos 1953).

The calculated relations referred to are shown in figure 3. They indicate the theoretical principle of temperature measurement with the Plumbicon camera tube. The calculated functions $L(\lambda_1, T)/L(\lambda_2, T) = f(T)$ for the wavelengths $\lambda_1 = 0.46$, $\lambda_2 = 0.64$ μm and tungsten ribbon are also given.

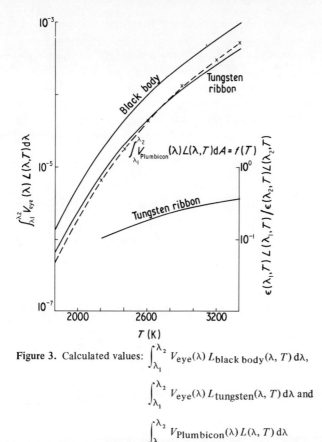

Figure 3. Calculated values: $\int_{\lambda_1}^{\lambda_2} V_{\text{eye}}(\lambda) L_{\text{black body}}(\lambda, T) \, d\lambda$,

$\int_{\lambda_1}^{\lambda_2} V_{\text{eye}}(\lambda) L_{\text{tungsten}}(\lambda, T) \, d\lambda$ and

$\int_{\lambda_1}^{\lambda_2} V_{\text{Plumbicon}}(\lambda) L(\lambda, T) \, d\lambda$

as a function of temperature T.

The fact that the slope of this last curve (for two-colour pyrometry) is less steep than of the video measurement indicates that the latter method is more sensitive and can theoretically detect smaller values of $\Delta T/T$.

4. The measuring device

The block diagram given in figure 4 illustrates the basic principle of the method of measurement (Kerkhof and Werner 1952, Jansen 1950).

We can distinguish three main parts: part I, the object to be measured; part II, the measuring instrument; and part III, the display and the analogue–digital registration.

The object to be measured (incandescent source) is projected by means of an optical system on to the photoelectric layer of the camera tube.

A preselected line $(n_{i=1-625})$ in the relevant area can be selected from the control unit. In the next stage the size of the measuring spot is adjusted and indicated (either manually or automatically). This stage is connected to two video amplifiers. The output of video amplifier 1 applies a signal with a suppressed measuring spot (see figure 4) to the input of the display (TV monitor) and indicates the measuring position on the image. The output of video amplifier 2 is connected to a peak detection memory circuit and the latter transmits to the recording instrument (and/or digital apparatus) a signal which is proportional to the temperature.

The optical system consists of a lens system (photo-optical system 1) and a suitable filter (the value of which will depend on the temperature range). Photo-optical system 2 (photoelectric layer) is a part of the camera tube.

For the system as a whole the following parameters are important for the operation of the measuring device.

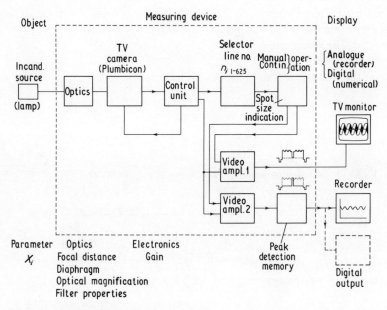

Figure 4. Basic principle of measurement.

Photo-optical system 1: diaphragm (D), focal distance (F), transmittance (τ), optical magnification (β), filter properties.
Photo-optical system 2: homogeneity of the (PbO) photoelectric layer in the relevant area, the spectral sensitivity.
Electronics: gain (A_v).

Introducing these parameters as correction factors into formula (2), we now arrive at the general formula:

$$E_{\text{signal}}(T) = c_{1\text{opt}}(\tau, \beta, f, D) \times c_{2\text{opt}}(x, y) \times c_{\text{electron}}(A_v)$$

$$\times \int_{\lambda_1}^{\lambda_2} V(\lambda)\, \epsilon(\lambda, T)\, L(\lambda, T)\, d\lambda. \tag{3}$$

5. Conditions for optimal operation. Experimental results

The above correction factors c_{1opt}, c_{2opt} and $c_{electron}$ are all less than unity. To obtain optimum conditions it is obviously necessary to choose parameter values which give maximum correction factors.

The following conclusions were arrived at on the basis of theory and experiment.

The correction factor $c_{1opt} = f(\tau, f, \beta, D)$ is a maximum for minimum focal distance, maximum transmittance and maximum diaphragm. Experimental results are in agreement.

For the measurement of temperature distributions in the area of measurement the 'homogeneity' of the photoelectric layer is the most important factor. Because of the lower sensitivity of the PbO layer for the penetration of 'shorter' wavelengths the homogeneity of the layer will be improved by using a BG-12 (Schott blue) filter. The 'sensitivity' will be somewhat decreased.

Measurements were made to estimate the sensitivity ($\Delta E/\Delta T$) of the device as a function of the true temperature (T). The influence of the gain is shown in figure 5.

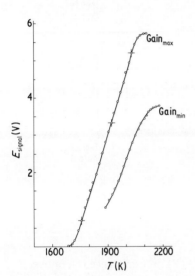

Figure 5. $E_{signal} = f(T)$ at optimum adjustment of gain.

$E_{signal} = f(T)$ for some values of the gain. Optimal adjustment of the gain is possible (this will also depend on the 'black-level contrast' adjustment).

Figure 6 gives $E_{signal} = f(T)$ for different values of the reduction filters. We note that the temperature range for which $E_{signal} = f(T)$ is a linear function can be increased by using a more severe reduction filter. It is clear that the sensitivity will then decrease. In the event of a reduction smaller than $\tau \leqslant 2\%$ $E_{signal} = f(T)$ becomes nonlinear.

The results are in agreement with the calculated values of figure 3.

A summary of the conditions for optimum operation is given in table 2.

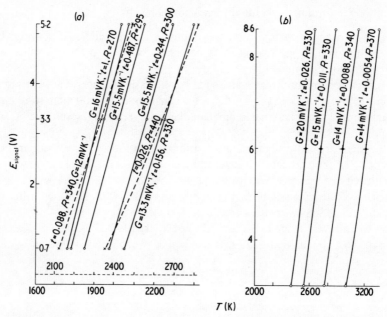

Figure 6. $E_{signal} = f(T)$ for several values of t. t = relative transmittance of the filter, R = temperature range, G = sensitivity ($\Delta E_{signal}/\Delta T$).

Table 2.

Optimum: $c_{1opt}(\tau, \beta, f, D)$	→ f_{min}
	D_{max}
	τ_{max}
$c_{2opt}(x, y)$	→ accuracy using BG-12 Schott filter $\Delta T/T = 0.3\%$ at $T = 2000$ K
$c_{electron}$	→ at maximum gain $\Delta E_{signal}/\Delta T = 16$ mV K^{-1} at $T = 1900$ K

6. Calibration

The following different types of lamp were used as standards for calibration purposes (Lechner and Schob 1975, de Vos 1953 and the 5th Temperature Symposium):

Tungsten ribbon lamp, Philips type 6002 R, temperature range 2000–2800 K (NPL calibrated).
Single-coiled halogen lamp 24 V/150 W, temperature range 2800–3200 K.
Comparative measurements were also made using this method and the colour pyrometer mentioned previously.

7. Measurement errors, internal calibration

Possible sources of measurement error (Lechner and Schob 1975, de Vos 1953) are contamination of the envelope and optical system and variations in surface roughness

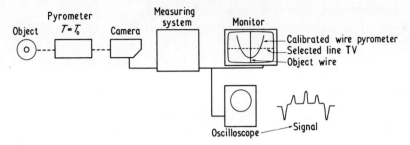

Figure 7. Calibrating during measurement.

(by etching), causing a change of emissivity. The latter could be observed on the image of the monitor.

A method of internal calibration which also reduces the error due to contamination of the optical system is shown in figure 7. The temperature of the object is measured at the same time as that of the calibrated wire of the optical pyrometer. The accuracy of calibration of this wire also determines the accuracy of the measuring system.

8. Some applications

In research on incandescent filaments, the study of the development of spots burning at temperatures higher than the mean level (hot spots $\Delta T/T_m$) is important (Balder and Meulders 1969, de Bie *et al* 1972, Hörster *et al* 1971).

Accurate registration of temperature differences is necessary. Transport phenomena such as the turn-to-turn diffusion, filament sag, mechanical deformation and the burn-out mechanism in relation to different parameters can be observed simultaneously.

Electrode temperature, the temperature distribution of discharge lamps under special conditions and the effect on arc properties can also be studied. The measuring system may be used for controlling temperature and temperature distribution in sintering processes.

Another field of application is the light measurements which have been normal practice for many years in the research and development of road-lighting luminaires.

9. Conclusions

A system of video temperature measurement using normal standard components and a Plumbicon camera tube has been developed.

This system has the following properties and advantages:

(i) It enables the temperature distribution of incandescent sources of small dimensions and complicated construction (coiled filaments with a wire diameter $< 100 \mu m$) to be measured. The accuracy of measurement of a tungsten ribbon lamp was $\Delta T/T \approx 0.3\%$ at $T = 2000$ K.
(ii) The maximum sensitivity for temperature measurement under the conditions stated is $<0.1\%$.
(iii) The temperature range with a linear signal temperature function is about 250 K at 1700–1950 K.

(iv) Reasonably fast measurements are possible.
(v) Facilities for the recording and analysis of information can be easily incorporated in the system.
(vi) Temperature measurement and visual observation can take place simultaneously.
(vii) The measuring spot is indicated electronically on the monitor. The spot size can be varied if required.

References

Balder J J and Meulders G J 1969 *Lichttechnik* 9 102A
de Bie J R, Klomp W G and Schob O 1972 *Light. Res. Technol.* 5 29
Dollekamp J, Schut Th J and Weyland W P 1970 *Electron. Appl.* 30 18
van Doorn A G 1966 *Philips Tech. Rev.* 27 1
van der Drift A G, de Haan E F and Schampers P 1963 *Philips Tech. Rev.* 25 25
—— 1964 *Philips Tech. Rev.* 25 133
Heyne L 1960 *Thesis* University of Amsterdam
Hörster W, Kauer E and Lechner W 1971 *Philips Tech. Rev.* 32 155
Jansen P C 1950 *Electron. Appl. Bull.* 11 61
Kerkhof F and Werner W 1952 *Television* Philips Technical Library
Lechner W and Schob O 1975 this volume
Levitt R S 1970 *J. SMPTE* 79 115
Rutgers G A W and de Vos J C 1954 *Physica* 20 715
de Vos J C 1953 *Thesis* Vrije Universiteit, Amsterdam

Time-resolved spectroscopic measurements on flashbulbs

L W van der Meer and J de Vries
Light Division, Philips' Gloeilampenfabrieken, Eindhoven, The Netherlands

Abstract. In combustion flashbulbs chemical energy is converted into radiant energy by high-temperature oxidation of thin metal shreds. The relative spectral intensity distribution in the visible range can be measured ten times during one flash of about 30 ms duration. The rapid-scanning spectrometer used is arranged to a fast measuring and data-handling system.

Combustion phenomena in flashbulbs were studied with different metal–gas systems. Some spectra of the combustion of Zr with O_2 and Al with O_2 are shown. The change in temperature during the flash is obtained from the change in the spectral intensity distribution with time by comparing an experimental spectrum with the best-fitting Planckian distribution. During the oxidation of zirconium in small flashbulbs a temperature of about 4600 K is reached which falls at a rate of approximately 35 K ms^{-1}. The spectrum of the combustion of aluminium in a large flashbulb shows pronounced emission bands of AlO which is an intermediate product in the ultimate formation of Al_2O_3.

1. Introduction

As the temperature of a thermal radiator rises, the luminous efficacy of the radiant flux increases. A maximum is reached at about 6000 K because the wavelength at which the eye's sensitivity is greatest coincides with the maximum in the radiant flux.

In combustion flashbulbs — used to replace daylight in photography — chemical energy is converted into radiant energy by high-temperature oxidation of a metal in the form of thin shreds. Until about 1960 the most-used metal was an Al–Mg alloy, but nowadays zirconium is mainly employed. In the study of the combustion processes it is of considerable importance to know how the temperature evolves during one flash of about 30 ms.

After ignition of a flashbulb the metal shreds are transformed into a collection of some hundreds of hot droplets. Most of the luminous flux from the lamp is generated by this cloud of liquid particles, as can be concluded from high-speed photography of the combustion process. A small proportion is due to selective thermal radiation by molecular species in the gaseous phase.

The spectral distribution of the radiant flux changes continuously with time and may deviate considerably from the Planckian distribution. Thus spectroscopic temperature determination during the combustion process necessarily involves time-resolved spectral measurements over a wide wavelength range.

The present paper discusses spectroscopic temperature determinations of combustion systems using a rapid-scanning spectrometer. A wavelength range from 340 to 760 nm is scanned in 1 ms with a dead time of 0·25 ms between two successive scans. The spectrometer is coupled to a mini-computer. This makes it possible to control the measuring process, to record a scanned spectrum instantly and to handle the data later on.

2. System organization

A flow chart of the complete system is shown in figure 1. During one scanning period of 1 ms, 200 spectral data points are transferred from the spectrometer to the computer. This relatively fast data flow is made possible by high-speed analogue–digital conversion of the spectrometer output and by a data channel which has direct access to the internal core memory (8 K) of the computer.

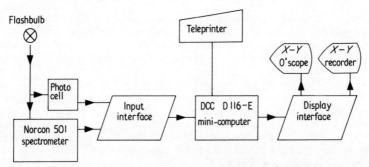

Figure 1. The system consists of the following main parts: Norcon rapid-scanning spectrometer, interfacing hardware, Digital Computer Controls mini-computer, output devices and a photocell.

Several scans, up to a maximum of ten, can be recorded in the memory. In the measurement set-up the software program enables these ten scans to be selected within a time interval, the shortest being 12·5 ms and the longest 80 000 ms.

Simultaneously with the spectral measurements, the relative luminous intensity of the flashbulb as a function of time is detected with a photocell and recorded. The data are used later to correct the measured spectral distribution for the change in the level of the radiant flux during a scanning period.

At the end of a measurement cycle, each recorded spectrum is processed in a number of steps by subroutines in the software program. The outcome of each step is displayed on an $X-Y$ oscilloscope and the calibrated result plotted by an $X-Y$ recorder. One axis represents intensity and the other wavelength. Communication with the computer is maintained via a teleprinter.

3. Wavelength scanning and data acquisition

A representative part of the radiant flux from the flashbulb is coupled into the spectrometer by means of a diffusing white screen of pressed $BaSO_4$ powder. The irradiated screen is imaged by collecting optics at the entrance slit of the spectrometer and it is actually irradiance which is measured. Imaging of the source itself is not employed because of the inhomogeneously radiating surface of the flashbulb.

In the Norcon model 501 spectrometer the grating is fixed and wavelength scanning is accomplished by sweeping a sequence of corner mirrors through an intermediate focal plane of the spectrometer (Dolin et al 1967). Twenty-four of these mirrors are mounted on the periphery of a scan wheel which is belt-driven at a speed of 33 rps. The passage of a single corner mirror through the focal plane results in simultaneous scanning across the two exit slits of two contiguous wavelength intervals in a time interval of one millisecond. By using a 300 lines/mm diffraction grating linear wavelength scanning is simultaneously obtained for ranges extending from 340 to 580 nm and from 520 to 760 nm (scanning speed 240 nm ms^{-1}).

Two photomultipliers detect the radiation through different cut-off filters at the exit slits. During one scanning period the currents of each multiplier are integrated over each of 100 successive time intervals of 9·5 μs, with a dead time of 0·5 μs during which the integrators are reset. The output of each integrator is sampled and held at the end of the integration period. After analogue–digital conversion the data channel provides for storage in the core memory.

A trigger pulse indicating the passage of each corner mirror is used to synchronize the scanning system with the recording system. A clock generating 1000 pulses between the start and finish of a wavelength-scanning period caters for the timing of the integrators and data transfer.

The resolution of a digital spectrum restricted by the integration time is 2·4 nm. The optical resolution is about 1 nm when slit widths of 200 μm are used. The spectral loss caused by the 0·5 μs reset time is 0·12 nm which is negligible in relation to the optical resolution.

The accuracy in the recorded spectral intensity is limited mainly by the photomultiplier signal-to-noise ratio which is determined by the integration time. Due to the intense radiant flux from a flashbulb an acceptable ratio is obtained in spite of the relative short integration time of 9·5 μs.

4. Calibration

Before spectral measurements on a lamp are carried out, the wavelength scale and the spectral sensitivity of the spectrometer are calibrated in accordance with procedures incorporated in subroutines of the software program.

The wavelength calibration is effected with a low-pressure mercury lamp. The wavelengths of the emission lines are known and their positions in relation to the 1000 clock periods between the start and finish of a scanning period are determined. This is done by differentiating the photomultiplier currents and counting the clock periods between the start pulse and each zero crossing. The overall reproducibility of the beginning of the wavelength scale is ±0·3 nm.

The intensity calibration at each wavelength is effected with a calibrated tungsten strip lamp. By imaging the strip at the entrance slit of the spectrometer a sufficiently high detection level is obtained. The spectral sensitivity is determined from the known spectral distribution of tungsten at a given temperature as compared with the recorded spectral distribution. A sequence of records is made for the 24 corner mirrors, resulting in an improved signal-to-noise ratio of the calibration spectrum and averaging of the mutual deviations of the mirrors. The deviations are determined separately and the associated factors – in the range from 0·95 to 1·05 – are applied to normalize actual spectral measurements.

The spectral sensitivity – described with 200 data points – is applied to the recorded spectral distributions measured via the screen. This is permissible since the reflectance of the pressed $BaSO_4$ powder is practically wavelength independent in the measured wavelength range.

5. Results and discussion

Combustion phenomena in different metal—gas systems have been studied. Spectroscopic temperature measurements for the combustion of Zr with O_2 in small bulbs and Al with O_2 in large bulbs are discussed in this section.

Approximately 15 mg of 20×40 μm zirconium shreds were combusted in a 0.25 cm^3 flashbulb with an initial oxygen pressure of about 15 atmospheres. Figure 2 (a–d) shows four spectra recorded at different moments during the combustion process. Each plot represents the relative spectral intensity distribution. The scanning periods are indicated in the curve representing the relative luminous intensity as a function of time (figure 2e).

The best fitting of a Planckian distribution to the experimental spectra is also indicated in figure 2. The fitting was performed by comparing an experimental spectrum with the Planckian distribution for temperature intervals of 25 K. The distribution temperature (Kaufman 1966) of the combustion phenomenon at a certain moment is thus obtained. It represents the true temperature of the hot droplets only if the collection behaves like a grey body radiator. The temperature measured 3 ms after ignition is 4625 K and falls at a rate of about 35 K ms^{-1}.

The temperature drop is caused by the depletion of oxygen in the vicinity of the burning droplets. This can be concluded from experiments in which one zirconium droplet was combusted in an abundant quantity of oxygen in which the temperature appeared to remain nearly constant for approximately 25 ms (Kettel 1973).

It can be seen in figure 2 (e and f) that the temperature is already decreasing while the luminous intensity is still increasing. This is due to the increasing number of zirconium droplets joining in the combustion process which can be actually seen from high-speed photography on flashbulbs (private communication). The temperature of the droplets may differ from one droplet to another. What is obtained is a weighted average temperature of all the radiating particles.

With experiments in small lamps no prominent emission bands of gaseous ZrO molecules in the ultimate formation of ZrO_2 are detected. The major part of the radiant flux apparently comes from the droplets the composition of which progresses from Zr to ZrO_2 (Harrison 1959, Levine and Nelson 1970, Maloney 1971).

In large lamps with low oxygen pressure transient molecular species are detected. This is demonstrated with the combustion of thin aluminium threads in a large flashbulb. The volume is 130 cm^3 and the initial oxygen pressure about 1 atmosphere. Figure 3 shows the relative spectral intensity distribution at a moment halfway between ignition and maximum luminous intensity.

In the green region figure 3 shows pronounced emission bands which must be attributed to gaseous AlO (Gaydon and Pearse 1965). This molecule is an intermediate product in the combustion process which ultimately results in the formation of Al_2O_3.

Since the spectral distribution of the radiant flux was considered to be a summation of the continuous radiation of droplets and the emission bands of the AlO molecules, the best fit of a Planckian distribution to the continuous part of the curve was employed to determine the temperature. Using this method a distribution temperature of

3875 K can be attributed to the hot droplets. Since the boiling points of Al and Al_2O_3 are reported to be about 2800 K (JANAF Thermochemical Tables 1971) and 3800 K (Brewer and Searcy 1951), respectively, the major part of the radiant flux here is apparently due to Al_2O_3 droplets (Johnson and Rautenberg 1960).

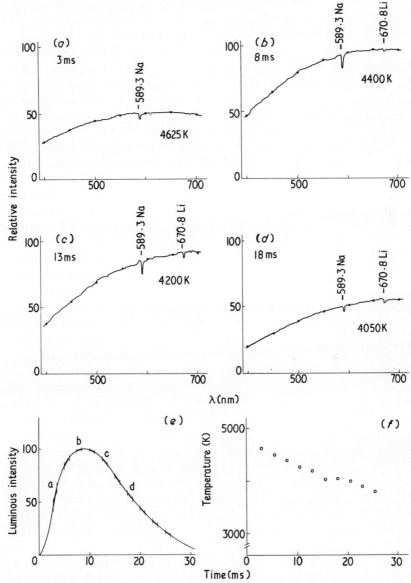

Figure 2. Measurements on the combustion of Zr with O_2 in a small flashbulb. (a)–(d): relative spectral intensity distribution at four different moments during the combustion process. Dots refer to black body radiators at the temperatures indicated. (e): the relative luminous intensity against time. Ten scanning periods are marked in the curve. (f): temperature as a function of time. Values are obtained from ten spectral distributions.

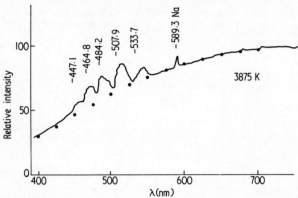

Figure 3. Measurements on the combustion of Al with O_2 in a large flashbulb. The relative spectral intensity distribution was scanned halfway between ignition and maximum luminous intensity. Dots refer to a black body radiator at the temperature indicated. Wavelength values of bandheads of AlO emission are also indicated.

Acknowledgments

The authors are greatly indebted to Mr Wijtvliet and Mr Gemmink, Techmation, Amsterdam, for their constructive cooperation and technical assistance.

References

Brewer L and Searcy A W 1951 *J. Am. Chem. Soc.* **73** 5308
Dolin S A, Kruegle H A and Penzias G J 1967 *Appl. Opt.* **6** 267
Gaydon A G and Pearse R W B 1965 *The Identification of Molecular Spectra* (London: Chapman and Hall)
Harrison P L 1959 *7th Int. Symp. on Combustion* (London: Butterworths) p913
JANAF Thermochemical Tables 1971 2nd edn (Washington: NBS)
Johnson P D and Rautenberg T H 1960 *J. Opt. Soc. Am.* **50** 602
Kaufman J E 1966 *IES Lighting Handbook* 4th edn (New York: Illum. Eng. Soc.)
Kettel F 1973 *Philips Res. Rep.* **28** 219
Levine H S and Nelson L S 1970 *High Temp. Sci.* **2** 343
Maloney K M 1971 *High Temp. Sci.* **3** 445

Heat flux pyrometer

I R Ashcroft and P A Norris
Corporate Engineering Laboratory, British Steel Corporation, London SW11 4LZ

1. Introduction

In the production of steel plate or heavy-gauge strip, large steel slabs (up to 0·3 m × 2 m × 8 m) are heated, prior to rolling, in a slab reheating furnace. Typically, a pusher type reheating furnace consists of two or more refractory-lined chambers heated by gas- or oil-fired burners mounted above and below the slabs. A series of water-cooled pipes with tough wearer bars (skid pipes) support the steel slabs as they are pushed through the heating chamber. A typical size furnace is 10 m wide, 6 m high and 30 m long (internal dimensions) capable of heating 200–280 tons/hour. A cross section of such a furnace is shown in figure 1. Heavy-duty rams push the adjacent slabs through the furnace such that, to get a hot (1240 °C) slab out at the discharge end, a cold (ambient) slab has to be pushed into the charge end.

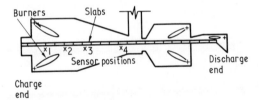

Figure 1. Furnace cross section.

In the control of such a furnace it is essential that the furnace operator knows the temperature profile of the stock in the furnace in order to maintain delivery of correctly heated slabs as demanded by the adjacent hot rolling mill. Overheating wastes valuable fuel, causes oxidation with subsequent loss of bulk material and, in the extreme, can melt the slab surfaces causing adjacent slabs to stick together. Underheating causes delays to the rolling programme while waiting for the slab to be heated to the correct temperature or, if the underheated slab is fed into the rolling mill, damage can be caused by overstressing the mill during rolling. In a conventional furnace system, thermocouples mounted in the refractory chamber lining enable operators to estimate slab temperatures at various places along the furnace length. Due to the large thermal mass of the refractory this is an inherently slow method and not entirely suitable for a fast-heating high-tonnage furnace.

Other methods of directly measuring the slab surface temperature at various points throughout the furnace length have not generally met with success. Radiation pyrometers are affected by reflected radiation from the furnace walls and, if not carefully sited, can sometimes measure the direct flame temperature.

Gold-cup total radiation pyrometers lowered on to the top surface of the slabs are unreliable because they are not robust enough for the hostile environment. Also, due to differing slab thicknesses, a level top surface is not always available. A retracting mechanism would therefore have to be provided, linked to the pusher, to withdraw the pyrometer, as occasionally slabs stack up on each other or turn upon their edges when pushed. A gold-cup pyrometer mounted below and pushed up against the underside of the slabs, which are maintained at a constant pass line, would rapidly fill with dust and scale, which would invalidate the reading.

The approach adopted by the Corporate Engineering Laboratory of the British Steel Corporation has been to develop a thermocouple temperature sensing device which is fixed in location beneath the slabs and supported on a water-cooled pipework system.

2. Theory

The original concept of the temperature sensor was a thin disc which was thermally insulated from its surroundings except for the slab whose temperature was to be monitored (figure 2a). In the ideal case the disc would obtain the temperature of the

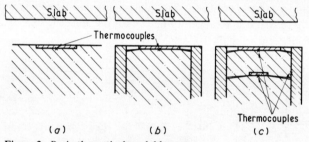

Figure 2. Basic theoretical model layouts.

slab by radiation exchange and the temperature would then be measured by the thermocouple. In practice the sensing disc was surrounded by solid refractory and connected to the main supporting body of the probe by three thin pins (figure 2b). This arrangement did not sufficiently isolate the disc and heat was conducted into the refractory behind the disc and also along the pins into the body of the probe. This caused the disc to register too low a temperature and also lengthened the time constant of the system due to the thermal capacity of the refractory. To compensate for this, the design was altered so that estimates could be made of the rate of flow of heat from the disc into the surroundings by measuring the surrounding temperatures with two extra thermocouples (figure 2c). This was done by attaching one thermocouple to the wall and the other to a second disc placed below the top sensing element. These measurements of the rate of heat flow from the centre of the disc could then be used to calculate the slab temperature. It was found, however, that the heat flow path was not sufficiently well defined and the time constant was too long.

To overcome these problems the disc was extended to cover the entire top of the probe thus establishing a conduction path along the radius of the disc and down the wall of the probe (figure 3). This design also allowed the use of an insulating material with a lower thermal mass and lower conductivity behind the disc, thus reducing the heat loss from the back of the disc and decreasing the time constant of the sensor. Two

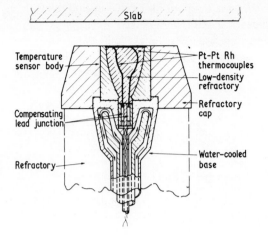

Figure 3. Sensor head unit.

thermocouples were then used (one at the centre of the disc and one at the edge) to measure the heat flow along a radius. These temperatures are then used in a mathematical model of the system to evaluate the slab temperature (see §4).

3. Design

The slab surface temperature sensor therefore consists of a thin metallic alloy disc continuously welded around its circumference on to the end of a cylindrical body of the same alloy. Thermocouples are welded on to the underside of the disc at its centre and, at a predetermined radial distance, very low-density refractory packing material is used to fill the remaining space in the cylinder (figure 3).

A refractory cylinder fits over the sensor cylindrical body to afford some protection from the environment. The rest of the pipework within the furnace is also refractory-coated. The encapsulated junction between the mineral-insulated thermocouple and the compensating lead is mounted in an aluminium plug, located within the water-cooled support, and is in intimate contact with it.

The sensor is mounted on a water-cooled support structure and is tack-welded in position. Heat conduction between the sensor base and the water-cooled support is maintained by bedding the sensor head down on a metallic-based epoxy resin layer prior to welding. For safety reasons this welding of the sensor on to the support structure is designed such that, if a distorted slab hits the sensor, the unit will break away leaving the water-cooled support system intact. Dirt and scale build-up on the sensor disc is largely prevented by the support skid pipe system vibrating each time a new slab is pushed into the furnace.

The water-cooled support pipe work is a triple concentric structure having a dry core tube carrying the thermocouple compensating signal leads to a suitable signal transfer point outside the furnace environment. Water inlet and outlet is via the two remaining concentric tubes. Cobalt–steel alloy brackets welded on to the skid pipe crossover supports are used to locate and support the sensors in position beneath the slabs.

It is envisaged that a modified support structure system mounted through the furnace floor would enable the sensor heads to be replaced and/or examined whilst the furnace is operational. Normal access to the furnace interior is limited to shut-down periods which typically occur only at three- or six-monthly intervals.

Results from sensors mounted in the slab reheating furnace at the British Steel Corporation South Teesside (Lackenby) Works are shown in figure 4. The three traces (numbers 1, 2 and 3) are the uncorrected disc centre thermocouple signals from sensors mounted at 4, 7 and 15 m from the furnace charge end.

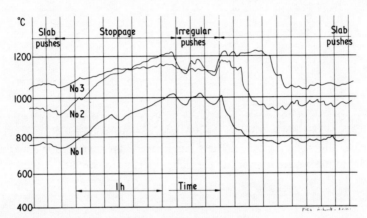

Figure 4. Typical temperature sensor output signals.

After the first few slab pushes the chart recording shows a steady increase in temperature of the slabs during a stoppage. The second part of the chart shows the effect of pushing slabs out after a stoppage; hot slabs are pushed through the furnace followed by cold slabs from the storage area. Combustion conditions within the furnace are then adjusted such that, with steady throughput of slabs, equilibrium heating and conditions are maintained at the three measuring locations.

4. Mathematical model

The diffusion equation in cylindrical polar coordinates is:

$$\frac{\partial^2 T}{\partial r^2} + \frac{1}{r}\frac{\partial T}{\partial r} + \frac{1}{r^2}\frac{\partial^2 T}{\partial \theta^2} + \frac{\partial^2 T}{\partial z^2} = \frac{1}{\alpha}\frac{\partial T}{\partial t}$$

where α = diffusivity = $k_D/s_D \rho_D$, and k_D = conductivity of the disc, s_D = specific heat of the disc, ρ_D = density of the disc and t = time.

The system is assumed to be symmetrical about the z axis and so the temperature at a given radius from the centre is constant, ie,

$$\frac{\partial T}{\partial \theta} = 0.$$

It is also assumed, since the disc is thin, that there is no variation in the z direction, ie,

$$\frac{\partial T}{\partial z} = 0.$$

Therefore

$$\frac{\partial^2 T_r}{\partial r^2} + \frac{1}{r}\frac{\partial T_r}{\partial r} = \frac{1}{\alpha}\frac{\partial T_r}{\partial t}$$

is the diffusion equation for the temperature of the disc at a distance r from the centre.

If a heat input term is included and it is assumed that the heat loss from the back of the disc is negligible, then

$$\frac{\partial^2 T_r}{\partial r^2} + \frac{1}{r}\frac{\partial T_r}{\partial r} + \frac{\sigma\epsilon}{k_D t_D}(T_s^4 - T_r^4) = \frac{1}{\alpha}\frac{\partial T_r}{\partial t} \tag{1}$$

where σ = Stefan's constant, ϵ = emissivity of the slab, t_D = thickness of disc and T_s = slab surface temperature.

Two main approaches have been taken in the solution of this equation: (*a*) a numerical technique – probably requiring the use of a digital computer on-line; and (*b*) an approximate empirical representative, which can be simulated by simple analogue circuits to give a real-time solution.

4.1. Numerical solution

A finite difference formulation of equation (1) is

$$\frac{(T_{i-1,j} - 2T_{i,j} + T_{i+1,j})}{h^2} + \frac{(T_{i+1,j} - T_{i-1,j})}{2ih^2} + K(T_{sj}^4 - T_{i,j}^4)$$
$$= \frac{1}{\alpha}(T_{i,j+1} - T_{i,j}) \tag{2}$$

where $i = 0:1:n$ and is the radial coordinate, $j = 0:1:\infty$ and is the time coordinate, and $h = R/n$, where R = radius of disc.

Associated with this equation there are the following boundary conditions:

T_0 known, ie, the temperature at the disc centre
T_n known, ie, the temperature at the disc edge

$$\left.\frac{\partial T}{\partial r}\right|_{r=0} = 0 \quad \text{ie, } T_{-1,j} = T_{1,j}.$$

Because the heat exchange between the slab and the disc are specified in terms of the slab and disc temperatures and the disc temperature is not constant across the width of the disc, the above boundary conditions are incomplete. The initial temperature profile across the disc at time $t = 0$ must therefore be calculated before we can proceed with the solution for all time increments.

Consider equation (2) for the first two time levels:

$$\frac{(T_{i-1,0} - 2T_{i,0} + T_{i+1,0})}{h^2} + \frac{(T_{i+1,0} - T_{i-1,0})}{2ih^2} + K(T_{s0}^4 - T_{i,0}^4)$$

$$= \frac{1}{\alpha}(T_{i,1} - T_{i,0}) \tag{3}$$

$$\frac{(T_{i-1,1} - 2T_{i,1} + T_{i+1,1})}{h^2} + \frac{(T_{i+1,1} - T_{i-1,1})}{2ih^2} + K(T_{s1}^4 - T_{i,1}^4)$$

$$= \frac{1}{\alpha}(T_{i,1} - T_{i,0}). \tag{4}$$

Note that both the right-hand sides are identical.

Assuming values for T_{s0} and T_{s1} and using our known values of $T_{0,0}$ and $T_{0,1}$ these equations can be solved alternately for $T_{i+1,0}$ and $T_{i+1,1}$ for $i = 0:1:n-1$. The values of $T_{n,0}$ and $T_{n,1}$ can then be compared with the known values $T_{n,0}$ and $T_{n,1}$. If they are different then T_{s0} and T_{s1} can be varied accordingly until $T_{n,0}$ and $T_{n,1}$ reach the required accuracy.

Having determined this initial temperature profile, equation (2) can now be used to determine the temperature profile for any future time coordinate.

Also by considering equation (2) with $i = 0$ and using the identity

$$\lim_{r \to 0} \frac{1}{r} \frac{\partial T}{\partial r} = \frac{\partial^2 T}{\partial r^2}$$

we can write

$$\frac{4}{h^2}(T_{1,j+1} - T_{0,j+1}) + K(T_{sj+1}^4 - T_{0,j+1}^4) = \frac{1}{\alpha}(T_{0,j+1} - T_{0,j})$$

for which the slab temperature T_{sj+1} can be extracted as all other temperatures are now known.

This solution was tried on a set of results obtained from an experimental laboratory furnace. Unfortunately the results suffered from noise which, because of the large time interval between successive sets of data (10 s), caused the solution to become unstable. On smoothing the input data, however, by fitting a fourth-order Chebyshev polynominal and interpolating to get a time interval of one second, it was possible to predict the slab temperature to better than $\pm 20\,°C$.

For practical purposes this smoothing process must be undertaken in real time. Therefore a recursive estimation procedure due to Kalman was chosen to do this. This technique is known as Kalman filtering. It uses a 'state space' approach which makes use of difference or differential equations rather than the integral equations of the more classical approach. The Kalman filter takes advantage of a system model relating the slab surface temperature to the disc centre and edge temperatures and the process by which the slab surface temperature is varied, in addition to the real-time measurements of disc centre and edge temperatures. Then, on the basis of all measurements up to the present time, it determines the most likely values of the states (ie, slab temperature, disc edge and centre temperatures).

values of the states (ie, slab temperature, disc edge and centre temperatures).

Again this technique has been applied to a limited set of data only, for which it predicted slab temperatures to within ±10 °C after eight iterations (ie, eight data inputs).

4.2. Empirical approach

By studying various experimental results it was found that the slab surface temperature could be related to the disc centre and edge temperatures by an equation of the form

$$T_s = aT_0 + b(T_0 - T_n) + c \frac{dT_0}{dt}.$$

An analogue circuit was designed to represent this equation and connected on-line to the centre and edge disc thermocouples of a probe which was mounted in the experimental furnace. The constants a, b and c were adjusted to give an output of T_s which was closest to the actual value of slab temperature. The values of the constants obtained were:

$$a = 1 \qquad b = 1 \cdot 3 \qquad c = 30.$$

Then, with these values held constant, a number of experiments were performed. These involved simulating pushes in the furnace by exposing the probe to a relatively cold slab at approximately 800 °C which was then heated to a temperature of approximately 1000 °C. When this temperature was reached a new cold slab was introduced and the process repeated. Predicted values of T_s were within ±20 °C and the time constant was in the order of 10 s. Because of the limited temperature range considered, it is impossible to draw any firm conclusions. The method looks promising, however, as a relatively cheap and simple system.

Some typical results obtained in the laboratory furnace are shown in figure 5.

T_0 = disc centre temperature $\qquad T_E$ = disc edge temperature
T_{s1} = measured slab temperature $\qquad T_{s2}$ = calculated slab temperature.

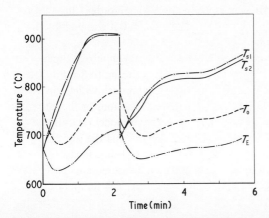

Figure 5. Laboratory furnace, corrected temperature signals.

5. Conclusions

From tests carried out both in real furnaces and for simulated furnace conditions within the laboratory, it would appear that a thermocouple temperature sensing device of this nature has many applications both within and outside the British Steel Corporation. Further work has to be done, however, before the sensor and associated electronics become a fully proven and tested temperature measuring device.

Acknowledgments

The authors wish to acknowledge the assistance of Messrs R Gray and M Dewshi in mathematical modelling and computation.

Inst. Phys. Conf. Ser. No. 26 © *1975: Chapter 6*

Plasma (arc) thermometry

J Richter

Institut für Experimentalphysik der Universität Kiel, Kiel, Germany

Abstract. The plasma temperature in an arc is defined only if the plasma is in a state of local thermodynamic equilibrium. This plasma state is analysed in detail and methods for checking the plasma state are discussed. A compilation is given of the experimental methods which yield information about the plasma and the plasma state. Some methods for measuring arc temperatures are discussed.

1. Introduction

In the last ten years our knowledge of arc physics has improved considerably mainly in two directions. We have learned to understand and to calculate the behaviour of arcs under the influence of gas flows and we have increased our information about the physical state in the arc plasma. The plasma state is closely connected with all questions concerning the arc temperature. We will thus focus our interest on the second direction of development.

The term 'temperature' is somewhat problematic if used in connection with electrical arcs. In its correct meaning, the word is reserved for systems in thermodynamic equilibrium (TE). It is well known that the TE state of a hot gas or plasma can only be obtained if the gas or plasma is enclosed in a cavity with a constant wall temperature and without fields of any kind or other distinguished directions, eg, in a black body cavity used as an high-temperature radiation standard.

After the cavity is removed, the plasma will lose energy by radiation, heat currents and diffusion processes. To compensate for these losses, the plasma has to be heated continuously; for the case of arcs, by ohmic heating. Gradients of nearly all quantities in the plasma thus result. Obviously, such systems are not in the TE state which can be described by a minimum number of variables. The situation is not hopeless, however, especially not in the case of electrical arcs at high or moderate electron densities. Many relations, which are applicable in the total volume of a TE plasma, can be locally realized in an arc column. The main properties and relationships of a TE plasma which are of interest in this connection are listed in table 1. (All symbols and designations are defined in the Appendix.)

In an arc, the homogeneity is not given and the radiation field is diluted at nearly all wavelengths

$$I_\lambda < B_\lambda(T)$$

because the reabsorption, necessary for the establishment of black body radiation, is very small. But in many types of arcs with a sufficiently high electron density, the

Table 1. Equilibrium properties and relations.

(i) Homogeneity: $\dfrac{d}{dx}, \dfrac{d}{dy}, \dfrac{d}{dz} = 0$

(ii) Black body radiation field:

$$I_\lambda = B_\lambda(T) = \frac{2hc^2}{\lambda^5} \frac{1}{\exp(h\nu/kT) - 1} \quad (1)$$

(iii) Maxwellian distribution of velocities:

$$\frac{dn}{n} = F(v, m, T)dv = \frac{4}{\sqrt{\pi}} v^2 \, dv \, \exp(-mv^2/2kT)/(2kT/m^3)^{1/2}. \quad (2)$$

(iv) Equation of state:

$$p = kTn_e + kT \sum_s (n_s + n_{s,+}). \quad (3)$$

(v) Boltzmann population of internal energies:

$$\frac{n_{s,i}}{n_s} = \frac{g_{s,i}}{Z_s} \exp(-E_i/kT). \quad (4)$$

(vi) Saha equation:

$$\frac{n_e n_{s,+}}{n_s} = S_s(T) = \frac{2Z_{s,+}}{Z_s} \left(\frac{2\pi mkT}{h^2}\right)^{3/2} \exp(-\chi/kT). \quad (5)$$

---

relationships (2) to (5) in table 1 are valid or, more correctly, approximately valid. This corresponds to a local temperature and a plasma state called local thermodynamic equilibrium (LTE). A volume element of such a plasma can be described locally by thermodynamic variables such as temperature, pressure, etc. Only the radiation field is reduced compared with the black body field of the local temperature.

In LTE arc plasmas, the electrons pick up energy from the electrical field and randomize this energy very quickly. The result is an electron gas with a kinetic temperature T_e which describes the velocity distribution $F(v, m_e, T_e) \, dv$. This electron gas and its temperature controls excitation and ionization, and the kinetic temperature of atoms, ions and molecules. The plasma state is dominated by electron collisions. It is relatively simple to understand how the Boltzmann population is established under the

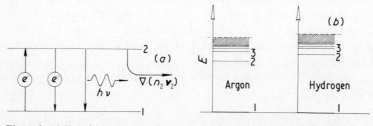

Figure 1. (*a*) Transitions in a two-level atom without reabsorption of radiation. $\nabla \cdot (n_2 \mathbf{v}_2)$ is the loss of excited particles by diffusion. (*b*) Simplified energy-level scheme of hydrogen and argon.

influence of these electron collisions. Let us now consider a two-level atom (see figure 1a). In the steady state case, the number of atoms entering the excited level '2' per second and cm^3 must be equal to the de-excitation rate. The population rate is given by inelastic electron collisions with the rate coefficient $R_{12}(T_e)$. The depopulation rate contains three terms namely, the super elastic collisions by electrons, the spontaneous emission (γ accounts for reabsorption, $\gamma < 1$), and the divergence of the particle flow $n_2 v_2$:

$$n_e n_1 R_{12}(T_e) = n_e n_2 R_{21}(T_e) + \gamma n_2 A_{21} + \nabla \cdot (n_2 v_2). \tag{6}$$

Using the relation

$$\frac{R_{21}}{R_{12}} = \frac{g_1}{g_2} \exp(E/kT_e) \tag{7}$$

which holds true if the electrons obey a maxwellian velocity distribution, one obtains

$$\frac{n_2}{n_1} = b \frac{g_2}{g_1} \exp(-E/kT_e) \tag{8}$$

with

$$b = \left(1 + \frac{\nabla \cdot (n_2 v_2) + \gamma n_2 A_{21}}{n_2 n_e R_{21}}\right)^{-1}. \tag{9}$$

With increasing electron density the underpopulation factor $b < 1$ converges towards unity, which means a population ratio according to TE.

In a more realistic case with many levels (including ionization) one has to replace (6) by a system of equations corresponding to the different levels and possible transitions. This is the system of 'rate equations' which can be solved with the help of a computer if the rate coefficients, the transition probabilities, and the flux of particles and radiation (determining γ) are known. The result of such a calculation is given in figure 2 for a homogeneous ($v = 0$) and optically thin ($\gamma = 1$) hydrogen plasma at $T_e = 16 \cdot 000$ K. In this figure, the logarithm of the factors b_i, defined by

$$\frac{n_i}{n_1} = b_i \frac{g_i}{g_1} \exp(-E_i/kT_e), \tag{10}$$

is plotted against the logarithm of the electron density, measured in cm^{-3}. For densities larger than 2×10^{17} cm^{-3}, one has $b_i = b = 1$ or LTE.

Very often the results of such complicated treatments can be approximated by a two-level system. Many elements have an energy-level scheme characterized by a large energy gap between the ground state and the first excited level (figure 1b). Because excitation and ionization are performed stepwise, the largest energy gap controls the establishment of the LTE state. Thus equation (9) can be used for an assessment whether LTE is obtained or not. In practice this criterion is only applicable if no particle flux and no reabsorption is present. But this last assumption is not correct, because, on the whole, the radiation field in the resonance lines is very strong ($\gamma < 1$). There are no simple and reliable criteria for deciding the very important LTE question. We will return to this problem later.

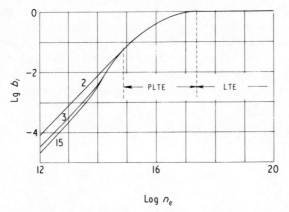

Figure 2. Underpopulation factors for the second, third and fifteenth quantum state of hydrogen at 16 K (McWhirter and Hearn 1963).

In figure 2 we can see that at electron densities not sufficient to establish LTE, a region of n_e exists where $b_i = b < 1$. This means an underpopulation independent of the level, or in other words, an overpopulation of the ground state by a factor $1/b$. This plasma state is called partial local thermodynamic equilibrium or PLTE. All excited levels and free electronic states are populated according to the electron temperature. We can describe this new state simply by equation (10) and

$$\frac{n_e n_{s,+}}{n_s} = b S_s(T_e). \tag{11}$$

It is reasonable to include into this PLTE state the possibility of different kinetic temperatures of electrons and heavy particles (T_e and T_g). The energy exchange between electrons on one side and atoms, ions and molecules on the other decreases with

Table 2. Equilibrium plasma models.

Model	Temperature	Boltzmann factor	Saha equation	Radiation
TE	T	$\exp(-E/kT)$	$S(T)$	$B_\lambda(T)$
LTE	T_e	$\exp(-E/kT_e)$	$S(T_e)$	$I_\lambda < B_\lambda(T_e)$
PLTE	$T_g = \beta T_e$	$b \exp(-E/kT_e)$	$b S(T_e)$	$I_\lambda < B_\lambda(T_e)$

Equation of state: $p = kT_e n_e + kT_g \sum (n_{s,} + n_{s,+})$

decreasing electron density. Thus another parameter $\beta = T_g/T_e < 1$ is introduced to describe the PLTE state. The three state models for the arc plasma are summarized in table 2. For all three models a combination of the Boltzmann factor and the Saha equation results in the following relation:

$$\frac{n_e n_{s,+}}{n_{s,i}} = \frac{2Z_{s,+}}{g_{s,i}} \left(\frac{2\pi m k T_e}{h^2}\right)^{3/2} \exp|-(\chi - E_i)/kT_e|. \tag{12}$$

Equation (12) is valid for all levels for the TE and LTE cases and for all levels $i > 1$ for the PLTE case ($i = 1$: ground state).

The first published literature on this subject seems to be the astrophysical investigations performed by Menzel (1937), Menzel and Cillie (1937) and Giovanelli (1948). A period of intensive investigations of equilibrium models (concerning laboratory plasmas) began after the publications of McWhirter (1961), Bates *et al* (1962) and McWhirter and Hearn (1963). A summary of the work in this field has been given by Richter (1971) and a detailed treatment of the non-LTE plasma state has been given by HW Drawin in Venugopalan (1971).

2. Optical measurements with arcs

In this section the problem of LTE will be left for the present and the methods for obtaining information about the plasma state by spectroscopic and optical measurements will be briefly discussed. The knowledge of these methods will help us to find possibilities to determine the plasma state and the plasma temperature.

The different techniques and methods will be listed and the quantities which can be obtained and the difficulties which may arise will be considered.

2.1. *Line intensities (from homogeneous, optically thin layers, eg 'end-on' measurements through the axis of the arc)*

The line intensity is given by

$$J_{ik} = \int I_\lambda \, d\lambda = \frac{h\nu}{4\pi} A_{ik} g_i n_i s. \tag{13}$$

where s is the length of the layer. The measurement of J yields n_i, the number density of particles in the upper quantum state of the line. For levels close to the ionization limit one has to account for a decrease of n_i by a factor

$$P_i = \exp[-\text{constant} \times n_e/(\chi - E_i)^3] \tag{14}$$

which has to be inserted on the right-hand side of equation (13) (Gündel 1970). For LTE plasmas one can calculate $n_i = f(T_e, p, c_s)$ and determine the temperature from n_i. If two or more lines with different excitation energies have been measured, a plot of $\lg(J/Ag\nu)$ against E yields a straight line with a slope $-1/kT_e$ (that is, a Boltzmann plot, proof of a Boltzmann population).

2.2. *Line profiles (measured from an optically thin layer)*

In arc plasmas line profiles are formed by Stark broadening (mostly dominant) and Doppler shift (observable only at small n_e and with inner shell transitions). The halfwidth of a Stark-broadened line is given by

$$\Delta\lambda_h = \begin{cases} \text{constant} \times n_e & \text{(quadratic Stark effect)} \\ \text{constant} \times n_e^{2/3} & \text{(linear Stark effect; eg, } H_\beta\text{).} \end{cases} \tag{15}$$

Thus the Stark width yields n_e if the constants in equation (15) are known (see Griem 1974).

The full halfwidth of a Doppler-broadened line yields the gas temperature

$$T_g = \frac{mc^2}{8 \lg 2} \left|\frac{\Delta\lambda}{\lambda}\right|^2 \tag{16}$$

where m is the mass of the radiating particle.

2.3. Continuous spectra

The recombination and Bremsstrahlung spectrum can be described by

$$I_{\lambda,c} = \text{constant} \times \xi(\lambda, T_e) n_e n_+ (T_e)^{-1/2}. \tag{17}$$

(Maxwellian distribution of electron velocities, not necessarily LTE!) The factor ξ describes the fine structure of the spectrum and the deviation from the hydrogenic continuum (Biberman and Norman 1960, Biberman and Ulyanov 1961). Those ξ factors have been calculated by Schlüter (1968).

Plasma impurities can increase the intensity considerably especially if their ionization potential is small. In a one-component plasma ($n_e = n_+$), n_e can be calculated from $I_{\lambda,c}$.

2.4. Scattering of laser radiation

In a typical arc plasma ($T_e \simeq 10^4$ K, $n_e \simeq 10^{16}$ cm^{-3}), the Debye shielding distance λ_D is of the order $5-10 \times 10^{-6}$ cm. Therefore the scattering spectrum is much more complicated than in the case of simple Thomson scattering which is valid for

$$\alpha = \lambda/(4\pi\lambda_D \sin\tfrac{1}{2}\theta) < 1. \tag{18}$$

(λ is the laser wavelength of the order 6×10^{-5} cm and θ the scattering angle). Detailed calculations for $\alpha \gtrsim 1$ lead to very complicated spectra from which, under special conditions, n_e, T_e and T_g can be obtained (Rosenbluth and Rostoker 1962).

2.5. Index of refractivity (interferometry with laser radiation)

The index of refractivity of a plasma is given by

$$N = 1 + 2\pi\alpha_0 n_0 + 2\pi\alpha_+ n_+ - \frac{e^2 \lambda^2}{2\pi m_e c^2} n_e \tag{19}$$

where α_0 and α_+ are the coefficients of polarization of neutrals and ions. The last term on the right-hand side of (19) dominates under arc conditions. One can determine N at two different wavelengths and eliminate the influence of atoms and ions. The result is a very exact value of n_e (eg, Meiners and Weiss 1973).

2.6. Radiation from optically non-thin layers (eg, in the centre of a strong emission line; arc end-on)

In this case the optical depth $\kappa(\lambda)s$ is approximately 1. The intensity is then given by the source function S:

$$I_{\lambda,0} = S_\lambda \{1 - \exp[-\kappa(\lambda)s]\}. \tag{20}$$

With a concave mirror (reflectance R) behind the arc which refocuses the light back into the arc column, one obtains an increased intensity

$$I_\lambda = I_{\lambda,0} + I_{\lambda,0} R \exp[-\kappa(\lambda)s]$$

and after elimination of $\kappa(\lambda)s$,

$$S_\lambda = I_{\lambda,0} R/(1 + R - I_\lambda/I_{\lambda,0}). \tag{21}$$

The source function in a spectral line is given by

$$S_\lambda = \frac{2hc^2}{\lambda^5} \frac{1}{n_i g_k/n_k g_i - 1} \to B_\lambda(T_e) \quad \text{for LTE}. \tag{22}$$

For $\kappa(\lambda)s \to \infty$, $I_\lambda = I_{\lambda,0} = S_\lambda$ follows. An analysis shows that this method yields accurate temperatures only if the emission is observed in the UV region ($h\nu/kT_e > 1$). Atomic constants are not necessary.

2.7. Gas temperature

Two relatively simple methods exist for measurement of the ratio $\beta = T_g/T_e$.

Finkelnburg and Maecker (1956) and others used the energy balance of the electron gas to determine this ratio:

$$\sigma E^2 = 4kT_e n_e f \frac{m_e}{m_a} (1 - T_g/T_e). \tag{23}$$

$\sigma = e^2 n_e/(m_e f)$ is the electrical conductivity and f the collision frequency of the electrons ($f = \bar{v}_e \Sigma n_k Q_{ek}$). With a measured electrical field strength and a calculated collision frequency, β can be evaluated.

Gurevich and Podmoshenskii (1965) have proposed another method which turned out to yield accurate values if $\beta \simeq 1$ (Richter et al 1975). After a sudden arc shortening one observes a line intensity increase which can be understood if n_i from equation (13) is introduced into equation (12):

$$J = \text{constant} \times n_e n_+ \exp[(\chi - E)/kT]/T^{3/2}. \tag{24}$$

With the assumptions of a constant n_e directly after the short circuit and a rapid fall in the electron temperature down to the gas temperature ($\sigma \cdot E = 0!$), this intensity jump can be calculated from equation (24) and T_g/T_e can be determined from the measured intensity increase.

3. Plasma diagnostics and temperature measurement

As we have seen from the previous sections, there is no simple and realistic LTE test, because radiation and diffusion fluxes, operating against LTE, cannot be determined locally. To answer the important question of whether a special arc LTE is obtained or not, one can use two general methods.

3.1. Calculations

The energy balance over the arc radius (the Ellenbaas–Heller differential equation) can be solved for a given arc current and a given radius and gas, taking into consideration the system of rate equations which include radiation and diffusion processes. This problem can be solved numerically if all cross sections and transition probabilities are available. Unfortunately, there is no gas which fulfils this condition and so one has either to make approximations and special assumptions (eg, PLTE), or include some measured quantities into the computations. Kruger (1970) and Uhlenbusch *et al* (1970) have carried out such an investigation. Here we will discuss some of the results of Uhlenbusch *et al*. They solved the problem for noble-gas arcs of different currents by measuring some quantities ($n_i(r)$) to avoid theoretical difficulties arising from unknown atomic data. Some interesting results of their work are given in figure 3,

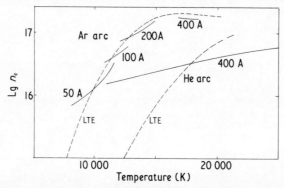

Figure 3. LTE in argon arcs and non-LTE in a helium arc. (Calculations by Uhlenbusch *et al* 1970.)

Table 3. Data on noble-gas atoms and plasmas.

	He	Ne	Ar	Kr	Xe
Ionization energy (eV)	24·6	21·6	15·6	14·0	12·1
E_2 (eV)	19·8	16·6	11·5	9·9	8·5
T_e (max n_e) (K)	24600	20400	15800	14300	12700
$n_{e,\,max}$ (10^{17} cm^{-3})	1·2	1·5	1·9	2·1	2·4
$n_{e,\,cr}$ (10^{17} cm^{-3})	12	6·5	1·9	1·2	0·7

where n_e is plotted against T_e for an argon and a helium arc. For the case of argon (and the heavier noble gases) the results closely follow the LTE curve and the helium (and neon) arcs show strong deviations from the LTE state. These results are understandable if one compares the different ionization energies and excitation potentials of the first excited levels (table 3). As the analysis of the detailed results show, the main sources for the deviations are the diffusion currents of neutral helium (or neon) atoms in the ground state towards the arc axis and diffusion currents of charged particles towards the arc edge (ambipolar diffusion).

From equation (9) (assuming no diffusion and no reabsorption ($v_2 = 0$, $\lambda = 1$), one obtains

$$b = 1/(1 + A_{21}/R_{21}n_e) \simeq 1/(1 + 10^{13} T_e^{1/2} E_2^3/n_e). \tag{25}$$

(with T_e in K and E_2 in eV). LTE is reached if $b \simeq 0.9$ or

$$n_e \geqslant 10^{12}\sqrt{T_e}\,E_2^3 = n_{e,\,cr} \tag{26}$$

(Wilson 1962). Equation (26) contains an overestimation of escaped radiation and an underestimation of diffusion losses. In the case of noble-gas arcs both effects seem to approximately compensate each other. In table 3 the values of $n_{e,\,cr}$ are compared with the maximum electron densities which can be obtained in LTE at 1 atmosphere pressure. For $n_{e,\,max} < n_{e,\,cr}$ no LTE is reached as the calculations of Uhlenbusch *et al* show.

3.2. Experimental LTE check

A straightforward LTE test by experiment requires:

(i) T_e determination via the Boltzmann plot,
(ii) n_e determination, eg, by interferometry,
(iii) n_i determination by line intensity,
(iv) measurement of total pressure, and
(v) proof that $T_g = T_e$.

These data yield the ground-state density n_1 (using the equilibrium relations of table 1) and the factor b can be calculated. Because T_e behaves exponentially (equation (4)), the resulting b factor is normally inaccurate (errors of 30 to 200%).

A second experimental method used by Richter (1965) and Shumaker (1974) is based on measurements at very different electron densities or temperatures. The idea of this method is that deviations from LTE will decrease with increasing electron densities (according to equation (9)) and a comparison of measured quantities with LTE values will indicate deviations from LTE. Figure 4 shows such an extended LTE test (Shumaker and Popenoe 1972). The plot of $\lg n_i(\text{Ar I})$ against $\lg n_k(\text{Ar II})$ contains

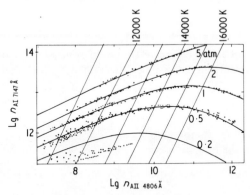

Figure 4. LTE test of argon arc plasmas at different pressures and arc currents. Solid curves correspond to LTE (Shumaker and Popenoe 1972).

the LTE curves and values measured from the line intensities given by equation (13). Uncertainties of atomic transition probabilities do not enter into the results because in the lg–lg plot a parallel shift of both curves against each other is allowed. Figure 4 shows that argon arcs are in LTE at atmospheric pressure and high currents as in the work of Uhlenbusch *et al* (1970).

An example of an extreme non-LTE state is shown in figure 5 where the intensity ratio of two copper lines emitted from an argon–copper arc are plotted over the

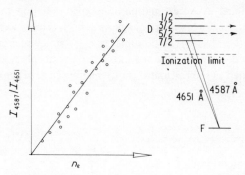

Figure 5. Non-LTE population of an auto-ionizing CuI level measured with a copper–argon arc (Keller 1973).

measured electron density. The upper level of the 4587 Å line shows auto-ionization with a very high transition probability into the continuum. This means (see equation (9)) that $A_{21} > R_{21} n_e$ and one obtains a line intensity proportional to n_e. The second line in figure 5 (4651 Å) does not show this effect (Keller 1973).

A rough conclusion of all these investigations is that the LTE state can be expected to occur in arcs burning in the heavier noble gases and other gases with similar excitation energies E_2. The electron density should be close to the maximum density and the arc diameter should be as large as possible to avoid strong gradients (which enlarge the diffusion), and to avoid the escape of radiation, especially resonance radiation.

3.3. Experimental temperatures

The temperature measurement of an arc plasma in LTE is, in principle, very simple. If plasma composition (c_s) and pressure are known, nearly all intensities or other optical quantities can be used to determine the temperature T_e (see §2). In practice one has to look for quantities which depend strongly on T_e. Here one has to distinguish between methods which require the knowledge of atomic constants and those which are independent of transition probabilities or other data.

Sensitive methods belonging to the first group are:

(i) Temperature from absolute line intensities (equations (4) and (13)). The relative error of temperature is given by

$$\frac{\Delta T}{T} = \frac{kT}{E} \left| \frac{\Delta A}{A} + \frac{\Delta J}{J} \right|.$$

kT/E is normally about 0·1 or less.

(ii) The electron density can be measured from the recombination continuum (equation (17)), from the halfwidth of H_β at small hydrogen admixtures (equation (15)), or by a two-wavelength interferometry (equation (19)). All measurements yield accurate temperatures if the degree of ionization is small. This means that n_e has a strong temperature dependence.

Without knowledge of special atomic data one can apply the following methods:

(i) Emission from optically non-thin layers (equation (22)). Here one has to be very careful to avoid reabsorption in the cooler layers of the plasma. The accuracy of the method depends on $\kappa(\lambda)s$ and $h\nu/kT$.

(ii) Fowler and Milne (1923, 1924) have introduced a general method based on the temperature dependence of spectroscopical quantities. Using figure 4 it can be demonstrated that if two spectroscopic quantities have been measured over a wide range of arc current for example, one can then plot the logarithm of the first quantity against the logarithm of the second; the parameter of this experimental curve is the arc current. Using the LTE relations one can draw the corresponding curve with the temperature as a parameter. If both quantities depend linearly on the (unknown) atomic constants, one can amalgamate both curves and obtain the temperature as a function of arc current. This procedure can yield very accurate temperatures but one has to measure a large amount of spectroscopic intensities.

(iii) Larenz (1951) has measured the intensity of a spectral line over a wide temperature range. This intensity curve has a maximum at a temperature which can be calculated from the LTE relations (the method of normal temperature).

Under optimum conditions it is possible to measure arc plasma temperatures with an accuracy of the order of 1%. But it appears that a reliable temperature value can only be obtained if different methods yield the same result. Further details on temperature measurements are given in the following books on plasma spectroscopy and plasma diagnostics: Griem (1964), Huddlestone and Leonard (1965), Lochte-Holtgreven (1968), Venugopalan (1971).

Appendix. List of symbols

A_{ik}	transition probability (s^{-1})	Q	cross section
b, b_i	underpopulation factor (equations (9) and (10))	r	arc radius
		s	length, designation of plasma component
c	velocity of light, concentration		
e	electron charge	T_e	electron temperature
E, E_i	excitation energy	T_g	gas temperature
E	electrical field strength	v	velocity
g	statistical weight	Z	partition function
h	Planck's constant	α	coefficient of polarization, characteristic parameter (equation (18))
I_λ	spectral intensity (radiance) (erg cm^{-2} s^{-1} cm^{-1} sr^{-1})		
J	line intensity = $\int I_\lambda \, d\lambda$ (erg cm^{-2} s^{-1} sr^{-1})	β	T_g/T_e
		γ	coefficient accounting for reabsorption
k	Boltzmann's constant		
m	mass	θ	scattering angle
n_s	number density of species 's'	$\kappa(\lambda)$	absorption coefficient (cm^{-1})
$n_{s,+}$	number density of ions	ν	light frequency
n_+	number density of ions	ξ	coefficient, describing continuous spectra
$n_{s,i}$	number of particles in quantum state i		
		σ	electrical conductivity
N	index of refraction	χ	ionization energy.

References

Bates D R, Kingston A E and McWhirter R W P 1962 *Proc. R. Soc.* A **267** 297
Biberman L M and Norman G E 1960 *Opt. Spektrosk.* **8** 230
Biberman L M and Ulyanov K N 1961 *Opt. Spektrosk.* **10** 297
Finkelnburg W and Maecker H 1956 *Handbuch der Physik* XXII ed S Flügge (Berlin: Springer-Verlag)
Fowler R H and Milne E A 1923 *R. Astr. Soc.* **83** 403
—— 1924 *R. Astr. Soc.* **84** 499
Giovanelli R G 1948 *Aust. J. Sci. Res.* **1** 275
Griem H R 1964 *Plasma Spectroscopy* (New York: McGraw-Hill)
—— 1974 *Spectral Line Broadening by Plasmas* (New York and London: Academic Press)
Gündel H 1970 *Beitr. Plasmaphys.* **10** 455
Gurevich D B and Podmoshenskii I V 1965 *Opt. Spectrosc.* **18** 319
Huddlestone R H and Leonard S L 1965 *Plasma Diagnostic Techniques* (New York and London: Academic Press)
Keller R 1973 *Dissertation* University of Kiel
Kruger C H 1970 *Phys. Fluids* **13** 1737
Larenz R W 1951 *Z. Phys.* **129** 327
Lochte-Holtgreven W 1968 *Plasma Diagnostics* (Amsterdam: North-Holland)
McWhirter R W P 1961 *Nature* **190** 902
McWhirter R W P and Hearn A G 1963 *Proc. Phys. Soc.* **82** 641
Meiners D and Weiss C O 1973 *Z. Naturf.* **28a** 1294
Menzel D H 1937 *Astrophys. J.* **85** 330
Menzel D H and Cillie G G 1937 *Astrophys. J.* **85** 88
Richter J 1965 *Z. Astrophys.* **61** 57
—— 1971 *10th Int. Conf. on Phenomena in Ionized Gases* (Oxford: Donald Parson)
Richter J *et al* 1975 to be published
Rosenbluth M N and Rostoker N 1962 *Phys. Fluids* **5** 776
Schlüter D 1968 *Z. Phys.* **210** 80
Shumaker J B 1974 *J. Quant. Spectrosc. Radiat. Transfer* **14** 19
Shumaker J B and Popenoe C H 1972 *J. Res. NBS* **76A** 71
Uhlenbusch J, Fischer E and Hackmann J 1970 *Z. Phys.* **238** 404
Venugopalan M 1971 *Reactions under Plasma Conditions* (New York: Wiley Interscience)
Wilson R 1962 *J. Quant. Spectrosc. Radiat. Transfer* **2** 477

A fully automatic system for determining temperature distributions in thermal plasmas

K C Lapworth and L A Allnutt
Division of Quantum Metrology, National Physical Laboratory, Teddington, Middlesex

Abstract. An automatic system has been developed for determining temperature distributions in thermal plasmas by spectroscopic techniques. The system has been used for measurements on an atmospheric pressure argon plasma produced by a radio frequency discharge. Temperatures of 9000 K to 10 000 K in the central region of the plasma have been determined with an estimated uncertainty of ±200 K.

1. Introduction

The temperature distribution throughout a plasma can be determined from the distribution of volume emission coefficient for suitable line or continuum radiation. The volume emission coefficient can be determined for a plasma of cylindrical symmetry from side-on measurements of the specific intensity made at a number of points across the plasma. In general this must be done at several axial positions along the plasma. In addition, so-called cylindrically symmetrical plasmas often present slight asymmetries and it is necessary to obtain more than one side-on view around the axis of the plasma in order to deal with the effects of such asymmetries. In the equipment described in this report, provision is made to obtain two orthogonal side-on views of the plasma. In order to provide adequate side-on intensity data it is necessary to obtain several hundred readings, and a fully automatic system has been developed in order to carry this out rapidly and efficiently.

2. Experimental arrangement

2.1. Plasma generator

The plasmas are produced by means of a 4·5 MHz radiofrequency discharge at atmospheric pressure. The plasma torch consists of a 40 mm outside diameter silica tube with a helical flow of gas led into the top of the tube and discharged to atmosphere at the lower end of the tube. The silica tube, which is 300 mm long, is placed inside a four-turn coil carrying RF power. A discharge is initiated with a tesla coil and, when the gas becomes sufficiently conducting, power couples inductively from the coil into the gas to form an intensely bright thermal plasma at atmospheric pressure. Plasmas of argon, air and nitrogen have been produced in this torch but, to date, temperature measurements have been carried out only in argon.

2.2. Optical arrangement

This is shown schematically in figure 1. Mirrors M1 and M3 can be turned automatically to view either the plasma or the standard lamps. Mirror M2 can swing between the two positions shown so that either a side view or front view of the plasma may be obtained.

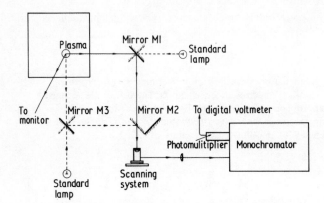

Figure 1. Optical arrangement.

The horizontal and vertical scanning of the plasma is achieved by means of an optical train consisting of a 1 m focal length collimating lens, followed by two plane mirrors which deflect the collimated radiation first downwards then horizontally, and finally a 0·5 m focal length lens which focuses the beam on to a horizontal field slit just in front of the monochromator slit. The collimating lens and plane mirrors are moved horizontally and vertically on slides in such a way that throughout the scanning the plasma stays focused on the field slit and the direction of the beam entering the monochromator remains unchanged.

Radiation from the plasma is also focused on to a second smaller monochromator which monitors the intensity of a spectrum line. This serves to check the constancy and steadiness of the plasma during an experiment.

2.3. Automatic control

The horizontal and vertical slides of the scanner are positioned by stepping motors driven from a commercial machine tool numerical control unit. The wavelength drive of the monochromator is similarly driven. Each of the mirrors M1, M2 and M3 can be rotated from one set position to another by signals derived from the numerical control unit. The control unit is commanded by pre-programmed punched tape so that any desired sequence of movements of slides, wavelength setting and mirror settings may be obtained. A block diagram of the automatic equipment is shown in figure 2. On completion of a slide movement or change of wavelength setting a 'stop' signal from the control unit commands the digital voltmeter to sample the signal from the photomultiplier over a selected integration time and, on completion of sampling, a print command signal serves to punch out the reading and also to initiate the next instruction on the program tape in the numerical control unit. Movements of the rotating mirrors

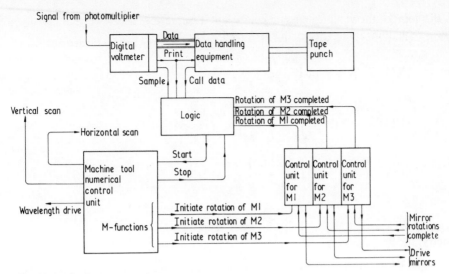

Figure 2. Block diagram of control system.

M1, M2 and M3 are initiated by switch closures in the control unit (M-functions, figure 2). The completion of a mirror movement is sensed by a microswitch which serves to initiate the next instruction to the numerical control unit.

3. Results

For a cylindrically symmetrical plasma, the volume emission coefficient, $\epsilon(r)$, is related to the side-on intensities, $I(x)$, across the plasma of radius R by the well known Abel-type integral equation:

$$I(x) = 2\int_{r=x}^{R} \frac{\epsilon(r)\,r\,\mathrm{d}r}{(r^2-x^2)^{1/2}}. \tag{1}$$

This can be inverted to yield

$$\epsilon(r) = -\frac{1}{\pi}\int_{x=r}^{R} \frac{\mathrm{d}I(x)/\mathrm{d}x}{(x^2-r^2)^{1/2}}\,\mathrm{d}x. \tag{2}$$

We have found slight departures from cylindrical symmetry so that the above equations do not strictly apply in our case. In order to deal with the asymmetrical case, the method of Freeman and Katz (1963) has been employed. In this method the volume emission coefficient is assumed to have radial and angular dependence of the form given by

$$\epsilon(r, \theta) = H(r) + K(r)\cos\theta + L(r)\sin\theta. \tag{3}$$

With this form of the volume emission coefficient it can readily be shown that the following equations result (Freeman and Katz 1963):

$$\frac{1}{2}\{I(x) + I(-x)\} = 2\int_x^R \frac{H(r)r\,dr}{(r^2 - x^2)^{1/2}}$$

$$\frac{1}{2x}\{I(x) - I(-x)\} = 2\int_x^R \frac{K(r)}{r} \frac{r\,dr}{(r^2 - x^2)^{1/2}}$$

$$\frac{1}{2}\{I(y) + I(-y)\} = 2\int_y^R \frac{H(r)r\,dr}{(r^2 - y^2)^{1/2}}$$

$$\frac{1}{2y}\{I(y) - I(-y)\} = 2\int_y^R \frac{L(r)}{r} \frac{r\,dr}{(r^2 - y^2)^{1/2}} \tag{4}$$

where x and y refer to coordinates in the two orthogonal views and $\theta = 0$ is taken along the x axis. Since these equations are analogous to the Abel-type equation for cylindrical symmetry, they may be inverted to give $H(r)$, $K(r)$ and $L(r)$.

The side-on intensity data $I(x)$ and $I(y)$ are fitted by Chebyshev polynomials up to twelfth order. An example of such a 'best-fit' profile is shown in figure 3. Using the best-fit Chebyshev polynomials, the integral equations (4) above are inverted analytically to yield $H(r)$, $K(r)$ and $L(r)$. The volume emission coefficient can then be calculated for any r and θ from equation (3). Values of $H(r)$, $K(r)$ and $L(r)$ have been used to calculate the volume emission coefficient for both the argon continuum radiation at 431·47 nm and for the argon line at 430·01 nm. Five horizontal planes at equally spaced intervals along the vertical axis of the plasma in the region of the coil were surveyed, requiring, in all, 820 side-on intensity measurements. The total time taken for the experiment was approximately half an hour. For the line measurement, the volume emission coefficient integrated over the line profile is required and, in order to avoid lengthy spectral scanning, the exit slit of the monochromator was opened to 200 μm and a correction was made for line wing loss using Stark width data determined by Schulz and Wende (1968). This correction never exceeded 7%. The volume emission coefficient derived from the continuum measurements gives the temperature through the relation

$$\epsilon(\lambda, T) = C_1 \frac{N_e^2}{\lambda^2 T^{1/2}} \xi(\lambda, T)$$

where

$$C_1 = \frac{8e^6}{3c^2}\left(\frac{2\pi}{3km_e^3}\right)^{1/2}$$

and

$$\xi(\lambda, T) = \frac{\gamma}{U_+(T)} \xi^{fb}(\lambda, T)[1 - \exp(-c_2/\lambda T)] + \xi^{ff}(\lambda, T)\exp(-c_2/\lambda T)$$

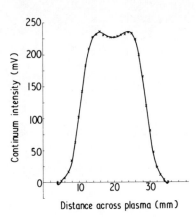
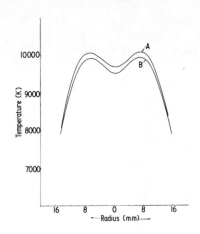

Figure 3. Data and fitted curve. ×, data points; ——, best-fit curve.

Figure 4. Temperature distribution along the x axis of plane 3. Curve A, from intensity of continuum at 431·47 nm; curve B, from intensity of line at 430·01 nm.

where ξ^{fb} is a factor related to the free–bound radiation calculated by Schlüter (1968), γ is the statistical weight of the parent ion (in the present case $\gamma = 6$ for argon), ξ^{ff} is a factor that takes into account free–free radiation and $U_+(T)$ is the partition function for the ion. The remaining symbols have their usual meaning.

The temperature is evaluated from spectral line measurements using the following formula for the integrated line volume emission coefficient

$$\int_{\text{line}} \epsilon(\lambda, T)\, d\lambda = \frac{1}{4\pi} \frac{N_0 \exp(-E_j/kT)}{U_0(T)} g_j A_{ji} \frac{hc}{\lambda_{ji}}.$$

Here N_0 is the neutral atom number density, $U_0(T)$ is the partition function for the neutral atom, E_j is the energy of level j relative to the ground state, g_j is the statistical weight of level j, A_{ji} is the transition probability for transition $j \to i$ and λ_{ji} is the wavelength of the line resulting from this transition.

Temperature distributions along the x axis of plane 3 derived from both continuum and line volume emission coefficients are shown in figure 4. The occurrence of peak temperatures off the axis is due to skin-depth heating associated with the high-frequency power. Temperature profiles were prepared for all the observation planes and these profiles were used to derive isotherms in the plasma. An isotherm map, derived from temperatures given by the continuum emission, is shown in figure 5. This shows clearly the presence of asymmetries and the off-axis temperature maxima due to skin-depth heating.

The uncertainties contributing to the determination of the volume emission coefficient are estimated as follows:

(i) uncertainty due to scattering of radiation in the monochromator, ±1% for line and ±1·5% for continuum;

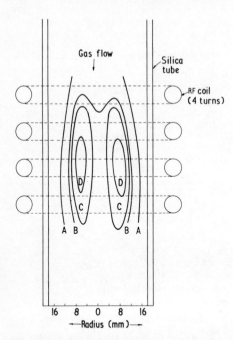

Figure 5. Isotherms in one diametral plane. Curves A, 8600 K; curves B, 9600 K; curves C, 9800 K; curves D, 10 000 K.

(ii) uncertainty in the high-temperature transmission of silica containing the plasma, ±0·5%;

(iii) uncertainty in intensity of calibration lamp due to uncertainty in its temperature, ±5%;

(iv) uncertainty in intensity of calibration lamp due to uncertainty in the emissivity of tungsten, ±3%;

(v) uncertainty in derivation of the volume emission coefficient in the central region of the plasma, ±3%;

(vi) uncertainty in exit slit width of monochromator (for line measurements only), ±1%;

(vii) uncertainty due to nonlinearity of radiation detector, ≪1%.

The most pessimistic compounding of these uncertainties gives, at worst, a total uncertainty of ±13% for the continuum volume emission coefficient and ±13·5% for the integrated volume emission coefficient for the line. From these figures it is reasonable to assign an estimated uncertainty in the volume emission coefficient for both line and continuum of ±10%.

The uncertainties contributing to the determination of temperature are estimated as follows:

(i) uncertainty due to uncertainty in volume emission coefficient, ±0·5%;
(ii) estimates of local thermodynamic equilibrium using Griem's criteria (Griem 1964) indicate that over the central regions of the plasma where the temperature is greater than about 9000 K, level populations associated with the spectrum line and with the free–bound contributions to the continuum radiation are probably within 10% of

equilibrium populations. This leads to an uncertainty in temperature of approximately ±0·5%;

(iii) the transition probability for the 430·01 nm line given by Wiese *et al* (1969) is assigned an uncertainty of ±25%. Schlüter (1968) assigns an uncertainty of ±30% to his theoretically calculated factor (ξ^{fb}) for the continuum radiation. Schulz-Gulde's (1970) measurements of ξ^{fb} agree very well with Schlüter's calculated factor over the wavelength range used in the present work. Schulz-Gulde assigns an uncertainty of ±20–25% to his measurements. Therefore, it is probably reasonable to assign an uncertainty of ±25% to the value of ξ^{fb} taken from Schlüter's paper. This uncertainty for both line and continuum leads to an uncertainty in temperature of ±1·5%.

Therefore, the most pessimistic assessment of uncertainty in the temperature determinations in the central region of the plasma ($T \gtrsim 9000$ K) would be ±2·5%. From the figures, it is reasonable to assign an uncertainty of ±2% to the temperature. That is the temperature uncertainty in the central regions of the plasma is estimated to be ±200 K. The temperature profiles determined from continuum and line measurements, as shown in figure 4, agree within the limits of this estimated uncertainty.

Acknowledgment

The authors would like to thank Mr G F Miller of the Division of Numerical Analysis and Computing for obtaining the analytic solution of the integral equations (4).

References

Freeman M P and Katz S 1963 *J. Opt. Soc. Am.* **53** 1172–9
Griem H R 1964 *Plasma Spectroscopy* (New York: McGraw-Hill) pp 148–52
Schlüter D 1968 *Z. Phys.* **210** 80–91
Schulz P and Wende B 1968 *Z. Phys.* **208** 116–28
Schulz-Gulde E 1970 *Z. Phys.* **230** 449–59
Wiese W L, Smith M W and Miles B M 1969 *Atomic Transition Probabilities* NSRDS-NBS vol II (Washington: NBS)

A plasma standard of temperature and vacuum ultraviolet radiation

R C Preston
Division of Quantum Metrology, National Physical Laboratory, Teddington, Middlesex

Abstract. A standard of temperature above 10^4 K and of vacuum ultraviolet radiation based on a wall-stabilized arc requires detailed knowledge of the behaviour of such a source and the existence of local thermodynamic equilibrium (LTE) in the plasma. This paper gives preliminary results of the axial temperature of a 3 mm channel arc in argon at a pressure of $1 \cdot 75 \times 10^5$ Pa based on eight different spectroscopic measuring techniques. Arc currents between 28 and 75 A were used yielding temperatures with a total scatter of 300 K at 12 000 K and 600 K at 14 400 K. The existence of small deviations from LTE in the lowest excited states of krypton when it is introduced at a concentration of 0·3% have been shown to exist. Perturbations to the plasma temperature on introduction of less than 1% of carbon dioxide have also been studied.

1. Introduction

A wall-stabilized arc is both a stable and reproducible plasma source which enables accurate temperature measurement techniques to be extended to well above 10^4 K. Eventually, it is hoped that such a source will also be theoretically predictable (Shawyer 1975) in which case it will be an ideal standard of temperature.

The existence of local thermodynamic equilibrium (LTE) is not a foregone conclusion in such a plasma where strong thermal gradients are present (Uhlenbusch *et al* 1970, Shumaker and Popenoe 1972). Non-LTE leads to differences between the plasma characteristic temperatures (Drawin 1970) and hence there is a need to determine temperature using as many different and accurate techniques as possible.

This paper presents the results of temperature measurements made by a number of well established spectroscopic techniques. It also describes a preliminary investigation into the use of optically thick vacuum ultraviolet lines as a means of temperature measurement (Boldt 1970, Stuck and Wende 1972).

2. The plasma source

The plasma source consists of a wall-stabilized arc operating in an atmosphere of argon with very small additions of trace gases. The arc has 17 insulated water-cooled copper sections 3 mm thick with a central discharge channel of 3 mm diameter as shown in figure 1. Electrodes consist of an offset thoriated tungsten cathode and polished copper anode.

The arc is attached to a vacuum system through a differential pumping system which, with a 0·5 mm diameter first-stage orifice, allows arc pressures up to 10×10^5 Pa to be used whilst maintaining a pressure below 4×10^{-3} Pa in the vacuum system.

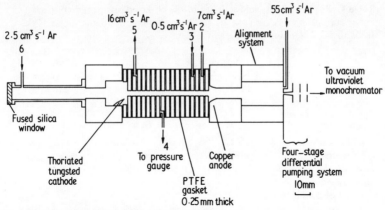

Figure 1. The wall-stabilized arc plasma source.

Main argon flow rates are set up as shown in figure 1 and the trace gases are present between inlet 3 and outlet 5. Vacuum ultraviolet resonance line emission from these gases can then be observed along the axis of the arc through the differential pumping orifices without intermediate absorption taking place.

The arc is run at constant current up to a maximum of 75 A with a stability of ±0·02 A over 4 hours after a warm-up drift of ±0·02 A in 15 minutes. The plasma is initiated by evacuating the channel and discharging a capacitor between the electrodes.

3. Optical arrangement for plasma diagnostics

Two separate optical systems are used to observe the radiation emitted end-on from the axis of the plasma source. The vacuum ultraviolet radiation from the anode end passes into an f/180 aperture system and then into a 3 m monochromator having a resolution of 0·005 nm. The radiation from the cathode end passes through an f/150 aperture system into a 3 m visible spectral region monochromator. An intensity calibration facility using a tungsten-ribbon lamp is provided for this optical system.

4. Reproducibility and stability of the plasma source

The arc described in §2 has been in operation for over 1 year although detailed measurements have only been undertaken in the last 6 months. The stability of the plasma source and measurement system combined is indicated in table 1.

5. Operation of the plasma source during the diagnostic experiments

All work described in this paper was performed at a pressure of $1·75 \times 10^5$ Pa. In addition to the argon flow, a flow of $0·04$ cm^3 s^{-1}† of krypton was introduced at inlet 3 for all the measurements. It was found that for krypton flows between 0·02 and 0·06 cm^3 s^{-1} the optically thick plateau level of the krypton $\lambda 123·5$ nm line was stable to ±0·2% with a perturbation to the argon continuum emission at $\lambda 431·4$ nm of less than ±0·2%.

† All flow rates are given in cm^3 s^{-1} at a pressure of 10^5 Pa and a temperature of 20 °C.

Table 1. Reproducibility and stability of the observed plasma radiation at an arc current of 60 A after the arc had been running for 30 minutes and for a further 5 minutes if the current is changed.

	Krypton λ123·5 nm black body line†	Argon λ430 nm line	Argon λ431·4 nm continuum
Stability of observed signal for a period of:			
(a) 5 minutes	±0·2%	±0·2%	±0·2%
(b) 4 hours	±1%	±2%	±2%
Reproducibility of the absolute emission for a period of 2 months‡	–	±3%	±3%

† No absolute calibration available in the vacuum ultraviolet.
‡ The diffraction grating in the visible monochromator and the calibration source were changed during this period.

6. Non-LTE in the krypton lower levels

The existence of LTE in the krypton lowest excited states was investigated by determining the slope of the straight-line plot of the logarithm of the peak intensities of the two optically thick lines at λ123·5 nm and λ116·4 nm. If Planck's law holds, then the slope should be equal to the ratio of the wavelengths. A systematic difference in slope of 0·5 ± 0·2% has been found, indicating a deficiency in emission at λ123·5 nm. This could mean that the radiation temperature associated with the upper energy level of the λ123·5 nm line is about 60 K lower than that for the λ116·4 nm line, which would be the case if collisional excitation rates were insufficient to prevent partial radiative depopulation of the excited state.

7. Temperature measurement

The plasma temperature was measured using eight different techniques. Some of these were more sensitive than others to the effect of temperature inhomogeneity near the anode and cathode. A factor indicating the sensitivity in terms of the resulting temperature error is presented in the last column of table 2, normalized to unity for the least sensitive methods.

Table 2 gives the results of two experimental sets of data, each of up to 6 hours duration. The two sets were obtained four weeks apart with the exception of the H_β measurement in the first set which was obtained approximately midway through the four-week period.

With the exception of the H_β measurements, we assume the existence of LTE and all methods rely on the solution of the partial pressure law, the Saha equation and the electrical neutrality equation (Lochte-Holtgreven 1968, p149) to give a unique relation between the absolute temperature, T, and the number densities of electrons, $N_e(T)$, or argon neutrals $N^0(T)$.

Plasma temperature and radiation standard

Table 2. Wall-stabilized arc axial temperature at an absolute pressure of 1.75×10^5 Pa for different arc currents.

Method	Temperature (K)												ΔT at 60 A	Requirement of the method†	End-effect factor
	28 amp	32 amp	36 amp	40 amp	45 amp	50 amp	55 amp	60 amp	65 amp	70 amp	75 amp				
Abs. ArI λ430 nm	11 845 / 11 830	12 163 / 12 107	12 440 / 12 378	12 695 / 12 604	13 002 / 12 888	13 240 / 13 143	13 485 / 13 382	13 730 / 13 610	13 935 / 13 816	14 178 / 14 044	14 398 / 14 254		±870	a, d, e, f, g, h, j, k	3
Abs. ArII λ481 nm	11 940 / 11 905	12 270 / 12 225	12 540 / 12 505	12 770 / 12 735	13 040 / 13 040	13 320 / 13 285	13 560 / 13 515	13 780 / 13 737	14 010 / 13 950	14 210 / 14 173	14 420 / 14 370		±210	a, d, e, f, g, h, j	1
Rel. ArI/ArII	12 002 / 11 970	12 354 / 12 328	12 608 / 12 598	12 820 / 12 834	13 062 / 13 135	13 360 / 13 370	13 610 / 13 580	13 815 / 13 796	14 043 / 14 015	14 235 / 14 227	14 435 / 14 392		±340	b, g, h, j, k	1
H_β plus ArI/ArII	12 030							13 844 / 13 775					±390	b, h, j, k, l	1
Abs. cont. λ431 nm	11 937 / 11 914	12 251 / 12 220	12 524 / 12 490	12 759 / 12 740	13 044 / 13 015	13 299 / 13 274	13 549 / 13 512	13 773 / 13 740	13 994 / 13 960	14 209 / 14 183	14 415 / 14 398		±640	c, d, e, f, g, h, k	2
Log ArI against log Kr	11 850 / 11 800	12 168 / 12 063	12 446 / 12 333	12 694 / 12 556	12 995 / 12 825	13 236 / 13 087	13 500 / 13 319	13 731 / 13 550	13 950 / 13 751	14 174 / 13 953	14 369 / 14 159		±140	g, h, j, k	3
Log ArI against log ArII	11 735 / 11 742	12 050 / 12 042	12 317 / 12 311	12 541 / 12 536	12 800 / 12 815	13 062 / 13 059	13 294 / 13 276	13 500 / 13 491	13 712 / 13 694	13 902 / 13 900	14 103 / 14 066		±220	e, g, h, j, k	4
Log (cont.) λ431 nm against log Kr	11 966 / 11 861	12 295 / 12 152	12 570 / 12 430	12 793 / 12 663	13 094 / 12 933	13 356 / 13 188	13 606 / 13 424	13 837 / 13 643	14 069 / 13 844	14 292 / 14 064	14 478 / 14 267		±320	g, h, k	5
Log (cont.) λ482 nm against log Kr	12 046 / 11 920	12 362 / 12 220	12 632 / 12 492	12 883 / 12 723	13 186 / 13 012	13 444 / 13 263	13 700 / 13 507	13 948 / 13 726	14 185 / 13 938	14 405 / 14 151	14 655 / 14 350		±320	g, h, k	5

† (a) Argon transition probability. (b) Relative argon transition probabilities. (c) Continuum Schlüter factor. (d) Absolute calibration source. (e) Linear detector over three decades. (f) Arc window transmission. (g) No de-mixing. (h) Self-absorption correction. (j) Correction for line wing loss. (k) Correction for interference from neighbouring lines. (l) H_β Stark width theory.

7.1. Absolute intensity of the argon neutral line at λ430 nm

The total intensity emitted in an argon neutral line of wavelength λ_{ki}^0 corresponding to a transition between the atomic levels k and i is given (for the optically thin approximation) in W m^{-2} sr^{-1} by

$$I_{ki}^0(T) = \frac{hc}{4\pi\lambda_{ki}^0} g_k^0 A_{ki}^0 \frac{N^0(T)}{Z^0(T)} l \exp(-E_k^0 c_2/T) \tag{1}$$

where l is the plasma column length, g_k^0 and E_k^0 are the statistical weight and energy respectively of the upper level k (Wiese et al 1969), and $Z^0(T)$ is the argon neutral partition function (Sparks and Fischel 1971).

The transition probability, A_{ki}^0, is taken from Wiese et al (1969) as 0.00394×10^8 s^{-1} for the argon line λ430 nm. This line is least perturbed by the Stark-broadened wings of neighbouring lines. The total intensity was obtained at each arc operating current by planimetry of the spectral scan over the line and the neighbouring spectrum. A theoretical model of the line profile was reconstructed using the Stark width measurements of Schulz and Wende (1968) for the λ430 nm line and relative Stark widths for neighbouring lines taken from Griem (1964). This model was used to correct the measured total intensities for line wing and perturbation losses (up to 40% correction). Correction was also made for line self-absorption by calculation and continuum self-absorption using data given by Morris and Yos (1971) (total correction was less than 7%). With an absolute intensity calibration, the total intensity I_{ki}^0 was determined from which the temperature was calculated using equation (1), knowing $N^0(T)$ as already mentioned in §7.

7.2. Absolute intensity of the argon singly ionized line at λ481 nm

For a singly ionized argon line, the total intensity $I'_{ki}(T)$ in the optically thin approximation is given by equation (1) where the parameters now refer to the singly ionized atom. The transition probability for the line at λ480·6 nm is taken from Wiese et al (1969) as 0.79×10^8 s^{-1} which is in excellent agreement with a later measurement by Shumaker and Popenoe (1972) of $0.786 \pm 0.01 \times 10^8$ s^{-1}. The total intensity was again obtained by planimetry, then corrected for line wing loss using the Stark widths of Morris and Yos (1971) and also for line self-absorption by calculation and continuum self-absorption using the data of Morris and Yos (1971). Using the standard lamp calibration, I'_{ki} was determined for a particular arc current and hence the corresponding temperature calculated using equation (1).

7.3. Relative intensities of the argon neutral line at λ430 nm to the singly ionized line at λ481 nm

The ratio of intensities as calculated from equation (1) combined with the experimental total intensity measurements yielded another result for the temperature.

Unlike the previous two methods, this method has the advantage that (i) it requires only relative transition probabilities, (ii) it does not need an absolute intensity calibration, (iii) it does not depend on the linearity of the detector because the two lines are of similar intensity, and (iv) it is insensitive to changes in the arc window transmission.

7.4. Electron number density from H_β width combined with the relative intensities of the argon neutral to singly ionized lines

The use of the partial pressure equation is not necessary if the electron density is obtained from a measurement of the Stark-broadened width of H_β combined with line-broadening theory. This is of considerable advantage in a multi-species plasma where de-mixing is important.

A small amount of hydrogen ($0 \cdot 3$ cm^3 s^{-1}) was introduced into the arc at the inlet 3 in figure 1 and the H_β line profile at $\lambda 486$ nm was scanned. The hydrogen caused an overall increase in the plasma axial temperature of 30 K, observed by monitoring the krypton $\lambda 123 \cdot 5$ nm black body line.

The asymmetrical double-peak profile was corrected for the argon continuum contribution and for changes in the monochromator and detector efficiency over the profile. Both the blue and red wing half maximum intensity widths were measured using the criterion for peak intensity illustrated by Locht-Holtgreven (1968, p159) and the electron density was obtained by application of the theoretical results of Kepple and Griem (1968). The mean value of the results for the two wings was taken because the theory does not predict the observed asymmetry.

By this method the first experimental data set gave an electron density of $2 \cdot 11 \times 10^{23}$ m^{-3}, which is in excellent agreement with the value of $2 \cdot 07 \times 10^{23}$ m^{-3} obtained by method §7.3. The corresponding temperature was determined by the use of the relative total intensity measurements of the argon neutral and singly ionized lines, the Saha equation and the electrical neutrality equation.

7.5. Absolute intensity of the argon continuum at $\lambda 431 \cdot 4$ nm

The volume emission coefficient, $\epsilon^c(\lambda, T)$, for the emission of free–bound and free–free radiation from an argon plasma (Schulz-Gulde 1970) is given in W m^{-3} sr^{-1} nm^{-1} by

$$\epsilon^c(\lambda, T) = \frac{C_1 N_e^2(T)}{\lambda^2 T^{1/2}} \left\{ G \exp(-c_2/\lambda T) + [1 - \exp(-c_2/\lambda T)] \frac{g_1' \xi(\lambda, T)}{Z'(T)} \right\} \quad (2)$$

with $C_1 = 1 \cdot 6318 \times 10^{-34}$ W m^4 K$^{1/2}$, where G is the Gaunt factor for free–free transitions ($G = 1 \cdot 7$) and g_1' is the argon singly charged ion ground state statistical weight ($g_1' = 6$).

The factor $\xi(\lambda, T)$, which is a weak function of λ and T, has been calculated by Schlüter (1968) with an uncertainty of ±30%. Measurements of the absolute volume emission coefficient were made at $\lambda 431 \cdot 4$ nm and corrected for (i) the line wing contribution from the two lines ArI $\lambda 430$ nm and ArI $\lambda 433$ nm and (ii) the continuum self-absorption using the absorption data of Morris and Yos (1971). Application of equation (2) yielded the temperature.

7.6. Log I(ArI λ430 nm) against log I(KrI λ123·5 nm)

Any two measurable spectroscopic parameters which have different known dependences on temperature can be used to determine temperature without the use of intensity calibrations or fundamental spectroscopic parameters which is the basic concept of this and the next two methods of temperature measurement (Richter 1965).

The logarithm of the Planck function $B(\lambda, T)$ (Griem 1964, p31) yields

$$\log B(\lambda, T) = -\log [\exp(c_2/\lambda T) - 1] + \text{constant}, \qquad (3)$$

where only temperature dependent terms have been considered. Likewise, the logarithm of the total intensity of an argon neutral line, from equation (1), yields

$$\log I_{ki}^0(T) = \log [N^0(T) \exp(-E_k^0 c_2/T)/Z^0(T)] + \text{constant}.$$

A theoretical plot of $\log B(\lambda, T)$ against $\log I_{ki}^0(T)$ as a function of temperature and an experimental plot of the observed intensity of the krypton black body line at λ123·5 nm against the corrected total intensity of the ArI λ430 nm line as a function of arc current, are produced. One of the curves is then translated without rotation until a best fit is obtained with the other curve. Each point on the experimental plot corresponding to a unique arc current can then be transposed to a unique temperature from the theoretical plot.

In fitting the two curves, a perfect fit could only be made over 50% of the curve. If the upper current points were fitted it was necessary to increase the effective radiation temperature for the λ123·5 nm line by about 50 K to bring the lowest current (28 A) point onto the theoretical curve. This is in agreement with the results found for the possible non-LTE in the krypton lower levels indicated in §6.

7.7. Log I (ArI λ430 nm) against log I (ArII λ481 nm)

A similar method as described in §7.6 was applied to the ArI λ430 nm and ArII λ481 nm lines. This resulted in an excellent fit of the experimental to the theoretical data. Temperatures were again read off directly.

7.8. Log I (argon continuum) against log I (KrI λ123·5 nm)

If the weak temperature dependence of the Schlüter factor and the exponential terms in equation (2) are ignored then the logarithm of the argon continuum volume emission coefficient is

$$\log \epsilon^c(\lambda, T) = \log [N_e^2(T)/T^{1/2}] + \text{constant}.$$

A theoretical plot of $\log \epsilon^c(\lambda, T)$ against $\log B(\lambda, T)$ from equation (3) was produced and compared with the experimentally determined logarithmic plot of the volume emission coefficient at λ431·4 nm and λ482 nm against the krypton λ123·5 nm line. As in

§7·6 a perfect fit could only be obtained over 50% of the curve. The magnitude of the difference was also identical and a similar explanation appears most likely.

7.9. Summary of the temperature determinations

The results presented in table 2 represent the two individual experimental determinations of the temperature for eleven different arc currents (except for the H_β measurement) using the eight different techniques.

The uncertainties quoted in the column headed ΔT represent the systematic uncertainty for each measurement at the current of 60 A. In the first five methods the main uncertainties contributing to ΔT are in the transition probabilities or the continuum Schlüter factor, and in methods §7.1, §7.2 and §7.5 there is an additional experimental uncertainty in the measurement of absolute intensity. End-effects and line wing corrections are also important in method §7.1 and represent the major contribution to the uncertainties in methods §7.6 and §7.7. Method §7.8 is the most dependent on end-effects and is also sensitive to contributions to the observed argon continuum signal which are not directly attributable to the true continuum emission (scattered light, impurity spectra). The last two columns in the table give the particular theoretical or experimental requirements for each method and the end-effect factor respectively.

Random uncertainties have not been quoted because of insufficient data but some indication of these is given by the reproducibility of the two sets of data and of the increments in temperature with current increase.

The log–log plots produce the lowest degree of reproducibility between the sets. This is probably caused by their reliance on a large amount of data collected over a period of a few hours and their sensitivity to the end-effects. However, these methods are extremely powerful because they do not require (i) knowledge of transition probabilities or continuum Schlüter factors, (ii) absolute or relative intensity calibrations, (iii) knowledge of the plasma emitting length or (iv) arc window transmission data. Non-fitting of the experimental and theoretical curves is also a strong indication of lack of LTE.

No attempt has been made to give a mean temperature. Further measurements must be made to investigate the validity of the line wing corrections and self-absorption corrections applied to the experimental results. In addition, more data on long-term reproducibility is essential for the plasma source to be used as a standard.

8. Perturbation to the plasma temperature on introduction of carbon dioxide

At present we are investigating the use of optically thick vacuum ultraviolet lines as a means of measuring temperature (Stuck and Wende 1972). The intention here is to measure the absolute intensity at the centre of such lines using emission standards which have been established by means of synchrotron radiation. Since the measured intensity will effectively be that of a black body radiator, the corresponding temperature will be given directly by the Planck radiation law. The lines we are using are those of the carbon I multiplet at $\lambda 165$ nm and the subject of this preliminary study has been the perturbation of the arc which results from the introduction of carbon in the form of carbon dioxide.

The carbon dioxide was introduced at inlet 3 in figure 1. Perturbations to the plasma temperature at a current of 60 A were monitored by using the krypton black body line at $\lambda 123 \cdot 5$ nm and the argon continuum at $\lambda 431 \cdot 4$ nm whilst varying the CO_2 concentration from 0·01 to 0·2 cm^3 s^{-1}.

The CI multiplet peaks were scanned and their peak levels determined as a function of the CO_2 flow rate. On increasing the flow each multiplet component was found to reach the same plateau level but at different flow rates. Further increase in flow was accompanied by a slower rate of increase in the peaks (+2% for 0·15 cm^3 s^{-1} increase in CO_2 flow). Meanwhile, the krypton line at $\lambda 123 \cdot 5$ nm also increased by a similar amount. Careful examination of the krypton and carbon optically thick peaks showed this increase to be a result of a superimposed 'bump' on a flat optically thick plateau. This is caused by a local increase in temperature at the trace gas inlet (Boldt 1970), the magnitude of which cannot be determined from the measurements because the radiation emitted from this hotter region is not optically thick.

The maximum flow of CO_2 resulted in a drop of 10% in the continuum level, probably caused by a de-mixing process which increases the axial carbon concentration and reduces the argon ion concentration (zero de-mixing would result in a 2·5% drop in the observed continuum level). An alternative explanation, proposed by Boldt (1970), Hofmann and Weissler (1971) and Müller (1970) using large CO_2 flow rates, is that the overall axial temperature decreases. If this is the case here, the magnitude of such a temperature change is masked by the perturbation at the inlet.

Despite this uncertainty about the cause of the observed perturbations, it is likely that a CO_2 flow rate of 0·03 cm^3 s^{-1} is just sufficient to bring the $\lambda 165 \cdot 7$ nm component of the multiplet up to the black body limit, the overall axial temperature will be within 30 K of its value prior to introducing the trace gas.

9. Conclusion

The results of applying several different spectroscopic temperature measuring techniques to a 3 mm channel wall-stabilized arc have been presented in this paper under the assumption of LTE. Small deviations from LTE have been shown to be present in the krypton lower excited levels and further deviations may well be encountered in the outer regions of the plasma when side-on measurements are made.

Detailed Stark width and self-absorption measurements on the $\lambda 430$ nm line of neutral argon should provide some improvement in these results together with side-on measurements to eliminate end-effects. Combined with an absolute calibration of the carbon optically thick line at $\lambda 165 \cdot 7$ nm, axial temperatures should be determined with uncertainties of ±0·5% but reproducible to ±0·2%.

Acknowledgments

I wish to express my appreciation to C Brookes for his help in setting up and operating the wall-stabilized arc and to Dr K C Lapworth for valuable discussions during the course of the work.

References

Boldt G 1970 *Space Sci. Rev.* **11** 728–72
Drawin H-W 1970 *High Temp. High Press.* **2** 359–409
Griem H R 1964 *Plasma Spectroscopy* (New York: McGraw-Hill)
Hofmann W and Weissler G L 1971 *J. Opt. Soc. Am.* **61** 223–30
Kepple P and Griem H R 1968 *Phys. Rev.* **173** 317–25
Lochte-Holtgreven W 1968 *Plasma Diagnostics* (Amsterdam: North-Holland)
Morris J C and Yos J M 1971 *Radiation Studies of Arc Heated Plasmas* ARL Final Report ARL 71-0317
Müller D 1970 *Proc. Int. Physics of Ionized Gases Summer School* ed B Navinšek (Ljubljana, Yugoslavia: Inst. 'J Stefan') pp694–705
Richter J 1965 *Z. Astrophys.* **61** 57–66
Schlüter D 1968 *Z. Phys.* **210** 80–91
Schulz P and Wende B 1968 *Z. Phys.* **208** 116–28
Schulz-Gulde E 1970 *Z. Phys.* **230** 449–59
Schumaker J B and Popenoe C H 1965 *J. Res. NBS* **69A** 495–509
—— 1972 *J. Res. NBS* **76A** 71–6
Shawyer R E 1975 this volume
Sparks W M and Fischel D 1971 *Partition Functions and Equations of State in Plasmas, NASA Report* SP-3066
Stuck D and Wende B 1972 *J. Opt. Soc. Am.* **62** 96–100
Uhlenbusch J, Fischer E and Hackmann J 1970 *Z. Phys.* **239** 120–32
Wiese W L, Smith M W and Miles B M 1969 *Atomic Transition Probabilities, Nat. Stand. Ref. Data Ser. (US)* 22

Temperature scale and thermal radiation standards in the visible and vacuum ultraviolet spectral region between 10 000 and 20 000 K

B Wende
Physikalisch-Technische Bundesanstalt (PTB) Institut, Berlin, Germany

Abstract. For the realization of a temperature scale between 10 000 and 20 000 K and for general requirements of high-temperature metrology thermal radiation standards in the visible and vacuum UV spectral region are being developed. These standards are based on wall-stabilized steady-state arc plasmas operated with argon under a pressure between 0.5×10^5 Pa and 4×10^5 Pa (0.5 atm and 4 atm). For electron number densities $n_e > 10^{23}$ m^{-3} temperatures are produced under the condition of a sufficient approximation to local thermal equilibrium. Temperature measurements by radiation thermometry in the visible and vacuum UV spectral region are discussed and the uncertainty components are analysed. Plasma black body radiation in the vacuum UV at the CI transitions $2s^2 2p^2\,{}^3P - 3s\,{}^3P^0$ ($\lambda \approx 165$ nm) and at Lyα ($\lambda = 121.6$ nm) is utilized to measure temperatures corresponding to the common pyrometric technique below 4000 K. The relative uncertainty $\Delta T/T$ of the established temperature scale is dependent upon the temperature and the plasma pressure. At 15 000 K and 10^5 Pa (number density of the electrons 1.8×10^{23} m^{-3}) the temperature determination uncertainty is $\Delta T/T = \pm 9 \times 10^{-3}$.

1. Introduction

The realization of a temperature scale above 4000 K by radiation thermometry requires the development of reliable thermal radiation standards, because conventional black body radiators cannot be applied in this temperature region. For temperatures between 10 000 and 20 000 K, steady-state wall-stabilized arc plasmas (Maecker 1956), operated with argon, are well known radiation sources for many applications in the field of quantitative spectroscopy. However, when using these arc sources as accurate thermal radiation standards in order to establish a temperature scale, several difficulties occur:

(i) The degree of approximation to the state of local thermal equilibrium (LTE) is not exactly known.
(ii) The essential atomic parameters in the radiation laws of the spectral line and continuous radiation of the plasma are uncertain (transition probabilities, Gaunt factor, $\xi(\lambda, T)$ functions).
(iii) High-density corrections have to be used to calculate partition functions and to apply the Boltzmann factors, the Saha equations, and Dalton's law. While the Debye high-density correction is now the generally accepted one, there is not yet any direct experimental confirmation.

(iv) The temperature is spatially (in an axial direction) not uniform, which is generally not considered.
(v) The arc column length is not well defined and an effective length has to be determined.
(vi) The radiation measurements require small apertures, which can produce errors by diffraction.
(vii) In several cases strong reabsorption effects of the measured plasma radiation have to be taken into account.

With respect to these problems, results obtained in our laboratory are reviewed, in particular those concerning the development of thermal radiation standards in the visible and vacuum ultraviolet (VUV) spectral region and their utilization for a realization of a temperature scale between 10 000 and 20 000 K.

2. Thermal radiation standards in the visible and vacuum UV spectral region

The thermal radiation is excited in wall-stabilized arc plasmas operating in argon (pressure 5×10^4 Pa to 4×10^5 Pa, typical argon flow rate, 20 cm^3 s^{-1} at room temperature). The arc chamber consists of water-cooled copper plates (4 mm thick) with central bores (4 mm in diameter) electrically insulated from each other by spacers (0·5 or 0·8 mm thick) made of Teflon and silicon rubber. Three water-cooled tungsten cathodes and three water-cooled tungsten anodes are used. The arc current varies between 10 and 250 A (time stability 10^{-3} s). The radiation passes through quartz windows and is measured in the direction of the arc column axis (end-on) or transversely (side-on) for a small solid angle that defines a plasma layer of approximately constant temperature in the 'end-on' direction (typical f numbers between 50:1 and 200:1).

To measure the vacuum UV radiation the arc source has to be modified (Boldt 1961). The anode side of the source and the VUV optical system are connected without a win-

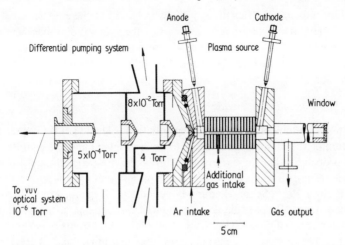

Figure 1. Thermal radiation standard in the VUV spectral region and differential pumping system (nominal arc channel diameter 4 mm, maximum power input 60 kW). A maximum pressure difference between 4×10^5 Pa in the plasma and 10^{-4} Pa in the VUV optical system is maintained.

dow by a differential pumping system consisting of three stages (figure 1) (K Grützmacher unpublished, Stuck and Wende 1972a). The radiation passes through small holes in the differential pumping system (minimum diameter 0·8 mm). A pressure difference between 4×10^5 Pa in the plasma and 10^{-4} Pa in the VUV optical system is maintained. The radiation must pass from the plasma to the differential pumping system without significant self-absorption in the layers of non-homogeneous temperature. Therefore, the plasma is operated with argon introduced through a gas intake near the anodes (typical total argon flow rate 50 cm^3 s^{-1} at room temperature, partial flow rate into the arc channel 8 cm^3 s^{-1}). A trace of an additional gas (CO$_2$ or H$_2$) is allowed to enter the arc channel through an additional gas intake. H$_2$ or CO$_2$ flow towards the cathode side only. Under these conditions the number density of the carbon or hydrogen atoms is sufficiently low in the plasma layers of non-homogeneous temperature near the anodes, thus allowing observation of the VUV resonance lines without critical self-absorption losses.

3. Temperature scale: investigation in the visible spectral region

During the last decade several spectroscopic methods in the visible spectral region have been applied in arc plasma thermometry. A well known method to determine arc temperatures is based on the measurement of the radiance L_{mn} of a well isolated spectral line and on the application of the following equations (1) and (2) (optically thin plasma at constant temperature).

$$L_{nm} = \epsilon_{nm} l \tag{1}$$

$$\epsilon_{nm}(T) = \frac{hc}{4\pi} \frac{g_m A_{nm}}{\lambda_{nm}} \frac{N(T)}{Z(T)} \exp\left(-\frac{E_m}{kT}\right). \tag{2}$$

ϵ_{nm} is the total line emission coefficient, λ_{nm} the wavelength of the emitted spectral line, l the plasma length at constant temperature, A_{nm} the transition probability, c the speed of light, h Planck's constant, k Boltzmann's constant, $N(T)$ the number density of the species, E_m the upper quantum level with the statistical weight g_m, and $Z(T)$ is the partition function.

3.1. LTE approximation and A_{nm} values

In order to establish a temperature scale using equations (1) and (2) the degree of LTE approximation of the plasma standard has to be determined and an accurate value of the transition probability has to be measured. Both of these problems are solved by simultaneously measuring different quantities of the plasma radiation as a function of the arc current (Richter 1965, Schumaker and Popenoe 1972), eg ϵ_{nm} of two lines of different ionization stages. By plotting the measured[†] values of ϵ_{nm} of the ArI line 714·7 nm against the ϵ_{nm} values of the ArII line 480·6 nm for currents between 25 and 200 A, the diagram shown in figure 2 is produced (nominal arc diameter 4 mm, three plasma pressures, $5·07 \times 10^4$ Pa, $1·01 \times 10^5$ Pa, $3·04 \times 10^5$ Pa). From this dia-

† See Stuck and Wende (1972a) for the spectroscopic arrangement.

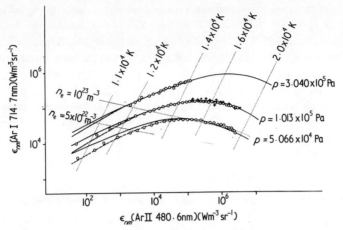

Figure 2. End-on (○) and side-on (+) measured line emission coefficients ϵ_{nm} of ArI 714·7 nm ($A_{nm} = 5·66 \times 10^5$ s^{-1}) against ϵ_{nm} of ArII 480·6 nm ($A_{nm} = 8·82 \times 10^7$ s^{-1}) for arc currents between 25 and 200 A (nominal arc diameter 4 mm, plasma pressures $5·07 \times 10^4$ Pa, $1·01 \times 10^5$ Pa and $3·04 \times 10^5$ Pa). The end-on measured ϵ_{nm} values have been corrected with respect to the temperature inhomogeneities along the arc axis, the arc column length, and reabsorption effects. The full curves represent calculated ϵ_{nm} values using the A_{nm} values listed in table 1 and assuming the LTE approximation (broken curves, PLTE).

Table 1. Two selected transition probabilities of argon according to figure 2 (Nubbemeyer 1974).

	Wavelength	Transition	A_{nm} value
ArI	714·7 nm	5p[5/2]2 − 4s[3/2]^{0}1	$5·66 \times 10^5$ s^{-1} ± 5%†
ArII	480·6 nm	4p ^{4}P$_{5/2}$ − 4s ^{4}P$_{5/2}$	$8·82 \times 10^7$ s^{-1} ± 7%†

† Because the measured A_{nm} values have been determined for the use of arc plasma thermometry, they are uncorrected with respect to the far wings of the line profile which are not measureable for a spectral line emitted by an arc plasma. For comparing the listed A_{nm} values with theoretical results or lifetime measurements, one has to correct them by +13% (see table 2).

gram the A_{nm} values (table 1) of the two spectral lines are determined with the comparatively small relative uncertainty† of ±5% (ArI) and ±7% (ArII) (Nubbemeyer 1974). For these two lines the energy difference between the upper quantum level and the ionization limit is $\gtrsim 2·5$ eV. Therefore high-density corrections of the Boltzmann factors (statistical weights) need not be taken into account. In figure 2 the full curves represent calculated ϵ_{nm} values using the A_{nm} values listed in table 1 and assuming LTE. For electron number densities $n_e > 10^{23}$ m^{-3} the calculated ϵ_{nm} values agree satisfactorily with the measured ϵ_{nm} values (agreement for ϵ_{nm} of ArII 480·6 nm for a range of about four orders of magnitude). Therefore we conclude: (i) for $n_e > 10^{23}$

† See table 2.

m^{-3} the LTE approximation is valid within a maximum relative uncertainty of the emission coefficient $\Delta\epsilon_{nm}/\epsilon_{nm} = \pm 0.07$ (ArII 480.6 nm, $T \lesssim 16\,500$ K); and (ii) the A_{nm} values listed in table 1 can be taken from the diagram of figure 2 with small uncertainties. In contrast to this, for $n_e < 10^{23}$ m^{-3} (pressure 5.07×10^4 Pa) the measured ϵ_{nm} values deviate from the ϵ_{nm} values calculated with the LTE model. These measured ϵ_{nm} values are represented sufficiently well by using the A_{nm} values listed in table 1 and by considering a plasma model in partial local thermal equilibrium (PLTE, broken curve). Further diagrams of the type of figure 2 measuring other radiation quantities confirm the results (Nubbemeyer 1974).

3.2. Temperature scale and uncertainties

When considering the discussed results, the well isolated ArII line 480.6 nm can be selected to establish a temperature scale between 10 000 and 20 000 K using equations (1) and (2). Difficulties occur at the limits of the temperature range: at 10 000 K the line is rather weak and deviation from the LTE approximation may occur; at 20 000 K optically thick emission in the line centre is observed. The accuracy of the determined temperatures depends essentially on the accuracy of the adopted A_{nm} value. Numerous measurements and calculations of this value are known and are compared in table 2. One can recognize a significant difference between the A_{nm} values based on lifetime measurements and those based on plasma radiation measurements. Taking into account the uncertainty $\Delta A_{nm}/A_{nm} = \pm 0.07$ of the measured A_{nm} value only (see table 2), the temperature is uncertain by $\Delta T/T = 3 \times 10^{-3}$ at 12 000 K, 5×10^{-3} at 15 000 K, and 7×10^{-3} at 18 000 K. However, the true uncertainty of the measured temperature is larger because several uncertainty components of the determined ϵ_{nm} values have to be considered and because the temperature is not constant along the arc column axis. The axis temperature is higher in the region of the cooled copper plates than in the region of the electrically insulating spacers. Therefore corrections have to be used in order to convert the end-on measured line radiance to accurate ϵ_{nm} values, which correspond to the maximum axis temperature in the arc regions of the cooled copper plates ('end-on' temperature, figure 3). These corrections depend upon the arc current and concern the temperature inhomogeneities along the arc axis, the arc column length, and reabsorption effects. The relative uncertainty components $\Delta y/y$ of the finally determined ϵ_{nm} values are listed in table 3. When they are combined in quadrature a relative uncertainty $\Delta\epsilon_{nm}/\epsilon_{nm} = \pm 0.055$ at $T = 15\,000$ K follows. When the uncertainties of A_{nm} and ϵ_{nm} are considered, the temperature is uncertain by $\Delta T/T = 9 \times 10^{-3}$ at $T = 15\,000$ K.

3.3. Comparative temperature measurements

The 'end-on' temperatures have been compared with the 'side-on' temperatures ($\Delta T/T = 9 \times 10^{-3}$ at 15 000 K)[†]. These temperatures (figure 3) have been calculated from the side-on measured line emission coefficient of ArII 480.6 nm, which corresponds to the minimum axis temperatures between the cooled copper plates. Above 13 000 K significant differences ΔT between both temperatures are observed: $\Delta T \approx 300$ K at

---

[†] Although the uncertainty components of the 'side-on' temperature differ from those of the 'end-on' temperature, the final uncertainty is approximately the same in both cases.

Table 2. Comparison of measured and calculated values of the transition probability ArII 480·6 nm ($4p\ {}^4P_{5/2} - 4s\ {}^4P_{5/2}$).

	Author	Method	A_{nm} ($10^7\ s^{-1}$)
Measured	Olsen (1963)	Arc plasma (side-on) Larenz method	7·86 ± 8%‡
	Richter (1965)	Arc plasma (end-on) method according to figure 2	6·0 ± 15%‡
	Popenoe and Schumaker (1965)	Arc plasma (side-on) H_β line-broadening diagnostic	13·1 ± 23%
	Berg and Ervens (1967)	θ pinch H and He line-broadening diagnostic	11·1 ± 40%‡
	Schnapauff (1968)	Arc plasma (end-on) branching ratios and lifetime	5·7 ± 10%
	Bues et al (1969)	Arc plasma (end-on) branching ratios and lifetime	5·9
	Schumaker and Popenoe (1969)	Arc plasma (side-on) method according to figure 2	8·72 ± 10%
	Schumaker and Popenoe (1972)	Arc plasma (side-on) method according to figure 2	7·86
	PTB Nubbemeyer (1974)	Arc plasma (end-on and side-on) method according to figure 2	8·82 ± 7%‡
		Branching ratios and lifetime	6·1 ± 25%
		Simultaneous emission and absorption	7·6 ± 26%‡
Calculated	Statz et al (1965)	Hartree–Fock–Slater	8·74
		Coulomb approximation	7·22
	Rudko and Tang (1967)		8·28
	Garstang and Odabasi (1971)	Coulomb approximation	8·36
	Luyken (1972)	Parametrized central field potential	9·08
Measured and adopted for T scale establishment	PTB		8·82 ± 7%‡

† For the lifetime the mean value $\tau = 10\cdot 1$ ns ± 20% was taken from four experiments performed between 1966 and 1970 (Assousa et al 1970, Bakos et al 1966, Denis and Gaillard 1970, Fink et al 1970).
‡ Without line wing correction.

15 000 K and $\Delta T \approx 600$ K at 18 000 K (nominal arc channel diameter 4 mm, pressure $1\cdot 01 \times 10^5$ Pa, copper plates 4 mm thick, insulating spacers 0·8 mm thick). Because of the observed temperature differences along the arc column axis, the geometrical state of the investigated plasma standard is not fully optimal and has to be improved. Therefore, the standard is now operated with 0·5 mm thick insulating spacers (see temperature measurements in the VUV spectral region).

In order to support the measured ArII $- A_{nm}$ value the 'end-on' temperatures have been compared with temperatures determined by simultaneous end-on absorption and emission measurements (Nubbemeyer 1974) in the line centre of ArII 480·6 nm and by

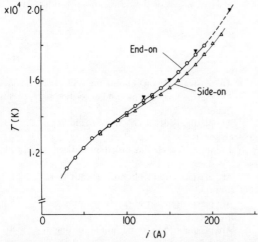

Figure 3. Temperature against arc current (nominal arc channel diameter 4 mm, copper plates 4 mm thick, Teflon spacers 0·8 mm thick, pressure $1·01 \times 10^5$ Pa, argon flow rate 20 cm³ s⁻¹ at room temperature). The 'end-on' temperatures (○) are calculated from the corrected end-on measured line radiance ArII 480·6 nm ($A_{nm} = 8·82 \times 10^7$ s⁻¹) and correspond to the maximum axis temperature in the arc regions of the cooled copper plates (for uncertainties see table 3). The side-on temperatures (△) are calculated from the side-on measured line emission coefficient ArII 480·6 nm and correspond to the minimum axis temperature in the arc region between the cooled copper plates. The end-on temperatures are compared with temperatures determined by simultaneous end-on absorption and emission measurements (▼) in the line centre of ArII 480·6 nm and by application of Kirchoff's law. At $T \approx 10\,000$ K deviation of the LTE approximation may occur; at $T \approx 20\,000$ K optically thick emission of ArII 480·6 nm is observed in the end-on direction.

application of Kirchhoff's law.† This method only requires the validity of the LTE approximation; no uncertain atomic parameters have to be used. Within the uncertainty of the absorption–emission method ($\Delta T/T = 3 \times 10^{-2}$ at $T = 18\,000$ K) the 'end-on' temperatures derived from the line radiance ArII 480·6 nm are confirmed (figure 3).

4. Temperature scale: investigation in the VUV spectral region

In contrast to the radiation measurements under approximately optical thin conditions in the visible spectral region, black body radiation is generated in the VUV region in the centre of the resonance lines in order to establish a temperature scale by a radiometric comparison between plasma black body radiation and synchrotron-calibrated transfer standards and by application of Planck's law.

4.1. Spectro-radiometric arrangement

To measure the VUV black body radiation the anode side of the arc column is imaged with a concave mirror (focal length 1·1 m, MgF₂ coating) enlarged by a factor

---

† In order to perform the simultaneous absorption and emission measurements we used a second plasma standard which was imaged end-on into the arc channel of the first standard. The absorption factor of the arc column (effective length 94 mm) in the line centre of ArII 480·6 nm is characterized by the ratio $L_\lambda/L_\lambda^H \approx 0·27$ at 15 000 K and $L_\lambda/L_\lambda^H \approx 0·96$ at 18 000 K (L_λ is the spectral radiance in the line centre and L_λ^H the black body spectral radiance).

Table 3. Relative uncertainty components $\Delta y/y$ of the line emission coefficient ϵ_{nm} and resulting temperature uncertainty (ArII 480·6 nm, end-on measurements, nominal arc channel diameter 4 mm, pressure $1\cdot01 \times 10^5$ Pa).

	Temperature		
	12 000 K $\Delta y/y$	15 000 K $\Delta y/y$	18 000 K $\Delta y/y$
[a]Spectral radiance transfer standard	±2%	±2%	±2%
Gold-point uncertainty			
Electronic nonlinearities			
Window transmissivity	±1%	±1%	±1%
f-number, diffraction	±3%	±3%	±3%
[b]Effective arc column length	±2%	±2%	±2%
Temperature inhomogeneities	±2%	±2%	±2%
Reabsorption	±1%	±2%	±14%
Plasma pressure, high-density corrections	±2%	±2%	±2%
[c]$\Delta\epsilon_{nm}/\epsilon_{nm} = [\Sigma(\Delta y/y)^2]^{1/2}$	±5·2%	±5·5%	±14·9%
$\Delta A_{nm}/A_{nm}$ (ArII 480·6 nm)	±7%	±7%	±7%
[d]$\Delta T/T$	±0·4%	±0·9%	±1·5%
ΔT	±50 K	±140 K	±260 K

(a) $\Delta y/y$ of the end-on measured spectral line radiance L_{nm}.
(b) $\Delta y/y$ of the necessary corrections to convert L_{nm} to ϵ_{nm} and T which correspond to the maximum axis temperature in the arc regions of the cooled copper plates.
(c) Resulting relative uncertainty of the determined ϵ_{nm} values and relative uncertainty of the transition probability $\Delta A_{nm}/A_{nm}$. The relative standard error of the mean of ϵ_{nm}, considering ten groups of measurement during two years, is ≈1%. The relative uncertainty of ϵ_{nm} calculated from the uncertainty of Boltzmann's constant is <0·3%.
(d) Temperature uncertainty from the combination of $\Delta\epsilon_{nm}/\epsilon_{nm}$ and $\Delta A_{nm}/A_{nm}$.

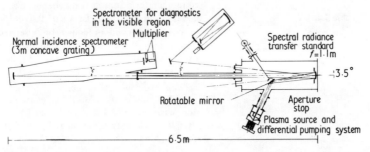

Figure 4. Spectro-radiometric arrangement for the measurement of the VUV radiation of arc plasmas. By turning the concave mirror (f = 1·1 m), the plasma source is imaged onto the entrance slit of a plane grating spectrometer to perform the plasma diagnostics in the visible spectral region.

of two onto the entrance slit of a 3 m normal incidence (Eagle mount) spectrometer (concave grating 1200 grooves/mm, plate factor 0·3 nm mm^{-1}, halfwidth of the apparatus profile $\Delta\lambda = 0\cdot0014$ nm at $\lambda = 185$ nm, figure 4) (K Grützmacher unpublished). The diaphragm in front of the concave mirror is the aperture stop. After turning the rotatable plane mirror through ≈120° the spectral radiance transfer standard is imaged onto the entrance slit under approximately the same conditions as the plasma standard.

4.2. Temperature scale and uncertainty

In previous papers (Stuck and Wende 1972a,b) the black body radiation of the longest wavelength CI resonance multiplet $2s^2\, 2p^2\, ^3P-3s\, ^3P^0$ ($\lambda \approx 165$ nm) had been selected for the temperature measurements. The relative uncertainty of the measured spectral radiance of the black body radiation at this wavelength using a synchrotron-calibrated deuterium lamp (Pitz 1969, Stuck and Wende 1972b) is estimated to be $\Delta L_\lambda/L_\lambda = \pm 0.075$, producing a relative uncertainty of the temperature of $\Delta T/T = \pm 8 \times 10^{-3}$ at 10^4 K.

To achieve a higher accuracy it was decided to generate black body radiation at shorter wavelengths, in particular at the wavelength of the hydrogen resonance line Lyα ($\lambda = 121.6$ nm). The use of Lyα is advantageous, because the Stark-broadening halfwidth and the transition probability are larger compared with the CI transitions at $\lambda \approx 165$ nm. Therefore, only a small number density of H atoms, added to the pure argon plasma, is required to generate black body radiation with a sufficient degree of approximation in a comparatively broad spectral bandwidth. Using the ratio of the flow rates of hydrogen to those of argon ($\lesssim 5 \times 10^{-3}$) the black body limited spectral bandwidth is of the order of 5×10^{-2} nm ($T \lesssim 17\,000$ K) (K Grützmacher unpublished). For the radiation measurements the spectral band pass of the spectrometer is considerably smaller ($\Delta\lambda = 6 \times 10^{-3}$ nm). When measuring the black body radiation at Lyα under these conditions the temperature of an approximately pure argon plasma is determined. As the development of synchrotron-calibrated transfer standards at Lyα has not been completed, at present temperatures cannot be measured using Lyα black body radiation. However, as in the common pyrometric technique below 4000 K, accurate temperature differences are determined between 10 000 and 20 000 K by measur-

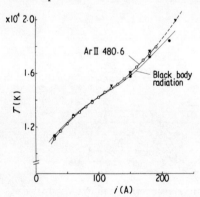

Figure 5. Comparison of the end-on measured temperatures (nominal arc channel diameter 4 mm, pressure 1.01×10^5 Pa) plotted in figure 3 with end-on temperatures derived from the measured black body radiation at Lyα (mean axis temperature using 0.5 mm thick insulating spacers in the arc chamber). O, ϵ_{nm}, ArII 480.6 nm ($A_{nm} = 8.82 \times 10^7$ s^{-1}); ▼, simultaneous absorption and emission measurements, 480 nm (Kirchoff's law); ●, black body radiation at Lyα (121.6 nm).

ing the ratios of the black body radiance at different arc currents and plasma pressures. For example: the temperature difference of 5000 K between $T_1 = 13\,000$ K and $T_2 = 18\,000$ K is determined with an uncertainty $\Delta T = \pm 80$ K, taking into account a relative uncertainty of the corresponding ratio of the black body radiance $L_2/L_1 = 12.6 \pm 4\%$[†] and an approximately known reference temperature. For the error estimation the un-

---

[†] Using the ratio of the flow rates of hydrogen to those of argon, up to 8×10^{-3} perturbations of the pure argon plasma were observed which were small compared with the uncertainty of L_2/L_1.

certainty of the second radiation constant c_2 need not be considered. The relative uncertainty of the black body radiation between 10 000 and 20 000 K calculated from the uncertainty of c_2 is $< 2 \times 10^{-3}$ at $\lambda = 100$ nm.

In figure 5 end-on measured temperatures, derived from the ϵ_{nm} value of the ArII line 480·6 nm and from the black body radiation at Lyα, are compared. Both temperature measurements are normalized at $T = 14\,000$ K. This temperature has been selected for the normalization because the temperature uncertainty calculated from the ϵ_{nm} uncertainty is sufficiently small and also because the number density of the electrons ($n_e = 1·5 \times 10^{23}$ m^{-3}) is sufficiently high with respect to the LTE approximation. For $T > 12\,000$ K both temperature measurements agree within the uncertainties discussed.

Acknowledgments

I gratefully acknowledge the support of my colleagues Dr H Nubbemeyer, Dr K Grützmacher and Dr D Stuck in preparing this paper.

References

Assousa G E, Brown L and Ford W K Jr 1970 *J. Opt. Soc. Am.* **60** 1311
Bakos J, Szigeti J and Varga L 1966 *Phys. Lett.* **20** 503
Berg H F and Ervens W 1967 *Z. Phys.* **206** 184
Boldt G 1961 *Proc. 5th Int. Conf. on Ionization Phenomena in Gases, München* (Amsterdam: North-Holland)
Bues I, Haag T and Richter J 1969 *Laboratoriumsbericht aus dem Institut für Experimentalphysik der Universität Kiel, 1966–67*
Denis A and Gaillard M 1970 *Phys. Lett.* **31A** 9
Fink U, Bashkin S and Bickel W S 1970 *J. Quant. Spectrosc. Radiat. Transfer* **10** 1241
Garstang R H and Odabasi H 1971 *Topics in Modern Physics, Boulder, Colorado* p261
Luyken B F J 1972 *Physica* **60** 432
Maecker H 1956 *Z. Naturf.* **11a** 457
Nubbemeyer H 1974 *Thesis* Freie Universität Berlin
Olsen H N 1963 *J. Quant. Spectrosc. Radiat. Transfer* **3** 59
Pitz E 1969 *Appl. Opt.* **8** 255
Popenoe C H and Schumaker J B Jr 1965 *J. Res. NBS* **69A** 495
Richter J 1965 *Z. Astrophys.* **61** 57
Rudko R I and Tang C L 1967 *J. Appl. Phys.* **38** 4371
Schley U, Wende B and Stuck D 1966 *PTB Mitt.* part 6 p509
Schnapauff R 1968 *Z. Astrophys.* **68** 431
Schumaker J B Jr and Popenoe C H 1969 *J. Opt. Soc. Am.* **59** 980
—— 1972 *J. Res. NBS* **76A** 71
Statz H, Horrigan F A, Koozekanani S H, Tang C L and Koster G F 1965 *J. Appl. Phys.* **36** 2278
Stuck D and Wende B 1972a *TMCSI* **4** part 1 105
—— 1972b *J. Opt. Soc. Am.* **62** 96

Theoretical prediction of temperature profiles in a wall-stabilized argon arc

R E Shawyer
Division of Quantum Metrology, National Physical Laboratory,
Teddington, Middlesex

Abstract. Temperature profiles have been evaluated for a wall-stabilized arc taking account of radiation losses. Inclusion of the radiation loss term has the effect of flattening the temperature profile across the arc compared with the sharper profile obtained when radiation losses are omitted from the calculation.

1. Introduction

The wall-stabilized arc has considerable potential as a standard of high temperatures and of vacuum ultraviolet radiation (Lapworth 1974). Such an arc is capable of operation in a very stable and reproducible manner and if the gas properties are sufficiently well known it should be possible to accurately predict the temperature profile across the arc, the operating characteristics of the arc, and the spectral distribution of radiation from the arc. In this paper a simple mathematical model of the arc is presented and the validity of the model is examined by comparing predictions from it with the results of some recent temperature measurements (Preston 1975).

2. Theory

The arc is assumed to be axially uniform and to be burning in a water-cooled tube of radius R whose walls are maintained at a constant temperature T_p. The temperature profile across the arc is determined by the steady-state energy balance equation which is

$$\sigma E^2 = u - \frac{1}{r}\frac{\mathrm{d}}{\mathrm{d}r}\left(rk\frac{\mathrm{d}T}{\mathrm{d}r}\right) \tag{1}$$

where σ is the electrical conductivity, E the electric field, u the net emission of radiation, r the radial coordinate, k the thermal conductivity and T the temperature. In equation (1) the input electrical energy due to the electrical field E is equated to the loss of energy due to radiation u plus the loss due to thermal conduction characterized by the thermal conductivity k. In describing the arc by equation (1) cylindrical symmetry has been assumed and the electrical field is taken to be spatially independent. A further fundamental assumption is that the gas must be in local thermodynamic equilibrium so that the properties σ and k exist as functions of the local electron temperature. The validity of this approximation has been assessed by a number of authors. For example, Uhlenbusch and Fischer (1970) have analysed measurements of temperatures, densities and field-strength/current characteristics of rare gas arcs in terms of a non-equilibrium model that includes diffusion, while Schumaker and Popenoe (1972) have detected departures from equilibrium in spectral intensities of optically thin lines measured as a function of pressure. The results of these investigations indicate that argon, krypton and xenon arcs at atmospheric pressure ($\sim 10^5 \text{ N m}^{-2}$) and above are substantially in local thermodynamic equilibrium but that the lighter rare gases show considerable departures from equilibrium. Under the assumption of local thermodynamic equilibrium u is related to the absorption coefficient K_ν at frequency ν and the temperature profile $T(r)$ by the following equations (Church et al 1966):

$$u = \nabla \cdot \mathbf{F} \tag{2}$$

$$\mathbf{F} = \int_{\nu=0}^{\infty} \int_{w} I_\nu(\mathbf{r}, \mathbf{n}) \, \mathbf{n} \, d\omega \, d\nu \tag{3}$$

$$\mathbf{n} \cdot \nabla I_\nu = B_\nu K_\nu - I_\nu K_\nu \tag{4}$$

where $\mathbf{F}$ is the radiation flux density, $I_\nu(\mathbf{r}, \mathbf{n})$ is the intensity of radiation of frequency ν at position $\mathbf{r}$ and in the direction defined by the unit vector $\mathbf{n}$, $\mathbf{n} \, d\omega$ is an element of solid angle about the direction $\mathbf{n}$ and B_ν is the Planck function. The equation of radiative transfer (4) may be integrated and the solution substituted into equation (3) to give for cylindrical arc geometry

$$F(r) = -4 \int_{\nu=0}^{\infty} \int_{\phi=0}^{\pi} \int_{s=a}^{b} K_\nu B_\nu G \left(\int_{s'=a}^{s} K_\nu \, ds \right) \cos \phi \, ds \, d\phi \, d\nu \tag{5}$$

where the function $G(x)$ is defined by

$$G(x) = \int_{0}^{\pi/2} \exp(-x/\sin\theta) \sin\theta \, d\theta. \tag{6}$$

The s integration in equation (5) is along a straight line in a plane perpendicular to the arc axis from a point a at radius r to a point b at the edge of the arc. This line makes an angle ϕ with the radius vector. The argument of the function G in equation (5) is an optical depth that determines the amount of self-absorption in the arc. We see that in general u is not just a function of the local temperature at r but due to self-absorption effects within the arc it is related to the entire temperature profile. To determine the temperature profile it is therefore necessary in general to solve a set of coupled integro–differential equations (equations (1), (2) and (5)). We shall assume that our arc is optically thin, ie, we neglect self-absorption of radiation. In this approximation u becomes

$$u = 4\pi \int_{0}^{\infty} K_\nu B_\nu \, d\nu \tag{7}$$

and equation (1) reduces to the optically thin Elenbaas–Heller equation (Elenbass 1951). In this equation σ, u and k are now all functions of the local temperature and the only unknown $T(r)$ may be determined by solving this equation subject to the two boundary conditions

$$\frac{dT}{dr} = 0 \text{ at } r = 0 \qquad T = T_p \text{ at the arc boundary}, r = R. \qquad (8)$$

It is convenient to cast equation (1) into a different form by introducing a transformation first used by Schmitz (1950). He defined the heat flux potential $S(T)$ by the relation

$$S(T) = \int_{T_p}^{T} k(T) \, dT. \qquad (9)$$

The energy equation (1) can then be rewritten in the form

$$\frac{d^2 S}{dr^2} + \frac{1}{r} \frac{dS}{dr} + \sigma E^2 - u = 0 \qquad (10)$$

and the boundary conditions (8) become

$$\frac{dS}{dr} = 0 \quad \text{at } r = 0 \qquad S = 0 \quad \text{at } r = R. \qquad (11)$$

Using S instead of temperature as the dependent variable removes the thermal conductivity k explicitly from the problem, but its influence is still present through the relation (9) and through σ and u which must now be expressed in terms of s. The temperature (or S) variation of the transport properties introduces nonlinearities into the problem and unless one is willing to make further simplifications (Maecker 1959) it is necessary to solve the energy balance equation (10) numerically.

A computer solution of equation (10) has been carried out by introducing finite difference approximations for the derivatives of S and numerically integrating the equation step by step from a given value of S, with zero slope, at the centre of the arc, $r = 0$ (corresponding to a given axial temperature) and continuing the process until S becomes zero at some arc radius r. A solution is usually required for given values of axial temperature and arc radius. The computer program finds such solutions by iterating on different values of the electric field strength E until for a particular value of the field the boundary condition $S = 0$ at $r = R$ is satisfied for the given axial temperature. Tables of values of the gas properties σ and u as functions of S are stored within the computer and their values at points required by the numerical integration scheme are found by interpolation on the tabulated values. The step length used in the integration process is reduced until a solution of any desired accuracy is found. Apart from this overall reduction of step length between iterations efficient use of computer time is ensured by using within any iteration a variable step length whose size depends upon the gradient of S. The step length is automatically varied in such a manner as to be reduced in value whenever S is varying rapidly. The end result of this process is to give S as a function of r. To find the temperature profile the inverse of the relation (9) is used to give $T(r)$.

Apart from the temperature profile other quantities of interest that are output by the computer program include the electric field strength E, the total arc current I, defined by

$$I = 2\pi E \int_0^R \sigma r \, dr \tag{12}$$

and the ratio ϕ of the power lost from the arc by radiation to the electrical power input, defined by

$$\phi = \frac{2\pi}{EI} \int_0^R u r \, dr. \tag{13}$$

3. Results

Equation (10) has been solved numerically, by the technique described in §2 of this paper, for a 3 mm diameter argon arc operating at a pressure of $1 \cdot 75 \times 10^5 \, \text{N m}^{-2}$. These arc parameters correspond to the dimensions and operating conditions of a wall-stabilized arc that is being investigated experimentally at the National Physical Laboratory as part of a research programme to develop a standard of temperature above 10^4 K and of vacuum ultraviolet radiation (Preston 1975). To be able to carry out the computations a knowledge of the material properties of argon is required. The values of thermal and electrical conductivity were taken from the recent computations by Devoto (1973), whose results are in reasonable agreement with experimental values. In addition the radiation absorption coefficient $K_\nu(T)$ as a function of frequency ν and temperature T is required to enable the function u to be evaluated from equation (7). However, due to an incomplete knowledge of line profiles and transition probabilities we have not attempted to make such a fundamental evaluation of this function. We chose to use in this preliminary study values of u based on the total radiation measurements reported by Bauder (1968).

Calculated temperature profiles across the arc using the above parameters and gas properties are shown in figure 1. These solutions were obtained for a constant temperature of 375 K on the plasma side of the channel wall. The temperature Tw of the cooling water is related to the temperature T_p of the wall on the plasma side by

$$\left.\frac{dT}{dr}\right|_{r=R} = -\frac{k_2(T_p - T_w)}{k_1 R \ln[(R+t)/R]} \tag{14}$$

where $dT/dr|_{r=R}$ is the temperature gradient in the plasma adjacent to the wall, k_1 the thermal conductivity of the plasma at the wall, k_2 the thermal conductivity of the copper wall and t the thickness of the channel wall. The relationship (14) follows from a solution of the equation of heat conduction within the channel wall under the assumption of constant thermal conductivity (ie, from equation (1) with $E = u = 0$ and $k = k_2$). Equation (14) together with the solutions depicted in figure 1 give a temperature difference $(T_p - T_w)$ across the channel wall of, for example, 4 K for an axial temperature of 10 000 K and 18 K for an axial temperature of 15 000 K. The temperature profiles of figure 1 are somewhat flatter than those obtained when the radiation terms in the energy equation are omitted from the calculations (Maxwell 1973), although the profiles do not have the nearly constant temperature form of a radiation dominated arc (Lowke 1970).

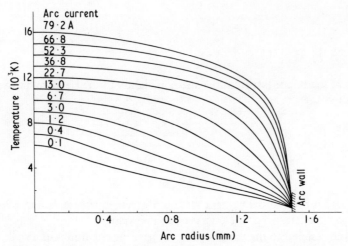

Figure 1. Temperature profiles across the arc.

In figure 2 the calculated axial temperatures are shown as a function of total arc current together with a set of recent measurements (Preston 1975). If we may assume that the approximations made in setting up our simple model are reasonable for the particular arc under consideration then the discrepancies between theory and experiment that can be seen in figure 2 could be explained by uncertainties in the values of the material properties of argon that we have used. In particular the total radiation measurements on which we base our values of the function u do not include all the radiation from the vacuum ultraviolet (Morris *et al* 1970, Krey and Morris 1970). In such experiments this radiation is almost completely absorbed by the cool argon gas between the arc and the detector. Lowke (1970) has shown that under certain conditions a dramatic lowering of the axial temperature can occur when ultraviolet radiation is included in calculations. The derived values of the field strength E and the arc current I for the temperature profiles of figure 1 were used to construct a field–current characteristic

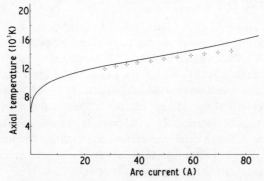

Figure 2. Axial temperature–arc current characteristic. The crosses depict the results of some recent measurements (Preston 1975).

for the 3 mm argon arc. This is shown in figure 3. The characteristic has negative slope for low currents and positive slope for high currents. The field values are higher than those found when radiation is neglected, ie, when conduction is the only energy loss mechanism (Maxwell 1973); this follows since additional power must be supplied to make up for that lost by radiation. To obtain a measure of the importance of the radiation term in the energy balance equation the ratio of energy lost by radiation to electrical energy input was evaluated using equation (13) and the temperature profiles of figure 1. The results are depicted in figure 4. It can be seen from this figure that radiation transport is significant for the arc currents, 28 to 75 A, used by Preston (1975). In the light of the above discussion on the role of neglected ultraviolet radiation it would appear that the results presented here give a lower limit to the effects of radiation.

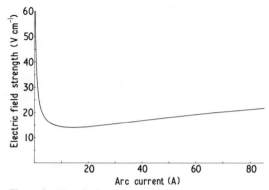

Figure 3. Electric field–arc current characteristic.

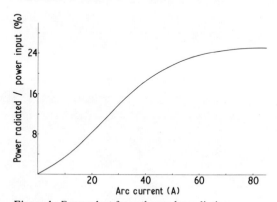

Figure 4. Energy lost from the arc by radiation.

References

Bauder U 1968 *J. Appl. Phys.* **39** 148–52
Church H C, Schelcht R G, Liberman I and Swanson B W 1966 *AIIA J.* **4** 1947–53
Devoto R S 1973 *Phys. Fluids* **16** 616–23
Elenbaas W 1951 *The High Pressure Mercury Vapour Discharge* (Amsterdam: North-Holland)

Krey R U and Morris J C 1970 *Phys. Fluids* **13** 1483–7
Lapworth K C 1974 *J. Phys.* E: *Sci. Instrum.* **7** 413–20
Lowke J J 1970 *J. Appl. Phys.* **41** 2588–600
Maecker H 1959 *Z. Phys.* **157** 1–29
Maxwell R W 1973 *NPL Quantum Metrology Report* No. 27
Morris J C, Rudis R P and Yos J M 1970 *Phys. Fluids* **13** 608–17
Preston R C 1975 this volume
Schmitz G 1950 *Z. Naturf.* **5a** 571
Schumaker J B and Popenoe C H 1972 *J. Res. NBS* **76A** 71–6
Uhlenbusch J F and Fischer E 1970 *Proc. IEEE* **59** 578–87

Thermal imaging, a technical review

Jens Agerskans
AGA Infrared Systems AB, Lidingo, Sweden

Abstract. After a short review of the fundamentals of IR theory, the performance of commercially available thermal imaging systems is discussed. The most common type of thermal-imaging system is the one based on opto-mechanical scanning. Different existing concepts and current developments of some functional blocks — optics, scanning devices, detectors, cooling techniques, data handling and presentation — are treated, with implications as to system performance. One type of non-mechanical scanning device under development (the pyroelectric vidicon) is also discussed. The paper concludes with a look at future thermal-imaging systems, where current developments in the data processing and military IR fields will play an important role.

1. Introduction

Thermal-imaging instruments have during the last 10–15 years found increasing use in areas such as industrial R & D and supervision, medical diagnosis, therapy and general research. New applications are continuously being developed and, along with this, new instruments and accessories. The most advanced technology in thermal imaging can be found in military equipment for tactical use, although the actual quantity of military units that have been produced so far does not seem to be very great.

Let us try to make clear what we mean by thermal imaging. One definition could read: 'A technique for converting the temperature pattern of a scene into a visible image'. However, thermal imaging is almost never used in such a wide sense because this definition would include contact as well as non-contact techniques. Generally, 'thermal imaging' is limited to non-contact techniques. This means that it is the *radiation pattern* of the scene that is converted into a visible image. The radiation properties of objects not only depend on the temperature of the objects but also on their emissivity. A preferred definition for thermal imaging would therefore be: 'A non-contact technique for converting the radiation pattern of a scene into a visible image'.

Two main categories of thermal imaging are distinguished. One is the measurement of radiation or a radiation-related parameter, of which temperature is the most important; the other is pure imaging without any need for quantitative information — examples being night vision for civil or military applications.

As will be discussed in the next section, the spectral region of prime interest is the infrared — although longer wavelengths, in the microwave region (Edrich and Hardee 1974), and shorter wavelengths, are used occasionally as well. This paper will concentrate on the basic techniques in use and under development for infrared (IR) thermal imaging equipment. The next section reviews the fundamentals of IR radiation theory. Succeeding sections describe system performance and the functional 'building blocks' which comprise IR thermal imaging equipment today, together with current technological trends. The paper concludes with a look at IR imaging in the future.

2. IR fundamentals

The differentiated Planck radiation law is illustrated in figure 1. As is well known, the Planck law gives the radiative power (or number of photons) emitted from a theoretical black body as a function of temperature and wavelength.

$$M(\lambda, T) = \frac{2\pi c}{\lambda^2} \frac{1}{\exp[(hc/\lambda kT) - 1]}$$

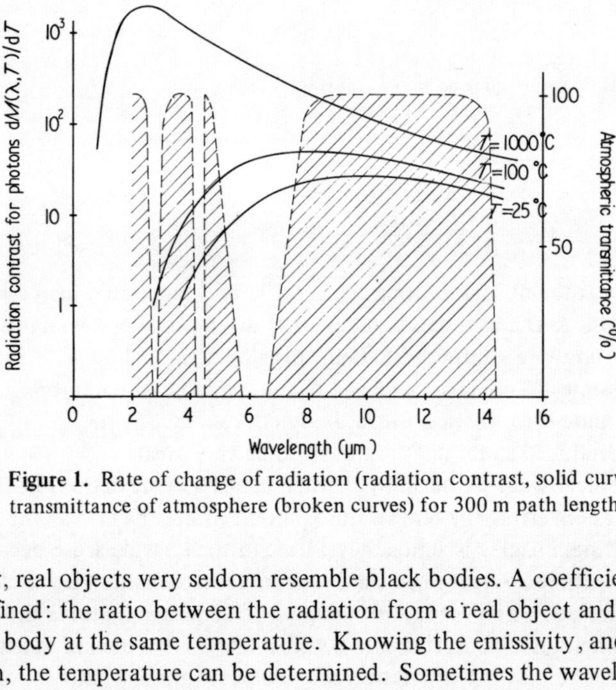

Figure 1. Rate of change of radiation (radiation contrast, solid curves) and transmittance transmittance of atmosphere (broken curves) for 300 m path length.

However, real objects very seldom resemble black bodies. A coefficient of emissivity has been defined: the ratio between the radiation from a real object and the radiation from a black body at the same temperature. Knowing the emissivity, and measuring the radiation, the temperature can be determined. Sometimes the wavelength and the temperature dependance of the emissivity have to be taken into account.

As can be seen from figure 1, the radiation contrast maximum is in the approximate range 1–10 μm for objects in the temperature range 0–1000 °C, which indicates that a spectral band in this range would be suitable for thermal imaging. Other factors that are important in the choice of optimum spectral band are the spectral sensitivity of the equipment and, for some applications, the spectral transmittance of the medium between the object and the equipment.

Some objects have a significant spectral signature which determines the optimum spectral band. An example is plastic film, whose spectral characteristic is shown in figure 2. Kirchhoff's law states that the absorptivity and the emissivity are equal. The absorption 'dip' at wavelength 3·4 μm can therefore be used for temperature measurement by narrow spectral filtering centred around 3·4 μm. If the filter is not narrow

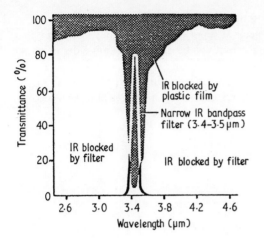

Figure 2. Narrow IR bandpass filter characteristic superimposed upon the carbon–hydrogen bond absorption band of polyethylene.

enough, the thermal imaging equipment will also see through the plastic film, with the result that radiation from the background will make the measurement inaccurate.

Similarly, for reflective (opaque or semitransparent) objects the radiation from the surroundings will be reflected from the object surfaces and introduce measurement errors.

3. Available equipment

Commercially available thermal-imaging equipments have typical thermal resolutions of the order of 0·1 °C, for 'black bodies' at room temperature. This has been found adequate in most measurement situations. For higher temperatures, the resolution increases. For lower, it decreases. The rate of increase or decrease depends on the spectral band to be used. Either the 3–5 μm or 8–12 μm bands are normally used. Figure 3 shows the thermal resolution of the two bands as a function of temperature

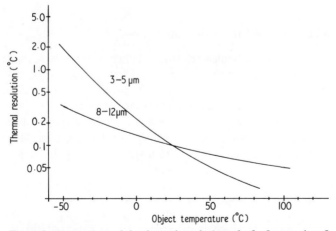

Figure 3. Comparison of the thermal resolution of a 3–5 μm and an 8–12 μm system.

for identical systems, except for those components that are significant for the spectral bands. It is interesting to note that for objects at room temperature, they give approximately the same performance. The spectral aspect will be discussed in more detail in the sections on optics and detectors.

For any specific system, the thermal resolution can be improved at the expense of other performance parameters. For a thermal resolution ΔT, a total field of view Ω, a field frequency f, and a spatial resolution ω, we have the relation

$$\Delta T = c \frac{\Omega f^{1/2}}{\omega} \tag{1}$$

where c is a constant.

One of the best known thermal-imaging systems in use today is AGA Thermovision 680†. This equipment has a thermal resolution of better than 0·2 °C, with a picture-field frequency of 16 Hz. The total field of view and spatial resolution is variable (by exchanging front lenses) with the ratio Ω/ω remaining constant. One example of its performance is a spatial resolution of 1·1 mrad, obtained with a lens attached which gives a total field of view of 8° × 8°.

From the formula, we see that for constant ΔT, Ω/ω (the number of resolution elements) can be increased at the expense of the field frequency. An example of a system with a low field frequency is the Spectrotherm 1000, which specifies a thermal resolution of 0·2 °C, a total field of view of 30° × 30°, and a spatial resolution of 525 lines per frame with 600 elements per line. It takes 2 s to scan the scene, during which the display is blanked and the information is read into a storage tube. This stored image is then presented 'frozen' for 2 s (or longer) during the flyback of the opto-mechanical scan. The effective picture-field frequency is thus 0·25 Hz, with a 50% duty cycle.

Since there still exists no standard for specifying thermal imaging equipment a direct comparison of performance is very difficult to make from the specifications in product brochures alone. The two systems are mentioned just to illustrate two different design philosophies: one employing real-time scanning and presentation, the other slow scanning and frozen presentation but better spatial resolution.

4. Basic function of an opto-mechanical scanning system

A block diagram of an opto-mechanical thermal imaging system is shown in figure 4. The image of the detector can be thought of as being projected on the scene via the optics and the scanner. The detector 'sees' only a small part of the scene at a time as the scanner translates the image of the detector across the scene, usually in a TV-like raster. The detector thus senses the radiation at each point in the scanned scene and converts it into a continuously varying electrical signal. The detector is usually cooled by liquid nitrogen in commercial systems. Detectors with no cooling give an inferior performance by a factor on the order of 100 or worse.

The detector signal is electronically processed and presented on a cathode-ray tube (CRT) screen. The sensitivity, or temperature contrast, on the screen and other para-

† Registered trademark

meters, such as temperature level, can be adjusted by special controls. To the CRT are also fed synchronization signals from the scanner via electronic shaping circuits.

In order to stabilize the video signal, a reference signal is often generated by letting the detector view a built-in stable radiation reference during the non-effective flyback portion of the opto-mechanical scan.

Another type of opto-mechanical scanning system is shown in figure 5. The block diagram is identical to the previous one in figure 4 for cooler, detector and electronics. The video signal is here used to modulate a light-emitting diode (LED) and together with an opto-mechanical arrangement for visible light similar to the one for IR, an observer can see a thermal image of the scene via an eye-piece. Both scanner blocks are often physically the same. The scheme shown in figure 5 is practical to use when the thermal-imaging system consists of a single unit. The system in figure 4 can, of course, also be built into one unit, although it usually consists of two units. Often it is desirable to have the camera (that is: optics, scanner, detector and some electronics) in one place, with the display presentation and control unit remote from the camera.

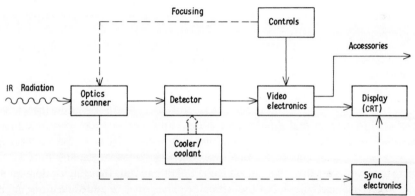

Figure 4. Opto-mechanical thermal-imaging system.

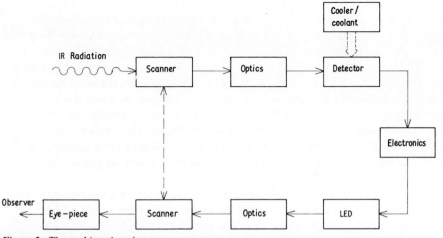

Figure 5. Thermal-imaging viewer.

5. Optics and scanning

The scanning is usually performed so as to generate a TV-type raster. Two basic types of scanning can be distinguished as far as the location of the scanner is concerned. One is generally referred to as 'object-space' scanning, the other as 'image-space' scanning. Figure 6 illustrates the basic principles of the two. *Object-space* and *image-space* refer respectively to the object and image sides of the objective lens.

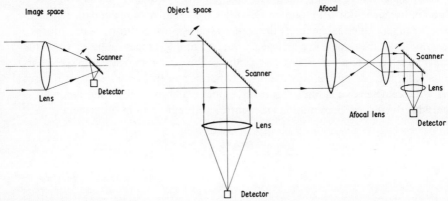

Figure 6. Scanning principles.

It can be understood from figure 6 that, especially for lenses with narrow fields of view (and thereby large diameters), the scanning mechanism of an object-space scanner has to be large. Fast and efficient scanning is therefore difficult to accomplish. (Note that in figure 6 a mechanism for only one scan direction is indicated.) Furthermore, there is no convenient way of changing the field of view of the system. The advantage of object-space scanning is that the optical requirements are quite low since the field of view of the lens itself is very narrow.

Image-space scanning makes fast and efficient scanning easier, although it may make the objective lens somewhat more complicated since the field of view of the lens has to be larger. The possibility of changing the field of view with a set of interchangeable lenses makes such a system very flexible. There is also the possibility of developing zoom lenses. So far, no commercially available zoom lens for a thermal imaging system exists, but it will certainly come.

An example of an image-space scanning system is AGA Thermovision, which is available with various lenses. An example of an object-space scanning system is Spectrotherm 1000. A third type of scanning system, which could be considered as an extension of the object-space scanning, is also illustrated in figure 6. In principle, an afocal lens is placed in front of the scanner, although the scanner has to be specially designed for this concept. Different fields of view can now be chosen by exchanging afocal lenses.

In figure 7, an example of an object-space scanning configuration is shown. Here the fast scanning in the horizontal direction (line scanning) is accomplished with a rotating mirror drum, and the slow vertical scanning (frame scanning) with a tilting mirror. An example of an image-space configuration is shown in figure 8. The rotating

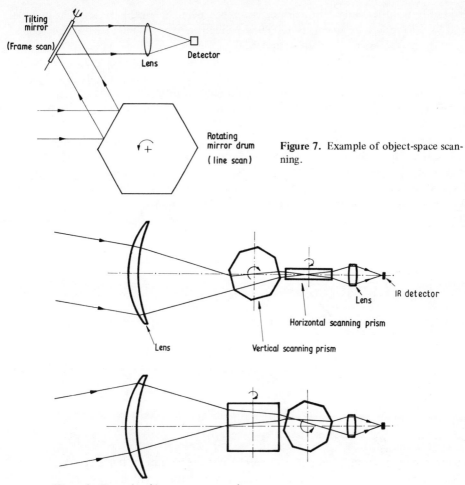

Figure 7. Example of object-space scanning.

Figure 8. Example of image-space scanning.

refractive polygon prisms scan by displacing the beam sideways. This scanning technique has been proven in serial production for three generations of AGA Thermovision development since 1965. The optical material used for the prisms is either silicon or germanium, which have high indices of refraction resulting in a large displacement and a favourable aberration effect. This prism-scanning concept, and other image-space scanning concepts are described by Agerskans *et al* (1974). An example of a system with an afocal objective lens is described by Anderson (1974).

The optics may consist of refractive and/or reflective elements, in other words lenses (prisms) or mirrors. Refractive elements generally provide a less complicated layout of the optical system, since no problem with obstruction (vignetting) exists, the elements are simply mounted in line (compare figure 8). For image-space scanning, entirely reflective collecting optics will not give sufficient optical resolution. A combination of lenses and mirrors (the so-called catadioptric system) may be advantageous, especially for narrow fields of view, since the collecting element has to be rather large for a narrow field of view optics, large mirrors being less expensive than large IR lenses. The material cost for IR refractive elements is quite high. Silicon and germanium have already been mentioned as suitable IR optical materials. Others exist, but silicon and germanium are the most common. Germanium can be used in the two bands 3–5 μm and 8–12 μm mentioned earlier, whereas silicon has a rather high absorption figure in the 8–12 μm band. Among other materials are: arsenic trisulphide, calcium fluoride, cadmium telluride, zinc selenide, zinc sulphide. (For further information see Hudson 1969, Wolfe 1965.)

Owing to the high indices of refraction for germanium and silicon (4·0 and 3·4, respectively) these materials require anti-reflective coatings. For uncoated thin elements of germanium and silicon the transmission is approximately 50% per element. Normally the coating consists of one layer on each surface, the thickness being a quarter of the wavelength where peak transmission is desired. The transmissions of an uncoated and a multilayer coated silicon plate are shown for comparison in figure 9. Multilayer

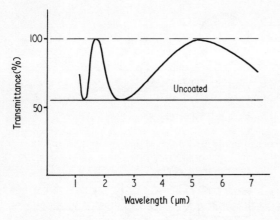

Figure 9. Transmittance of uncoated and anti-reflective coated silicon sample.

coatings are not normally used in commercial systems because of the high cost although it can be expected that the costs will go down in the future, with a resulting increase in their use.

Another way to increase the transmission is to use fewer lenses, through the use of aspheric surfaces. This is not usually the main reason for going to aspherics, however, but rather the cost/performance aspect. As mentioned earlier, the IR optical raw material is quite expensive, and this has a dramatic effect on the large lenses. Aspheric surfaces are therefore of particular interest for this type of lens.

6. Improving performance by multi-element detectors

Equation (1) gives the relation between some of the system design parameters. There are, of course, more parameters to consider in an actual design situation. For instance, by scaling the overall size a factor 2 upwards, the thermal resolution is improved by a factor of 2 if all other parameters are held constant. Another way to achieve improvement is to add more detector elements. Today the most widely used concepts in commercial systems operate on the principle of a single detector element with two scanning axes: horizontal and vertical. By adding more detector elements, the thermal resolution is improved by a factor which is the square root of the number of elements. Of course, if the thermal resolution is acceptable the gain in improvement can be used for some other parameter, such as field of view, field frequency, camera dimension, etc.

The conventional way to arrange a multi-element detector is shown in figure 10. The elements are placed in a row perpendicular to the line-scan direction. The line-scanning rate can thus be reduced by a factor equal to the number of elements. Each element

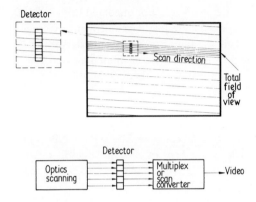

Figure 10. Multi-element detector arrangement for parallel scanning: (*a*) scanning raster; (*b*) block diagram.

has its own preamplifier and video processing, up to a point where multiplexing to a single video signal channel is done. Since line-scanning can be slower, the electrical bandwidth can be decreased accordingly, thereby reducing noise (whose square is proportional to the bandwidth).

Some concepts using this multi-element, *parallel* scanning technique (the different detector elements scan parallel lines) scan mechanically in only one direction. The obvious case is where the number of elements is equal to the number of scanning lines. Another concept is exemplified by a rotating mirror-drum, where each mirror is tilted a certain amount with respect to each other. The difference in tilt between two adjacent mirrors corresponds to the length of the detector row so as to achieve a contiguous line pattern.

A newer approach to arrange a multi-element detector is shown in figure 11. In this so-called '*serial*' scanning, the elements scan the same line with a short time lag. The signal from each element is delayed correspondingly and then added. The signals add linearly but the noise adds as root sum square. For instance, with 10 elements, the signal will be 10 times greater than for a single element, but the noise will be only $10^{1/2}$ times greater. The signal-to-noise ratio will therefore be improved by a factor of $10^{1/2}$. The theoretical sensitivity improvement in this serial arrangement is thus the same as in the parallel arrangement, when compared to the single-element detector arrangement.

The serial concept requires faster opto-mechanical line scanning and detectors with a higher bandwidth than the parallel concept. However, non-uniformities between elements are averaged out in the serial concept, while the same defects in the parallel concept may cause disturbances in the displayed picture, which then have to be corrected.

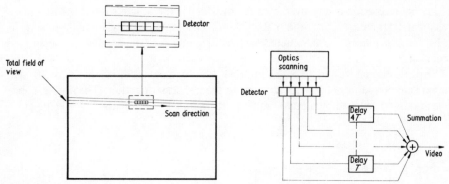

Figure 11. Multi-element detector arrangement for serial scanning: (*a*) scanning raster; (*b*) block diagram.

7. Detectors

There are basically two types of IR detectors: *photon sensitive* and *thermal sensitive*. Photon detectors are solid state devices whose electrical characteristics are functions of the number of photons absorbed by the detector. One type of photon detector is *photovoltaic*, consisting basically of a diode that generates either a current or a voltage, depending on mode of operation. Another type is *photoconductive*, consisting of a homogenous electrically biased chip with a resistance that is a function of the photon radiation. In today's thermal-imaging equipment, the use of photon detectors predominates. Thermal detectors are far inferior in performance to photon detectors by a sensitivity ratio of approximately two orders of magnitude. In order that they give not too poor a system performance, multi-element detectors have to be used. Furthermore, a rather low spatial resolution and low field rate have to be accepted in order to obtain a reasonable temperature resolution with thermal detectors. Among the thermal detectors are thermistors, thermocouples (thermopiles) and pyroelectric detectors. They are not expected to be of much importance in opto-mechanical thermal imaging, unless, of course, a real breakthrough should occur. A unique application of the pyroelectric principle will be discussed in the section on the pyroelectric vidicon.

There are today two outstanding photon detectors for IR equipment. The most highly developed is the photovoltaic (PV) indium antimonide (InSb). To obtain reasonable performance it has to be cooled, usually down to 77 K (the boiling point of liquid nitrogen). It has an upper cut off wavelength of $5{\cdot}35\,\mu\text{m}$ which is associated with the bandgap of the semiconductor material used for the detector. The spectral response is shown in figure 12. The fundamental sensitivity limit of a photon detector is set by the photon noise. For those photon radiation levels that normally are at hand in a thermal-imaging system, the sensitivity achieved in production equipment today is very close to this limit.

The other outstanding detector is the photoconductive (PC) mercury cadmium telluride (MCT). This semiconductor is actually an alloy of two semiconductors, mercury telluride (HgTe) and cadmium telluride (CdTe). It is sometimes written as $(\text{HgTe})_x(\text{CdTe})_{1-x}$, where the fraction x determines the spectral response. A MCT detector with a cut-off around $12\,\mu\text{m}$ is shown in figure 12. The MCT detector is of

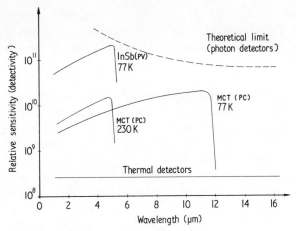

Figure 12. Spectral sensitivity of some IR detectors.

most importance for the 8–12 μm wavelength band, and is then generally cooled by liquid nitrogen. Today's MCT detectors are still some way from the theoretical sensitivity limit, but it is likely this situation will be improved in the future. With another value for x, this material is also available for the 3–5 μm band, and has a rather good sensitivity at the (higher) detector temperatures which can be achieved by thermoelectric cooling.

A few other detector materials are worth mentioning. An older material, lead selenide (PbSe) has a cut-off around 5 μm. One advantage with PbSe detectors is their relatively high sensitivity at detector temperatures in the range of 200–300 K. A rather new material is lead tin telluride (LTT, PbSnTe), similar to the MCT material in that it is an alloy of the two semiconductors PbTe and SnTe. Some disadvantages with respect to the MCT detector (which will not be discussed here) remain to be solved. There are also advantages which may make LTT a strong competitor to MCT. But, for the moment, MCT is used almost exclusively in systems operating in the 8–12 μm band.

Looking at figure 1, we see that there is more radiation contrast (dM/dT) in the 8–12 μm band for an object at 300 K than there is in the 3–5 μm band. However, figure 12 shows that the typical detector sensitivity is higher in the 3–5 μm band. These two factors just about cancel out, so that (for 300 K black bodies) roughly the same performance is achieved for two similar thermal imaging systems working in the two bands. There is, however, a potential for improvement of the 8–12 μm detector. For the moment this kind of detector has a considerably higher price, so a shift to the 8–12 μm band is not motivated except for special applications. Examples of such applications are: better sensitivity for low temperature objects, better transmittance in atmosphere for long ranges, and imaging of CO_2 laser radiation (10·6 μm).

8. Cooling techniques

For systems operating in real time, and with a picture-field scanning time down to a few seconds it is necessary to use cooled detectors, if reasonable thermal and spatial resolutions are to be obtained.

The dominant cooling technique in commercial systems is simply to use *liquid nitrogen*. Due to the heat load, the liquid boils and thereby maintains a constant temperature level. The storage vessel, the dewar, normally has a volume corresponding to 2–5 hours system operation. Sometimes the system is equipped with an automatic liquid nitrogen filling system, to achieve considerably longer operating times.

The *Joule–Thomson refrigerator* is a small gas liquefier which is placed in a special detector dewar. It is based upon the principle that a high-pressure gas cools down and liquefies as it expands when leaving a throttle valve. Nitrogen (77 K) and argon (87 K) are gases commonly used. The method has found very little use in thermal-imaging equipment, however, the prime reason being that the throttle valve can become plugged after a while since water vapour contained in the gas will freeze. This cooling technique is therefore not to be recommended when long, unattended operating times are required, unless very thorough precautions can be taken to purify the gas.

The *expansion-engine* principle is well known, the best known of which are the Stirling, Solvay, Gifford–McMahon and Vuilleumier cycles. The technical aspects will not be discussed here. It suffices to say that cooling to lower than 77 K is fairly easy to achieve. The expansion engine is rarely used in commercial systems because most applications cannot justify the higher price. However, in the military field, they are used to a greater extent. The potential exists for reducing costs, volume and weight, which will then increase the attractiveness of the expansion-engine principle for commercial systems.

Thermoelectric refrigerators are still very rarely used in commercial thermal-imaging equipment. Again, for military systems they seem to be of more interest, and then together with a multi-element detector. These refrigerators are based on the Peltier effect, which says that when an electric current flows through a circuit of two dissimilar metals, heat is absorbed at one junction and released at the other. Detector coolers with a cold end at 200 K and the warm end at 300 K are commercially available. Temperatures below 200 K are possible to achieve — and have been achieved — but at very low power efficiencies. It is still unclear as to what extent these low-temperature coolers can be improved, in terms of dimension, weight and power requirements. It is not unlikely, though, that for moderate performance systems, we will see thermoelectrically cooled detectors quite common for cooling detectors to 200 K and above.

9. Presentation and data handling

The ordinary picture presentation device is a black-and-white CRT which is intensity modulated, so that 'white' corresponds to heat and 'black' to cold, or vice versa. In some applications a colour CRT display is a useful accessory by providing a temperature-quantized picture. Another quantitative presentation technique is a CRT with γ deflection for temperature. When more than one line is presented in this way a so-called 'relief' picture is obtained.

Usually, permanent recording is done by ordinary photographic techniques using oscilloscope cameras. Polaroid[†] film, as well as regular 35 mm or 70 mm film types are used. For dynamic events, analogue magnetic tape recording is a convenient technique.

† Registered trademark

For special dynamic studies, where higher flexibility is required, a special magnetic disc memory has been developed for AGA Thermovision. In this memory, approximately 50 picture fields can be recorded and selected for play-back at any desired rate, or the fields can be presented 'frozen', one by one.

Some work has been attempted in which computers are used with thermal-imaging equipment. However, no real breakthrough advantages seem to have occurred yet. It can be expected, though, that during the next few years mini-computers will be used more frequently in various research, medical and industrial applications utilizing thermal-imaging equipment.

10. Pyroelectric vidicons

The pyroelectric vidicon is a thermal-imaging camera tube based on electron-beam scanning, as opposed to the opto-mechanical scanning treated so far. The thermal scene is projected via an IR lens onto a thin, disc-shaped pyroelectric crystal. A thermal *change* in the image on the crystal changes the spontaneous polarization, and produces a charge pattern with the same spatial and intensity variations as the scene. The charge pattern of the crystal on the rear surface of the disc is read out as a variable voltage signal resulting from ordinary vidicon electron-beam scanning techniques. For the signal processing and picture presentation, normal TV techniques are used.

It should be emphasized that only *changes in the thermal pattern* can be detected by the pyroelectric vidicon. Three possible modes of operation are feasible for the imaging equipment: (i) the incident radiation is chopped; (ii) the camera is panned, or the optical path is moved, for example, by an orbiting prism or mirror; (ii) only moving objects are detected. A typical performance figure for the pyroelectric vidicon is: $0.5\,^{\circ}C$ thermal resolution at a spatial resolution of 7 mrad, with a field of view of $20^{\circ} \times 20^{\circ}$.

Although there are important advantages, inherent in the pyroelectric vidicon, (like no cooling, and TV compatibility) the performance is today clearly inferior to what is required in most thermal-imaging applications. While improvements can be expected gradually in the future, a great deal of effort still has to be made in order to turn this type of device into an adequate measurement instrument system, with application-oriented accessories.

11. The thermal-imaging equipment of the future

Looking at the military IR field, we find that developments involving multi-element detectors and advanced coolers (expansion engine and thermoelectric) have been under way for a long time. A review from 1973 concerning military systems can be found in Miller (1973). Provided that military interest in IR imaging persists the cost of critical IR components, especially detectors and coolers, will drop as the number of military imaging systems produced increases. The corresponding civil market will certainly benefit from this in terms of improved cost/performance.

Development in the field of mini- (and now micro-) computers is very rapid and this can be expected to have a great influence on future commercial thermal-imaging systems insofar as improving the data-handling capability are concerned.

Opto-mechanical scanning techniques, as opposed to electron-beam scanning (as utilized in the pyroelectric vidicon), will continue to be superior for a long time yet. For some specific applications, where low-performance thermal imaging can be accepted, the pyroelectric vidicon may find a market in the near future. A relatively simple, opto-mechanical scanning system might still be competitive, however, through the utilization of a thermoelectrically cooled multi-element detector.

References

Agerskans J, Holm L and Lindberg P 1974 *Proc. 4th European Symp. on Military Infrared*
Anderson R F 1974 *Opt. Eng.* **13** 4 335–8
Edrich J and Hardee P C 1974 *Proc. IEEE* 1391–2
Hudson R D Jr 1969 *Infrared System Engineering* (London:Wiley)
Miller B 1973 *Aviation Week and Space Technol.* May 7, May 21
Wolfe W L (ed) 1965 *Handbook of Military Infrared Technology* (Washington, DC: US Govt Printing Office)

The freezing point of aluminium as a temperature standard

George T Furukawa, William R Bigge, John L Riddle and Martin L Reilly

Institute for Basic Standards, National Bureau of Standards, Washington, DC 20234, USA

Abstract. Six platinum resistance thermometers were 'calibrated' at the triple point of water and at the freezing points of tin, zinc and aluminium. By extrapolating the 'quadratic relation' the freezing point of pure aluminium was found to be 660·407 ± 0·005 'degrees C'. The advantages of having a platinum resistance thermometer calibrated at a fixed point (eg aluminium point) near the upper temperature limit are demonstrated.

1. Introduction

The platinum resistance thermometer (PRT) is the standard interpolation instrument for the International Practical Temperature Scale of 1968 (IPTS-68) from 13·81 K (−259·34 °C) to 903·89 K (630·74 °C). The lower temperature limit of the PRT range of the IPTS-68 is defined by the triple point (13·81 K) of equilibrium hydrogen. On the other hand, the upper temperature limit (630·74 °C) is not defined by a fixed point. Moreover, this temperature as defined by a PRT is one of the reference temperatures for calibrating standard platinum–rhodium alloy and platinum thermocouples that are used in realizing the IPTS-68 at higher temperatures. The temperature is based on the calibration of a PRT at the triple point of water (TP), the steam or the tin point, and the zinc point; at 630·74 °C an error of calibration of the PRT at the tin point or at the zinc point is amplified by about 3·5 (Riddle et al 1973).

McLaren and Murdock (1968) compared the freezing points of seven antimony samples; for a given sample they found the standard deviation of the observed freezing point to be ⩽ ±0·3 mK. A variation of 2·6 mK in the freezing points was found among five of the purest samples. Among eight PRTs they found the standard deviation and the range to be ±3·3 mK and 12 mK, respectively, for the value of temperature obtained for the antimony point. The measurements of Evans and Wood (1971) and Chattle (1972) show similar variations among PRTs for the value of temperature for the antimony point. Therefore, considering the high reproducibility of the antimony point (McLaren and Murdock 1968), above the zinc point the accuracy of temperature measurements with the PRT could be improved by including this fixed point in the calibration of the PRT.

Although the antimony point has been found to be highly reproducible (McLaren and Murdock 1968), a large supercool (up to 20 to 40 °C) has been experienced in freezing antimony. Liquid aluminium, on the other hand, supercools only 1 or 2 °C

(Furukawa 1974, McAllan and Ammar 1972); therefore, the preparation of a suitable freeze for measurements should be more convenient than that of antimony.

Aluminium is considered superior to antimony for a number of reasons besides supercooling much less during the freezing process. Aluminium is much more abundant than antimony and, because of the greater technological importance of aluminium, samples of high purity are available in larger amounts and at lower costs. Research to obtain purer aluminium is actively in progress. At the freezing point the vapour pressure of aluminium is only about 2×10^{-8} Torr (1 Torr = 133·3224 Pa) whereas that of antimony at its freezing point is much higher, being about 0·2 Torr (Nesmeyanov 1963); also, aluminium is less toxic than antimony. The thermal diffusivity of liquid and solid aluminium at its freezing point was estimated to be many times greater than that of liquid and solid antimony at its freezing point (Brandt 1967, Furukawa *et al* 1972a, *Gmelins Handbuch der Anorganischen Chemie* 1943, Powell and Childs 1972).

In an earlier paper (Furukawa 1974) we reported the reproducibility of the freezing point of aluminium samples from two 'batches' of similar purity. The comparison in terms of the resistance ratio $R(Al)/R(TP)$, the ratio of the PRT resistance at the aluminium point to that at the TP, showed that the range of the mean $R(Al)/R(TP)$ of five out of six specimens corresponded to 0·51 mK; the deviation of the mean $R(Al)/R(TP)$ of the sixth specimen from the mean $R(Al)/R(TP)$ of the other five specimens corresponded to $-1·31$ mK. (The sixth specimen may have been contaminated during the assembly of the freezing-point cell or the original sample bar was inhomogeneous.) The 'pooled' standard deviation of the ratio $R(Al)/R(TP)$ observed with the six specimens corresponded to ±0·43 mK, while the standard deviation of $R(Al)$ and $R(TP)$ corresponded to ±0·19 mK and ±0·14 mK respectively. (The errors in the observations of $R(0\,°C)$ are amplified by $-3·4$ in the ratio $R(Al)/R(TP)$.)

This paper presents the results of the investigation on the possible application of the freezing point of aluminium for improving the definition of the PRT temperature scale at the higher temperatures. The 'calibration' of the PRTs at the freezing points of tin, zinc and aluminium and the formulation of the results are discussed.

2. Thermometer design

Six PRTs of two designs were employed in the present investigation. They were kindly loaned to us by J P Evans and Sharrill D Wood of the Temperature Section of the United States National Bureau of Standards (NBS) who designed and assembled the thermometers for their research in high-temperature PRTs (Wood 1973). Three of the PRTs (HTSS type) were of the bird-cage design (Evans and Burns 1962); the sensors of the other three PRTs (HTFQ type) were of the bifilar helix construction and the platinum wire was wound on a cross-shaped fused quartz support (Wood 1973). The sensors and lead assemblies of all of the PRTs were encased in fused quartz sheaths which were then thoroughly evacuated and filled with dry air to a pressure of one atmosphere at 1065 °C. To reduce radiation losses by 'piping', the outer surface of each fused quartz sheath was given a matte finish by blowing alumina abrasive at high velocity against the surface. Table 1 gives a summary of the characteristics of the PRTs that were employed. The PRTs had been exposed to temperatures of 960 °C and 1100 °C for many hundreds of hours prior to the present investigation (Wood 1973).

Table 1. Thermometer characteristics and resistances at the triple point of water and at the tin, zinc, and aluminium points.

PRT serial number =	HTSS-15	HTSS-21	HTSS-22	HTFQ-23	HTFQ-24	HTFQ-25
Sensor support	Synthetic sapphire	Synthetic sapphire	Synthetic sapphire	Fused quartz cross	Fused quartz cross	Fused quartz cross
Lead insulation	Synthetic sapphire disks and rods	Fused quartz disks and tubes	Alumina disks and tubes	Fused quartz disks and tubes	Fused quartz disks and tubes	Fused quartz disks and tubes
Sensor style	Bird cage	Bird cage	Bird cage	Bifilar helix	Bifilar helix	Bifilar helix
$R(TP)$†	0·2698459	0·2977924	0·2595219	0·2292180	0·2224857	0·2179040
$R(Zn)$	0·6928562	0·6619855	0·6666472	0·5886329	0·5713772	0·5596194
$R(TP)$	0·2698460	0·2577927	0·2595220	0·2292180	0·2224858	0·2179044
$R(Sn)$	0·5105622	0·4878000	0·4911980	0·4337438	0·4210238	0·4123592
$R(TP)$	0·2698464	0·2577927	0·2595239	0·2292182	0·2224860	0·2179036
$R(TP)$	0·2698460	0·2577933	0·2595240	0·2292178	0·2224855	0·2179037
$R(Al)$	0·9104580	0·8699084	0·8760910	0·7735216	0·7508533	0·7354032
$R(TP)$	0·2698460	0·2577925	0·2595238	0·2292186	0·2224868	0·2179045
$W(Zn)$‡	$2·567697_{56}$	$2·567998_{58}$	$2·568849_{30}$	$2·568103_{23}$	$2·568250_{44}$	$2·568288_{60}$
$W(Sn)$	$1·892121_{28}$	$1·892290_{46}$	$1·892768_{24}$	$1·892347_{93}$	$1·892434_{13}$	$1·892461_{73}$
$W(Al)$	$3·374121_{89}$	$3·374578_{47}$	$3·375894_{46}$	$3·374740_{87}$	$3·374962_{58}$	$3·375026_{20}$
$t'(Al)$ 'degrees C'	660·4036	660·4014	660·4028	660·4003	660·4026	660·4112 §
α (10^{-3} °C^{-1})	3·92398	3·92472	3·92680	3·92497	3·92535	3·92549
δ (°C)	1·4963	1·4961	1·4958	1·4960	1·4962	1·4965

Mean $t'(Al)$ = 660·4021 'degrees C'; standard deviation = ±0·0013 'degrees C'

† The values of resistances that are given correspond to that at zero current obtained by extrapolation of observations made at two currents (3·0 and 5·0 mA); the resistances correspond to the temperature at the point of immersion in each of the fixed-point cells. (The depth of immersion of the thermometers in both the tin-point and zinc-point cells during the measurements was estimated to be 17·5 cm. The depth of immersion of the PRTs in the TP cell was 27·5 cm and in the aluminium-point cell was 16·7 cm.)
‡ The values of W are given for the fixed points at the 'standard' conditions, ie, at 1 atmosphere pressure for the tin, zinc, and aluminium points; $R(TP)$ was converted to $R(0\,°C)$.
§ This value is not included in the mean $t'(Al)$ or in the determination of the standard deviation.

To avoid quenching in defects in the PRT wire by cooling rapidly from the aluminium point, the PRT was withdrawn from the aluminium-point cell in six steps over a period of 30 min (Furukawa 1974). It was then annealed at 480 °C in a tube furnace for 30 min and removed to cool at room temperature. This treatment was found to yield resistance readings at the TP (using an AC bridge (Cutkosky 1970)) that were repeatable to ±0·1 mK or better. Berry (1966), McLaren and Murdock (1968), and Chattle (1972) have reported similar procedures for treating PRTs to obtain reproducible readings at the TP. No precaution was taken in cooling the PRTs from the tin point or the zinc point.

3. Resistance measurements

The Mueller bridge was used in the measurements. However, the Mueller bridge method is at a disadvantage with PRTs of low resistances, eg, the changes in the bridge zero and in switch contact resistances become relatively large uncertainties. When the thermometer resistance is smaller than the lead resistance, the variations in the lead resistance become a significant fraction of the PRT resistance. To minimize its uncertainty, the bridge zero was determined frequently during the PRT resistance measurements. All resistance values of the PRTs were reduced to zero current by extrapolation of the measurements at 3·0 and 5·0 mA.

For the results to be significant measurements with the Mueller bridge approaching the level of $10^{-7}\Omega$ were necessary. The PRTs were prepared to reduce temperature gradients or changes that would interfere with the measurements. Since the PRTs were designed for high-temperature research the fused quartz sheaths were exceptionally long (about 76 cm). The fixed-point cells employed in the present work were designed with a total thermometer immersion depth of about 42 cm. To reduce the short-term variations in the differences between the PRT lead resistances that can be caused by temperature changes along the exposed section of the thermometer sheath, a tube, made by rolling together sheets of polyethylene film and aluminium foil into a cylinder, was fitted around the exposed portion of the PRT sheath. Around this was placed a cylinder of insulation made by rolling a sheet of polyethylene foam. The 'head' of the PRT was protected in a similar manner with a cylinder of insulation made by rolling together sheets of aluminium foil and polyethylene foam. The thermometer lead cable was also protected from sudden changes in temperature or temperature gradient across the cable. Ribbons of aluminium foil were first wound helically over the lead cable and then sheets of polyurethane foam were wrapped around the aluminium foil covered cable.

4. Freezing-point cells of tin, zinc, and aluminium and furnaces

The designs of the freezing-point cells of tin, zinc, and aluminium, the purity of the metal samples that were used to assemble the cells, the furnaces and the procedures for the preparation of the freezes with the cells have been previously described (Furukawa 1974, Furukawa *et al* 1972b, Riddle *et al* 1973). With the tin-point and zinc-point cells the control of the furnace temperature was set to obtain freeze duration of about 12 to 14 h; during the first 50% of the freeze the temperature of the tin-point cell changed by less than 0·08 mK and the temperature of the zinc-point cell changed by less than 0·2 mK. With the aluminium-point cell (designated A-2, see Furukawa 1974) the freeze duration was about 12 to 16 h, of which during the first 50% the temperature change was less than 1 mK.

The immersion characteristics of the two PRT designs were checked in the four fixed-point cells that were employed. An AC bridge (Cutkosky 1970) was used for these tests. The results showed that when fully inserted the PRTs should be within 0·1 mK of the equilibrium temperature of the cell at the point of immersion.

5. 'Calibration' of the thermometers at the tin, zinc, and aluminium points

5.1. Procedure

The tin-point and zinc-point furnaces used for this work are employed in the routine calibration of PRTs received at NBS and were located three floors away from the aluminium-point furnace. It was decided to use the same Mueller bridge in the 'calibration' of the six PRTs. After measurements of the PRTs at the tin and the zinc points the Mueller bridge was moved to the location of the aluminium-point furnace. The good reproducibility of the measurements of $R(\mathrm{TP})$ obtained at the two locations indicates that the Mueller bridge was not affected significantly by the moving process.

Before inserting a PRT into the tin-point or zinc-point cell the PRT was preheated in an auxiliary furnace maintained close to the fixed-point temperatures. The PRT was extracted from the auxiliary furnace and quickly inserted into the tin-point or the zinc-point cell. In the case of the aluminium-point cell the PRT was preheated in the section of the cell above the graphite crucible over a period of about 15 min. The cold PRT was initially inserted so that the sensor end was about 2 cm above the graphite crucible (see Furukawa 1974). The PRT was inserted an additional 5 to 6 cm at the end of each 5-min interval until it was fully inserted.

For each PRT, after annealing at 480 °C for about 4 h, the sequence of measurements was $R(\mathrm{TP}), R(\mathrm{Zn}), R(\mathrm{TP}), R(\mathrm{Sn})$ and $R(\mathrm{TP})$ and, after moving the Mueller bridge to the location of the aluminium-point furnace, $R(\mathrm{TP}), R(\mathrm{Al})$ and $R(\mathrm{TP})$. The zero current values of resistance were calculated from measurements at 3·0 and 5·0 mA, and the hydrostatic head correction was applied to obtain the temperature at the location of the middle of the PRT sensor in the cells. The value of $R(0\,°\mathrm{C})$ was obtained by converting the observed values of $R(\mathrm{TP})$ employing the relation

$$R(t')/R(0\,°\mathrm{C}) = 1 + At' + Bt'^2. \tag{1}$$

Because of the small value of t' at the TP, the term containing t'^2 was neglected and the average value of $A(3\cdot98485 \times 10^{-3}\,°\mathrm{C}^{-1})$, found for PRTs calibrated at the NBS in the past years, was employed in the conversion. The uncertainty in the adjustment of the value from $R(\mathrm{TP})$ to $R(0\,°\mathrm{C})$ in this manner is less than $\pm 2 \times 10^{-8} R(0\,°\mathrm{C})$, which is negligible for the measurements obtained in this work. The ratio of the PRT resistance at each of the various metal fixed points to its resistance at 0 °C was calculated using the average of the values of $R(\mathrm{TP})$ obtained before and after the resistance measurement at the metal fixed point. (Henceforth, any reference to the PRT calibration at a metal fixed point includes the measurements at the TP before and after the measurement at the metal fixed point.)

5.2. Results and discussion

The results of the observations are summarized in table 1.

From 0 °C to 630·74 °C the values of temperature t_{68} on the IPTS-68 is defined by the equation:

$$t_{68} = t' + 0\cdot045 \left(\frac{t'}{100\,°\mathrm{C}}\right)\left(\frac{t'}{100\,°\mathrm{C}} - 1\right)\left(\frac{t'}{419\cdot58\,°\mathrm{C}} - 1\right)\left(\frac{t'}{630\cdot74\,°\mathrm{C}} - 1\right)\,°\mathrm{C}, \tag{2}$$

where t' is defined by

$$W(t') = R(t')/R(0\,°C) = 1 + At' + Bt'^2$$

given earlier. Equation (1) is equivalent to

$$t' = \frac{1}{\alpha}[W(t') - 1] + \delta\left(\frac{t'}{100\,°C}\right)\left(\frac{t'}{100\,°C} - 1\right), \qquad (3)$$

where

$$\alpha = A + B \times 100\,°C \qquad (4)$$

and

$$\delta = -\frac{B(100\,°C)^2}{A + B \times 100\,°C}. \qquad (5)$$

Except for the effect of possible experimental variations the coefficients A and B of equation (1) should be nearly independent of whether the PRT is calibrated at the TP, steam point and the zinc point or at the TP, tin point and the zinc point.

For the comparison of the measurements on the six PRTs at the tin, zinc, and aluminium points, the coefficients A and B of equation (1) were obtained from the values of $W(t')$ at the tin and zinc points for each of the PRTs. The values of temperature $t'(\text{Al})$ corresponding to the $W(\text{Al})$ were then calculated for each of the PRTs using equation (1) (extrapolated beyond the IPTS-68 limit of 630·74 °C) and the corresponding coefficients A and B. (Henceforth, $t'(\text{Al})$ refers to the value of temperature at the aluminium point obtained by the extrapolation of the IPTS-68 equation 1.) The results are summarized in table 1. The range of the values of $t'(\text{Al})$ obtained for five of the PRTs is 3·3 mK; the value of $t'(\text{Al})$ of the sixth PRT is somewhat higher, 9·1 mK higher than the mean value of $t'(\text{Al})$ (660·4021 'degrees C') of the other five PRTs.† The variations arise from the intrinsic differences unaccounted for among the PRTs and from the variations in the measurements. The fairly close agreement of the values of $t'(\text{Al})$ shows that the PRTs employed in the present work can be calibrated and used with less than ±2 mK scatter at the aluminium point (except for one of the PRTs).

To include the aluminium point in the PRT scale formulation the following relation was chosen:‡

$$W(t^*) = 1 + At^* + Bt^{*2} + Dt^{*2}(t^* - 231\cdot9292\,°C)(t^* - 419\cdot58\,°C). \qquad (6)$$

The coefficients A and B of equation (6) are the same as those of equation (1). A suitable value for t^* at the aluminium point can be assigned from the results of extrapolation of the quadratic equation (1) for a group of typical PRTs, similar to the procedure that was employed to obtain the value for the antimony point (McLaren and Murdock 1968) or (by interpolation) to obtain the value for the tin point (McLaren

---

† The value of $t'(\text{Al})$ given is not on the IPTS-68. To avoid confusion 'degrees C' is used instead of °C.

‡ The symbol D was selected instead of C to avoid confusion with the symbol °C for degrees Celsius.

and Murdock 1960)†. For a given PRT the magnitude of the coefficient D (normally about 2×10^{-15} or smaller) depends on how closely the observed $W(\text{Al})$ and the assigned $t^*(\text{Al})$ agree with the quadratic relation that is determined by the coefficients A and B based on the measurements at the tin and zinc points. Hence the numerical difference between t' (equation 1) and t^* (equation 6) at a given resistance ratio W depends upon the magnitude of the coefficient D. (At the fixed points the values of t' and t^* are the same. At the intermediate temperatures tabular differences between t^* and t' or t^* and t_{68} may be devised for a given PRT.) For temperature measurements above the zinc point equation (6), in conjunction with calibrations at the tin, zinc, and aluminium points and the assigned values of t^* at these fixed points, should reduce the uncertainty that arises from the extrapolation of equation (1) which is based only on the calibrations at the tin and zinc points. Any abnormally large value for D would suggest calibration errors.

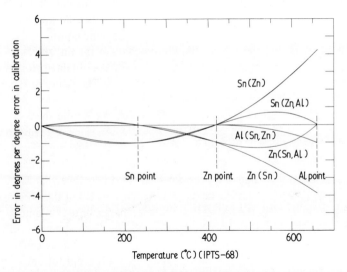

Figure 1. The error at various temperatures propagated from errors made in the calibration of a platinum resistance thermometer at the tin, zinc, and aluminium points. The curves show the error in the value of temperature caused by an error corresponding to 1 °C in $R(t)/R(0\,°\text{C})$ at the fixed points. The fixed point for which the error was made is indicated on the curve; the fixed point for which zero error is assumed is indicated in parentheses.

Employing the results obtained with PRT HTSS-15 the effect of an error corresponding to +1 °C in the determination of $W(t^*)$ at the tin, zinc, and aluminium points were analysed. (The analysis is applicable to any PRT.) Figure 1 summarizes the results of the analysis. The fixed point at which the error in $W(t^*)$ was made is indicated on the curve; fixed points at which the $W(t^*)$ was assumed to have been determined without

---

† In the cases of antimony and tin points the values of the temperature were obtained initially in terms of the IPTS-48 (Stimson 1948, 1961) and later converted to IPTS-68.

error are indicated in parentheses. As mentioned previously, errors of calibration at the tin and zinc points become amplified at the higher temperatures when equation (1) is employed and extrapolated. When the aluminium-point calibration is used with equation (6), the errors of calibration at the zinc and tin points are attenuated above the zinc point. Below the zinc point, the errors are shown to be attenuated to about 1 or 2% of the error at the aluminium point.

In some laboratories the antimony point is realized. Equation (6) could be used with calibrations at the tin, zinc, and antimony points; $W(Sb)$ and $t^*(Sb)$ would be employed instead of $W(Al)$ and $t^*(Al)$.†

McAllan and Ammar (1972) reported $660 \cdot 462 \pm 0 \cdot 004$ °C for the freezing point of pure aluminium. To compare the value of $t'(Al)$ ($660 \cdot 4021$ 'degrees C', the mean of values for five PRTs) obtained in this work with the above value, adjustments were made for the depression of the freezing point by the impurities present in the aluminium sample of cell A-2 and for what the above authors referred to as the 'γ' term, which corresponds to the last term of equation (2). The freezing points of aluminium samples of various purities obtained by McAllan and Ammar (1972) suggest that the freezing point of the aluminium sample in cell A-2 (purity of about 99·9993 to 99·9998%) could be 0·002 to 0·008 °C low. Assuming the freezing-point depression to be 0·005 °C, the freezing point of pure aluminium becomes 660·407 'degrees C' (in terms of the value of temperature obtained by extrapolating the IPTS-68 equation 1) for this work. The extrapolation of the γ term of equation (2) to $t'(Al)$ gives 0·045 °C which when added to 660·407 'degrees C' gives 660·452 'degrees C'.‡ This value is 0·010 °C lower than the value reported by McAllan and Ammar (1972). The difference may be associated with the prior treatment of the PRT. As noted earlier the PRT used in the present work had been exposed to temperatures of 960 °C and 1065 °C for many hundreds of hours prior to this investigation. Further work is necessary to resolve the difference.

Considering the possible errors of calibration (noting particularly the low resistance of the PRTs that were used in this work and the disadvantage of the Mueller bridge method at low resistances) and the amplification of the errors by the extrapolation to the aluminium point, the value 660·452 'degrees C' obtained for the freezing point of pure aluminium in the present work and the value 660·462 °C reported by McAllan and Ammar are closely consistent. To avoid confusion as a value of the aluminium point on the IPTS-68, for this work the value 660·407 'degrees C' ($t'(Al)$ adjusted for the impurity but not adjusted for the γ term) is being reported. The estimated uncertainty of the figure is ±0·005 'degrees C'. Half of the uncertainty estimate is assigned to the adjustment for the impurity.

The $W(Al)$ obtained previously with cell A-2 using the AC bridge and PRT HTSS-15 was $3 \cdot 374\,124_{85}$ (see Furukawa 1974) while the value obtained in the present work using the Mueller bridge is $3 \cdot 374\,121_{89}$ (see table 1). The difference $0 \cdot 000\,002_{96}$ corresponds

---

† For a given PRT, the coefficient D of equation (6) based on the Sn, Zn, and Al points, may be significantly different from that based on Sn, Zn, and Sb points, depending on how closely the PRT follows the quadratic relation given by equation (1) and on how closely the original PRTs, on which the assigned values of t^* are based, follow the quadratic relation.

‡ The value of the γ term was added for comparison purposes only. The values of t_{68} near the aluminium point are defined by the Pt–Pt–Rh thermocouple on the IPTS-68.

to 0·9 mK; the agreement is better than expected, perhaps accidental. More work is in progress to determine the relation between the DC and AC measurements.

6. Conclusions

The results on aluminium presented previously (Furukawa 1974) show that aluminium can provide a freezing point (near 660 °C) that is at least as reproducible as the freezing point of antimony (near 631 °C). The PRT temperature measurements above the zinc point can have large uncertainties because of possible variations in the calibration at the tin and zinc points. The use of a fixed point at the upper temperature limit of the PRT scale will reduce these uncertainties. The aluminium point is a suitable fixed point for the purpose.

Freezing-point measurements on aluminium samples of higher purity are planned for the future. PRTs of higher resistance and better immersion characteristics would help improve the accuracy and ease with which the aluminium point can be determined.

References

Berry R J 1966 *Metrologia* 2 80–90
Brandt J L 1967 *Aluminium. Properties, Physical Metallurgy and Phase Diagrams* ed K R Van Horn (Metals Park, Ohio: American Society for Metals) pp1–30
Chattle M V 1972 *TMCSI* 4 part 1 907–18
Cutkosky R D 1970 *J. Res. NBS* 74C 15–8
Evans J P and Burns G W 1962 *TMCSI* 3 part 1 313–8
Evans J P and Wood S D 1971 *Metrologia* 7 108–30
Furukawa G T 1974 *J. Res. NBS* 78A 477–95
Furukawa G T, Douglas T B and Pearlman N 1972a *American Institute of Physics Handbook* 3rd edn (New York: McGraw-Hill) pp4-105–4-118
Furukawa G T, Riddle J L and Bigge W R 1972b *TMCSI* 4 part 1 247–63
Gmelins Handbuch der Anorganischen Chemie 1943 (Berlin: Verlag Chemie) system-nummer 18
McAllan J V and Ammar M M 1972 *TMCSI* 4 part 1 273–85
McLaren E H and Murdock E G 1960 *Can. J. Phys.* 38 100–18
—— 1968 *Can. J. Phys.* 46 369–400, 401–44
Nesmeyanov A N 1963 *Vapor Pressure of the Elements* (New York: Academic Press)
Powell R L and Childs G E 1972 *American Institute of Physics Handbook* 3rd edn (New York: McGraw-Hill) pp4-142–4-162
Riddle J L, Furukawa G T and Plumb H H 1973 *NBS Monograph* 126
Stimson H F 1948 *J. Res. NBS* 42 209–17
—— 1961 *J. Res. NBS* 65A 139–45
Wood S D 1973 *NBS Tech. Note* 764

Systematic errors in high-temperature noise thermometry

L Crovini and A Actis
Istituto di Metrologia 'G Colonnetti', Torino, Italy

Abstract. The proposals for applying Johnson noise thermometry to the determination of high temperatures are becoming more frequent. A variety of methods have been presented but a systematic assessment of their relative merits is not yet available.

In the present paper the proposed methods are systematically compared with special regard to the factors that produce systematic errors. The influence of the dielectric properties of the connecting cables and the sensor design are throughly examined both theoretically and experimentally.

1. Introduction

Several proposals for the use of noise thermometry in practical measurements have been presented in the past five years (Storm 1970, Actis *et al* 1972, Kamper 1972, Brixy *et al* 1973, Borkowski and Blalock 1974). Some of these works were devoted to achieving moderate accuracies (±0·1% in general) which nevertheless could be maintained without recalibrations for long periods of time. In-pile temperature measurements (Brixy *et al* 1973) and temperature determinations in very high pressure environments (Wentorf 1971) are two possible areas where practical noise thermometry could be profitably applied.

A variety of methods has been introduced in order to overcome the usual difficulties of noise thermometry, such as the effect of the amplifier noise, the problems deriving from the gain stability and the definition of the measuring bandwidth, and the complexity of the measurement procedure. However, none of these works simultaneously takes into sufficient consideration all the factors which limit the final accuracy for high-temperature measurements ($T > 700$ K). A more detailed examination may reveal that the final accuracy is determined to a large extent by the sensor design.

The purpose of this paper is to evaluate and summarize all the error sources in order to obtain a realistic estimate of the final accuracy. This evaluation is supported as far as possible by the direct experience gained with the IMGC noise thermometer.

2. The sensor

In the case of thermal noise it is well known that the internal structure of the sensing resistor (eg, chemical composition, physical state, etc) does not influence the general relationship

$$\overline{v^2} = 4kRT\Delta f \tag{1}$$

which is known as the Nyquist equation. It relates the variance of EMF fluctuations at the leads of a resistance R to its thermodynamic temperature T where Δf is the measurement bandwidth and k Boltzmann's constant.

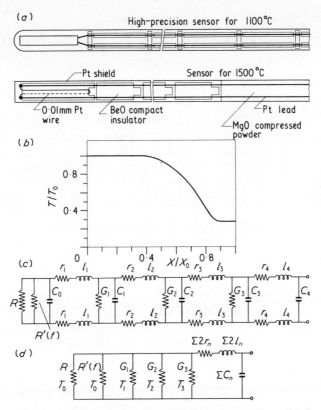

Figure 1. (a) Schematic diagrams of high-temperature thermal noise sensors. (b) Temperature profile in a comparison furnace: T_0 is the measurement temperature and x_0 is the maximum immersion depth in the furnace; T is the temperature at a particular position X in the furnace. (c) Lumped constant equivalent circuit of a high-temperature thermal noise sensor. (d) Simplified equivalent circuit.

A high-temperature environment makes it difficult to obtain R perfectly isothermal because lead resistances and isolator conductances at different temperatures contribute to the output voltage.

A conventional two-lead arrangement† may be represented as shown in figure 1(a). The usual temperature distribution over the sensor and its connecting cable for a conventional three-winding electrical furnace with metal equalizing block is shown in figure 1(b). Part of the cable-insulating material is exposed to high temperatures. High-purity metal oxide insulators exhibit noticeable electrical leakages above 700 °C. In addition insulating resistances may be frequency-dependent, as in the case of alumina and berillia. The non-isothermal circuit of figure 1(c) closely represents the actual behaviour of the sensor when a short length of the cable (less than 50 cm) is exposed to high temperatures. The circuit of figure 1(d) adequately replaces the more complex one if, as happens in practical cases, C_4 far exceeds the other capacitances and the lead

---

† Four-lead connections are possible in principle with the aid of the cross correlation technique, but require an excessively complex apparatus with two separate common points at the input. Since nobody has attempted to implement it, we shall ignore it.

impedances are negligible with respect to the shunting resistances. The reactive terms in combination with the resistances define the sensor bandwidth and will be considered at a later stage. The sensor measurable resistance is expressed as follows:

$$R''(f) = \Sigma 2r_n + \frac{1}{(1/R) + [1/R'(f)] + \Sigma G_n}. \qquad (2)$$

Since the contribution of $R'(f)$ is relatively small (but still significant for accurate measurements) the sensor resistance R_s may be introduced according to the following definition:

$$R_s = \frac{1}{(f_2 - f_1)} \int_{f_1}^{f_2} \frac{RR'(f)}{R + R'(f)} df \qquad (3)$$

where f_1 and f_2 are the limits of a rectangular-shaped measurement bandwidth. The Johnson noise across R'' may be related to a temperature T_a which is generally lower than the sensor temperature T_0:

$$T_a = \frac{1}{R''}\left(\Sigma 2r_n T_n + \frac{(T_0/R_s) + \Sigma G_n T_n}{[(1/R_s) + \Sigma G_n]^2}\right). \qquad (4)$$

Depending on the choice of R_s there is a trade-off between the effect of lead resistances and that of insulator conductances with two limiting cases, namely

$$T_a = \frac{R_s T_0 + \Sigma 2r_n T_n}{R_s + \Sigma 2r_n} \qquad (4a)$$

and

$$T_a = \frac{(T_0/R_s) + \Sigma G_n T_n}{(1/R_s) + \Sigma G_n}. \qquad (4b)$$

Equation (4a) holds for very low sensor resistances and 4(b) applies to values of R_s which are very high with respect to Σr_n.

Therefore it is evident that in the first stage of the noise thermometer an error is generated which is common to all measuring methods proposed so far. We may express $\epsilon_s = T_a - T_0$ in the following way:

$$\epsilon_s = \frac{\Sigma 2r_n(T_n - T_0)}{R''} + \frac{\Sigma G_n(T_n - T_0)}{[(1/R_s) + \Sigma G_n]^2 R''} + \delta \qquad (5)$$

where δ in the case of noise voltage measurements is given by

$$\delta = K\left(\frac{1}{(f_2 - f_1)} \int_{f_1}^{f_2} \frac{RR'}{R + R'} df - \frac{RR'(f_M)}{R + R'(f_M)}\right). \qquad (6)$$

In equation (6) we wish to measure R_s at a frequency f_M. When the temperature is derived by means of a noise power determination (Borkowski and Blalock 1974), δ vanishes provided that the noise voltage and noise current determinations are made in exactly the same frequency band.

2.1. Frequency dependence of the sensor resistance

Even if it may not be practical to perform an on-line determination of R_s in the appropriate frequency band, the deviation of R_s from a DC measured resistance can be easily obtained with an appropriate impedance comparator and the correction δ already calculated.

Some resistance against frequency determinations were obtained at IMGC by comparing the sensor at 780 °C with a standard resistor of the same DC resistance and the same time constant. The sensing resistor was a commercial alumina-insulated platinum resistance of about 750 Ω at the measurement temperature; the comparisons were made by means of a General Radio type 1654 impedance comparator in the frequency range 1–100 kHz. The average deviation of R_s in the frequency interval 10–100 kHz was 5×10^{-4} of the DC value. Figure 2 (full line) shows the isolator shunting resistance R' as a function of frequency.

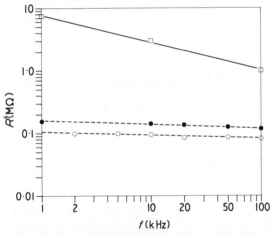

Figure 2. Frequency dependence of insulation resistance. The full line represents the behaviour of the insulation resistance for an alumina insulated 750 Ω platinum sensor at 780 °C. Broken lines represent the insulation resistance of a 99·5% alumina twin-bore isolator at 1300 °C (full circles) and 1400 °C (open circles).

Similar measurements were performed at 1300 °C and 1400 °C on a twin-bore 99·5% alumina insulator (4 cm long, 0·6 cm OD). The insulator was previously baked at 1300 °C for 1 hour to eliminate as far as possible any surface contaminant. Insulating resistances were determined by means of a sensitive capacitance bridge. The results are shown by the broken lines in figure 2 where a dramatic resistance drop without a significant frequency dependence occurs.

Measurements were repeated on a fused silica isolator (tetrasil grade transparent fused silica) from 600 to 1100 °C without detecting either a significant leakage or a frequency dependence.

2.2. Effect of non-isothermal shunting conductances

Isolator conductances which are located in the gradient zone of figure 1(b) are generators of the second term on the right-hand side of equation (5). A reliable determination of these systematic errors is possible only in laboratory experiments where materials, physical size and temperature distributions are reproducible with sufficient accuracy. At the same time a definite estimate in practical applications is impossible because of the large variations in both materials and physical arrangements. A first approximation evaluation is proposed in table 1 for alumina and berillia insulators. (When using alumina errors as large as 30 K at 1773 K are possible.) Errors are approximately proportional to the inverse of the temperature gradient on the cable. As a general rule, very steep temperature distributions must be provided in the high-temperature zone (20–40 K cm^{-1}). At the same time the sensor resistance should not exceed 500 Ω for temperature measurements in the range 1300–1800 K.

Table 1. Errors in temperature measurements at 1500 °C according to the second term on the right-hand side of equation (5).

Sensor resistance (Ω)	100	200	500
Temperature error Al_2O_3 † ‡	7·4 K	14 K	30 K
Temperature error BeO † §	1·7 K	3·4 K	8·2 K

† The error is contributed by an homogeneous cable having one extreme at 1500 °C and the other at 700 °C and submitted to a constant temperature gradient of 10 K cm^{-1}.
‡ The case of a compact Al_2O_3 insulator yielding 1 μS m^{-1} at 800 °C between the two leads is considered here. Calculations are accomplished by assuming the resistivity–temperature characteristics as given by Touloukian (1967).
§ The same initial condition of ‡ with the assumption of the resistivity–temperature characteristics of BeO (Touloukian 1967).

2.3. Lead resistance effect

Non-isothermal lead resistances yield temperature errors as given by the first term on the right-hand side of equation (5). For instance, in the IMGC noise thermometer, a platinum lead of 40 cm in length and 0·05 cm OD requires a 0·26 K correction when the 750 Ω sensor is held at 1050 K. This correction is expected to rise by approximately 0·24 K at 1300 K.

It is not practical to reduce lead resistance below 0·5–1 Ω for measurements in the range 1300–1800 K. These errors may be decreased by increasing the temperature gradient in the high-temperature side of the leads. Lead thermal conductivity is the ultimate limit for achieving very steep temperature distributions.

The worst estimate of such a limit is performed without taking into account heat transfer by radiation. Assuming the validity of Wiedeman and Franz's law for the lead metal and considering that the temperature distribution over the lead is uniquely determined by its thermal conductivity, the temperature relative error η may be expressed as the ratio of the lead Johnson noise to that contributed by the sensor, namely:

$$\eta = \frac{2}{R_s T_0} \int_{T_i}^{T_0} \frac{L_e}{\dot{Q}} T^2 \, dT \tag{7}$$

which yields

$$\eta = (2L_e T_0^2/3\dot{Q}R_s)[1-(T_i/T_0)^3] \tag{8}$$

where L_e is the Lorentz number, $\dot{Q}$ the heat flux through each lead and T_i the temperature of the 'cold' side of the lead. For instance, when T_0 is equal to 1500 K and T_i/T_0 is 0·5, then if $R_s\dot{Q} = 1$ (eg, 100 Ω and 10 mW per lead or 500 Ω and 2 mW), η amounts to 3·2%. Of course $\dot{Q}$ cannot be increased indefinitely without causing large temperature gradients in the sensor itself.

Heat transfer by radiation may be used to provide a certain degree of thermal anchoring of the lead high-temperature side. In this case the error will be somewhat reduced with respect to equation (8). More effective thermal anchoring systems are not suitable for the insulation limitations.

2.4. Sensor bandwidth

The equivalent resistance, inductance and shunting capacitance of figure 1(d) show the frequency response of the sensor and its time constant τ defined as

$$\tau = R''\left(\Sigma C_n - \frac{\Sigma l_n}{R''^2}\right). \tag{9}$$

If the measuring system presents a sharp-edged frequency band of corner frequencies f_1 and f_2, the effective sensor resistance R_{eff} differs from R'' in the following way:

$$\frac{R''-R_{\text{eff}}}{R''} = \frac{4}{3}\pi^2\tau^2(f_1^2+f_1f_2+f_2^2). \tag{10}$$

In comparative methods a time constant balance is provided between the measuring and reference sensors so that the effect of equation (10) is cancelled out.

Direct reading methods which use only one sensor to provide a linear output with respect to the input thermodynamic temperature may be influenced by equation (10). The time constant must be high enough to avoid any effect in the measurement bandwidth. However, if the sensor is remote with respect to the measuring equipment, connecting cables increase the capacitance and the time constant may exceed 20 ns, yielding a 1% systematic error for $f_2 \geq 100$ kHz. A one-point correction eliminates most of the systematic errors but resistance and capacitance changes may cause more than ±0·1% uncertainty.

3. Measurement systems

3.1. Amplifier

The amplifier performance is fundamental for the choice of the measurement method. In recent developments of noise thermometer apparatus, FET input solid state amplifiers have been generally adopted. FET amplifiers, though rather noisy at subaudio frequencies, exhibit low noise levels above 10 kHz. They approach the theoretical limit

$$R_e = 0.7 g_{fs}^{-1} \tag{11}$$

where R_e is the equivalent input noise resistance and g_{fs} is the transconductance in the common-source configuration (Van der Ziel 1962). The amplifier used at IMGC reaches 95 Ω for R_e in a twin-cascode configuration and with drain currents close to the saturation limit. It employs four selected 2N 5245; the two input FETs produce $g_{fs} = 13$ mS and $I_D = 16$ mA. The channel of plastic-case FETs is appreciably overheated with respect to the ambient temperature; this fact may partially explain the disagreement between experimental and theoretical values of R_e.

The stability of R_e is primarily affected by amplifier temperature fluctuations. In the IMGC realization the temperature coefficient $(1/R_e)(\Delta R_e/\Delta t)$ is 0.8% per degree; a ±0.5% supply stability and ±0.5 °C temperature control are provided. The resulting long-term stability was tested by means of six isotherm determinations over a two-month period as explained in the following section. The resulting standard deviation was better than 0.37%.

FET amplifiers provide very high input resistance and gate currents below 50 pA. The shot-noise produced by such a low current does not appreciably contribute to the output noise of the amplifier for sensor resistances not exceeding 10 kΩ.

3.2. Comparative methods

Garrison and Lawson (1949) proposed a comparative method where two resistances, R_1 and R_2, were switched at the amplifier input. The noise balance is achieved when

$$R_1 T_1 = R_2 T_2 \tag{12}$$

as is readily derived from equation (1) by assuming Δf to be constant. This condition corresponds to the assumptions that there is no change in the measurement apparatus and that the sensor bandwidth remains constant, that is, $\tau_1 = \tau_2$. The latter condition is usually achieved by implementing equation (12) in two different bandwidths (for instance, 10–100 kHz and 100–200 kHz). Short-term comparisons are not affected by the gain and the noise of the amplifier, in so far as R_e is independent of the sensor impedance.

The latter condition was seriously questioned by Hogue (1954), Pursey and Pyatt (1959) and Actis et al (1972) when using vacuum tube amplifiers. The shot-noise in the control-grid current and the capacitive feedback from output to input could be responsible for a slight dependence of R_e on the second power of the input resistance.

Calculations for FET input cascode amplifiers indicate that the effect should be negligible. Nevertheless a direct experimental confirmation was obtained at IMGC by comparing five resistors (with resistances in the range 300–1100 Ω) to a sixth one (750 Ω). All resistors were kept at the ice point within ±5 mK and sensor bandwidths were carefully equalized in advance by means of trimming capacitors. The five balances were obtained in less than three hours by means of a calibrated voltage divider, according to a procedure described in Actis et al (1972). The isotherm power attenuation ratio against sensor resistance was fitted either to a straight line or to a quadratic. The standard deviation of the experimental attenuation ratios from the best straight line was less than 1.9×10^{-4}, which may be reported as $0.2 \, \Omega$ in terms of noise resistance. The quadratic fitting did not give reproducible coefficients, while the improvement in the resulting standard deviation was not relevant. Therefore R_e was obtained by extrapolating the straight line for a vanishing sensor resistance.

The isotherms were repeated at the tin point with comparable results. Discrepancies may be found, however, when using a very low sensor resistance or a short circuit at the input. Spectrum distortions were detected in this case by means of a sensitive spectrum analyser. The effect, still unexplained, results in a gain reduction for very low source resistances and was detected on the IMGC apparatus when the grounding network had been slightly altered. It is therefore advisable to perform spectrum analyses before operating any noise thermometer.

The precise determination of the amplifier noise allows the IMGC noise thermometer to compare two equal resistors for absolute measurements at high temperatures, as initially proposed by Pursey and Pyatt at NPL (1959). The advantages of this comparative method are an easier time constant balance, regardless of frequency-dependent sensor resistances, and the possibility of determining temperature ratios with respect to high-temperature references, since small resistance imbalances can be tolerated and the results corrected for.

3.3. One-resistor method

Storm (1970) proposed the simultaneous amplification of the noise voltage of a resistor with two separate amplifiers and send the output signals to a correlator. The noise generated in the amplifiers would be almost completely eliminated and the output signal would be proportional to the resistor thermodynamic temperature. The advantage of this method lies in its simplicity. There are, however, the following drawbacks:

(i) gain instabilities may affect the calibration;
(ii) analogue correlators (multipliers) do not afford more than ±0.2% accuracy;
(iii) resistance changes in the sensor ought to be accounted for;
(iv) sensor bandwidth changes must be accounted for;
(v) correlated noise in the two input amplifiers produce an off-set in the output signal.

A two-point calibration may eliminate or reduce the errors of points (ii) and (iii). Gain instabilities may be reduced below the 0.1% limit. To eliminate the resistance dependence Borkowski and Blalock (1974) have recently proposed the successive measurement of both the RMS noise voltage and the RMS noise current of a resistor. By taking the product of the resulting signal the output is proportional to the noise power of the sensor and to its thermodynamic temperature regardless of resistance variations. The amplifier's noise is kept at a minimum to have less than 0.1% effect on the output voltage at full scale. It may be noted that this method is not influenced by the sensor frequency dependence so long as the frequency responses of the voltage amplifier and the current amplifier are the same. However, it requires the simultaneous use of two nonlinear transducers, in other words a true RMS meter and a multiplier, with the consequent penalty in the final accuracy.

A comparative summary of the measurement method performances is given in table 2.

Table 2. Comparison of measurement features of proposed methods for noise thermometry.

Measurement method	Amplifier noise dependence	Sensor resistance dependence	Sensor bandwidth dependence	Apparatus gain dependence	Correction for frequency-dependent sensor resistance	Bandwidth balance is temperature-dependent	Calibration points	Accuracy limit†
Garrison and Lawson (1949)	no	yes	no	no	yes	yes	1	±0·03%
IMGC	yes	yes	no	small‡	yes	no	1	±0·03%
Storm (1970)	small	yes	yes	yes	yes	–	2	±0·2%
Borkowski and Blalock (1974)	yes	no	no	yes	no	–	2	±0·2%

† The accuracy limit is estimated for a 10 minute integration on the output reading and a 100 kHz bandwidth. Only those factors which limit the electrical and electronic apparatus accuracy are considered here.

‡ This method is affected only by the preamplifier gain. Its influence on the final accuracy is inversely proportional to the ratio (R_e/R_s).

3.4. Integration time and external interference

The deviation of any single measurement from the average of an infinite ensemble in the case of noise-temperature measurement can be expressed as follows (Van der Ziel 1954):

$$\frac{\Delta T}{T} = \frac{K}{(\Delta f \tau_0)^{1/2}} \tag{13}$$

where τ_0 is the integration time, or twice the time constant of the reading instrument, and K is close to two for any method which implies two successive determinations and for which $R_e \ll R_s$. Here K reduces to unity for the direct reading method. Therefore ±0·1% accuracy is obtained with 100 kHz bandwidth and 40 s integration time in one case and with 10 s in the direct reading method of equal bandwidth.

Such a long time, when it is acceptable in practical measurements, may cause trouble on account of electromagnetic interferences (EMI). Power switches in the surroundings of the apparatus may introduce bursts of noise which usually saturate the preamplifier. Electromagnetic shields, EMI filters and a proper arrangement of the grounding network greatly help in reducing the EMI effect, but an individual measurement can still deviate by more than 0·1% from the average as a consequence of EMI.

A substantial improvement, in this respect, has been achieved by interrupting both the input signal and the time base of the integrator when a burst of EMI appears in the noise thermometer apparatus (Cibrario 1973). Figure 3 shows the IMGC noise thermometer where the integration is accomplished by means of a voltage-to-frequency converter and an up–down counter; the effective integration time is 8 minutes. The interruption of the signal and the time-base occurs when the measured noise exceeds a threshold which is set to 5 V_{RMS} by more than 1 ms. None of the input resistor noise spikes is able to produce a transition, while almost any burst of EMI noise can trigger the input gate which isolates the counter and interrupts its time base for a sufficient time interval to allow the effect of the EMI to fade down. To demonstrate the reproducibility of the IMGC noise thermometer the temperature of an inconel block was

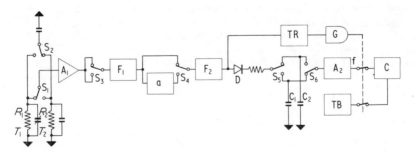

Figure 3. Schematic diagram of the IMGC noise thermometer: A_1, low-noise amplifier; F_1, F_2 pass-band filters; a, attenuator; C_1, C_2, memory capacitors; A_2, low-frequency amplifier, demodulator and voltage-to-frequency converter; C, up–down counter; TB, time-base generator; TR, EMI detecting circuit and gate trigger; G, gate; S_1 to S_6, mechanically driven synchronous switches.

repeatedly determined during three days (nine measurements). The temperature of the block was kept constant within ±0·03 K at approximately 1053 K by means of a standard platinum resistance thermometer. The noise thermometer determinations exhibited a standard deviation of 0·20 K. It was observed that the standard deviation could be significantly reduced by improving the amplifier temperature control.

4. Conclusions

All measurement methods considered here are adequate to provide a ±0·2% accuracy. A stronger limitation in accuracy may derive from the sensor design. In absolute measurements, such as those underway at IMGC, the sensor and the furnace are designed to allow a reliable determination of lead and insulator contribution to the output noise. In field applications it is not always possible to obtain reliable information of this kind and therefore systematic errors ranging from 0·1 to several per cent are introduced in temperature measurements from 1000 to 1800 K. Applications above 1800 K are still further degraded for the collapse of insulators and the necessity of using metals with a higher resistivity than platinum as lead materials.

We may therefore conclude by saying that high-temperature application of noise thermometry is primarily a matter of sensor design. Different measurement methods are applicable but not all of them can provide sufficient accuracy. The final choice will be oriented by a trade-off between accuracy and system complexity.

References

Actis A, Cibrario A and Crovini L 1972 *TMCSI* **4** part 1 355
Borkowski C J and Blalock T V 1974 *Rev. Sci. Instrum.* **2** 151
Brixy H, Hecker R and Overhoff T 1973 *Nuclear Power Plant Control and Instrumentation* (Vienna: IAEA) pp 637–48
Cibrario A 1973 *IMGC Technical Report* S/114
Garrison J B and Lawson A W 1949 *Rev. Sci. Instrum.* **11** 785
Hogue E W 1954 *NBS Report* 3471
Kamper R A 1972 *TMCSI* **4** part 1 349
Pursey H and Pyatt E C 1959 *J. Sci. Instrum.* **36** 264
Storm L 1970 *Z. Angew. Phys.* **6** 331
Touloukian Y S (ed) 1967 *Thermophysical Properties of High Temperature Solid Materials* (New York: Macmillan) vol 4, part 1
Van der Ziel 1954 *Noise* (Englewood Cliffs, NJ: Prentice-Hall)
—— 1962 *Proc. IRE* **50** 1808
Wentorf R H Jr 1971 *NBS Special Publication* 326 p81

Non-contact temperature measurement using forced air convection

I R Fothergill
Alcan International Limited, Banbury, Oxon

Abstract. This method exploits the phenomenon of heat transfer between a surface and air flowing over it when a temperature differential exists between the two. The air is delivered to the specimen surface by means of two or more jets. The device may be operated in one of two possible modes. The first involves passing initially unheated air over the surface and predicting the temperature of the latter from the temperature differential in the air before and after it has passed over the surface. The alternative mode involves preheating the air until this differential disappears and measuring the corresponding air temperature which will, under these zero net heat transfer conditions, be equal to the surface temperature.

The principle has been applied to the temperature measurement of both plane surfaces and surfaces of cylindrical geometry, ie a rod. A different device geometry is necessary, however, in each case.

1. Introduction

The most common method of non-contact temperature measurement is radiation pyrometry which exploits the electromagnetic radiation emitted from a surface. This radiation is however not uniquely dependent upon temperature (unless a black body emitter is involved) since the energy flux and spectral distribution are also functions of the spectral emissivity of the surface. The spectral emissivity is mainly responsible for errors associated with this technique.

The discussion to follow will describe a non-contact temperature-measuring technique employing forced air convection in an attempt to avoid these emissivity errors.

2. Theory of forced air convection method

If a jet of cold air is blown across a hot body and the jet temperature is measured before and after it has passed over the body giving respective temperatures T_1 and T_2, a difference between the two measurements will be found due to the transfer of heat from the body to the air with T_2 being greater than T_1.

If precautions are taken to minimize the temperature gradients in the jet arising from entrainment of cold air, then as the jet temperature is increased this temperature differential will decrease, becoming zero when the jet and body temperatures are equal. Thus, by measuring the air temperature under these balance conditions the body temperature can be determined without making contact.

This phenomenon has been exploited in the temperature measurement of surfaces of cylindrical geometry, ie a rod as well as two-dimensional plane surfaces.

3. Temperature measurement of 9 mm diameter aluminium rod

The device constructed for this purpose is shown in figure 1. It was fabricated from aluminium (because of its high thermal conductivity) and consisted of an annular cavity sealed at both ends with an inner diameter of 40 mm, an outer diameter of 60 mm and a length of 230 mm (Fothergill 1974). Lying close to the axis of the device was the rod, the temperature of which was to be measured.

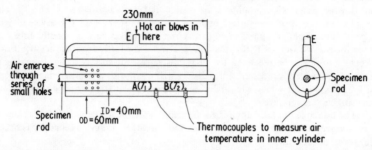

Figure 1. Temperature measurement of a 9 mm diameter rod.

Air was introduced into either end of the cavity via tube E and forced to circulate round the cavity before finally emerging through a series of 6 mm diameter holes in the inner thin aluminium wall.

On passing into the inner tube the air circulated round the rod and in so doing extracted heat from the latter (when the rod was hotter than the air). Two thermocouples, A and B, placed at different locations along the length of the tube, indicated respective temperatures T_1 and T_2. Since the air temperature was proportional to both the temperature difference between itself and the rod as well as the time it had been exposed to the rod, T_2 was greater than T_1.

On reducing the temperature differential between the rod and air, a smaller value of $(T_2 - T_1)$ was found until a state of zero net heat transfer between the two resulted when the rod and air temperatures became equal. Under these conditions T_1 equalled T_2. The purpose of the annular cavity was to maintain the walls of the inner tube at the temperature of the air in this tube, when zero net heat transfer conditions between the rod and air prevailed.

In order to measure the rod temperature, therefore, the in-going air temperature was adjusted until T_1 equalled T_2 and under these balance conditions the rod temperature was inferred directly from a measurement of either T_1 or T_2. It is estimated that measurements with accuracies of better than 2% can be made, the technique being inherently independent of the cross sectional geometry and surface emissivity of the specimen.

An alternative mode of operation where a rod of high thermal capacity is employed, as would be the case for a long length of continuously moving rod, is to dispense with hot air and calibrate the temperature differential $(T_2 - T_1)$ against a static calibration rod (externally heated). Such a procedure has certain practical advantages in that it avoids the necessity for an air heater and is also more responsive than the preheated air mode because of the thermal inertia of the air heater. The tolerance in location of the rod with respect to the measuring device is smaller for the initially unheated air mode than for the preheated air mode, however.

For temperature measurement of continuous rod during production, the device was split in a longitudinal direction into two halves (the exposed parts of the cavity being sealed) which were brought together round the rod during operation. In order to minimize any errors arising from large radial displacements of the moving rod the single thermocouples A and B were each replaced by two sets of four thermocouples symmetrically displaced at the same axial locations as A and B.

Errors arising from surface emissivity are again likely to be minimal for the initially unheated air mode since the radiation contribution to each set of thermocouples will be approximately the same.

4. Temperature measurement of a plane surface

4.1. Wall-jet method

When a round free jet impinges normally onto a plane surface the fluid (air in this case) moves out radially from the origin (see figure 2) forming a so-called 'wall-jet'. Glauert (1956) and Bakke (1957) have analysed the flow characteristics of the wall jet theoretically and experimentally respectively and have shown it to possess characteristics similar to a boundary layer in the inner wall region and characteristics similar to a free jet in the outer regions.

Figure 3 shows the radial variations of mean wall-jet temperature for two assumed heat transfer coefficients but neglecting both entrainment and radiation effects. These curves have been derived theoretically by Fothergill (1974).

One possible way of exploiting these characteristics would be to place two thermal sensors at different locations along a radius in the wall-jet and to adjust the free-jet temperature until the temperature differential between the sensors disappears. If entrainment and radiation effect were insignificant the balance would correspond to an

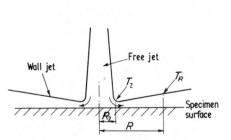

Figure 2. A wall jet resulting from the impingement of a free jet onto a surface.

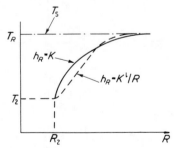

Figure 3. Typical variations of mean wall-jet temperature when the surface is hotter than the impinging free jet. h_R = heat transfer coefficient of wall jet, K and K^1 are constants.

equality between the wall-jet and surface temperatures, the latter thus being inferred from the wall-jet temperature. Alternatively, if an unheated free jet were employed, the temperature differential between the two sensors could be used as a parameter from which to predict the surface temperature. Unfortunately, the wall-jet thickness is very small (~mm) for practical values of R and when a temperature differential exists between the surface and wall-jet there is a very steep temperature gradient normal to the surface in the boundary layer region. Both these characteristics would make sensor location extremely critical and a different flow characteristic was therefore developed.

It was found that by directing two free jets displaced from each other normally onto a surface the two corresponding wall-jets, when they impinged, produced a flow condition which moved away from the surface (see figure 4). This new flow condition was called a WIS-jet which is an acronym for wall-jet impingement synthesis. It was also found that the WIS-jet had characteristics far more amenable to exploitation in temperature measurement.

Two modes of operation may be employed, one involving unheated free jets and the other heated free jets. In the former case, the surface temperature T_S is predicted from a measurement of the temperature differential $(T_W - T_F)$ where T_W is the WIS-jet temperature and T_F the free-jet temperature, $(T_W - T_F)$ being proportional to $T_S - T_F$. If the free-jet temperature is increased, then $(T_W - T_F)$ will decrease until, when $T_F = T_S$, $T_W - T_F$ will equal zero. Thus by measuring the temperature of either the free or WIS-jets under these balance conditions the surface temperature may be predicted.

It has been shown theoretically and experimentally (Fothergill 1974) that if $T_S > T_F$ then the temperature will increase along y +ve and y −ve directions in the WIS-jet (figure 4). Thus by placing two thermal sensors at two different y displacements the temperature differential will be proportional to $T_S - T_F$ and thus provides a parameter from which to predict T_S. If T_F is again heated until it equals T_S then the temperature differential will disappear and the surface temperature may be inferred from a measurement of either sensor.

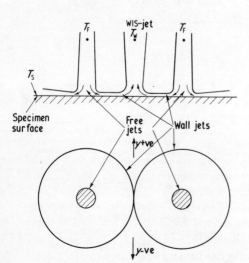

Figure 4. The formation of a WIS-jet by the impingement of two identical radial wall jets.

4.2. Method involving three collinear asymmetrically displaced free jets

In this case three free jets are employed resulting in the production of two WIS-jets (see figure 5).

If the free jets are unheated (ie T_F = ambient) then, because they are asymmetrically displaced, the WIS-jet W_x corresponding to the close free-jet spacing will be cooler than the WIS-jet W_X corresponding to the wider free-jet spacing. T_S is predicted from the

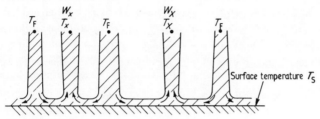

Figure 5. Temperature measurement using three free jets.

difference between the large and small WIS-jet temperatures $T_X - T_x$ since $T_X - T_x$ is proportional to $T_S - T_F$ and if T_F is constant $T_X - T_x = \alpha T_S - \beta$ where α and β are constants.

Figure 6 shows the device employing the three free-jet method. The free jets are produced by forcing air through the three asymmetrically displaced orifices A, B and C. The thermocouples T_x and T_X are placed 6 mm from the orifice plate and are supported by four posts, P.

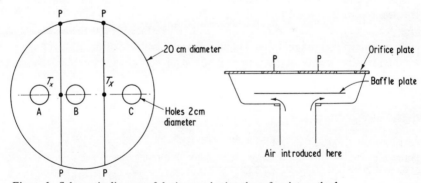

Figure 6. Schematic diagram of device employing three free-jet method.

It can be shown that the effects of surface velocity, surface radiation to the measuring thermocouples, surface cooling by the jets and surface roughness are insignificant for both the two and three free jet modes when preheated free jets are employed.

For unheated free jets, however, the three free-jet mode gives optimum performance. A device based on this mode has been constructed and installed on a pilot line for the temperature measurement of continuously annealed 0·25 mm aluminium strip moving at 90 m min^{-1}. It has been estimated that with this strip velocity and a free-jet velocity of approximately 750 m min^{-1}, the indicated temperature was approximately 1% higher than if the strip had been stationary. Since this error was always positive it was easily corrected for.

With a surface emissivity of 0·1 and a temperature of 300 °C the indicated surface temperature contained a 2% radiation contribution with the above free-jet velocity and air temperature measuring thermocouples of diameter 0·64 mm and emissivity 0·5. It has been estimated that the strip temperature was reduced by 0·5% by the cooling influence of the jets.

The significance of surface roughness on the heat transfer coefficients of the free and wall jets (hence the indicated temperature) depends upon the viscosity of the air as well as its velocity. It has been shown that with free-jet velocities of 750 m min^{-1}, as long as the surface roughness does not exceed 0·25 mm, its effect on temperature measurement will be insignificant.

It is found that by using unheated free jets the device will operate satisfactorily up to a surface–device displacement of about 3 cm and will tolerate a variation in this distance of about ±0·5 cm without upsetting the calibration.

The heated free-jet mode however requires a closer surface–device displacement and will operate satisfactorily up to a displacement of about 1·5 cm with no sacrifice in accuracy.

5. Conclusion

It has been seen that the forced air convection method lends itself to various modes of operation and two possible specimen geometries have been considered.

The preheated air mode has several advantages over the initially unheated air mode. It is less sensitive to the effects of radiation, surface velocity and surface roughness and also produces negligible cooling of the specimen surface. It has the disadvantage, however, of requiring more peripheral equipment and because of the necessity to preheat the air the measurement response time will be longer than for the initially unheated air mode.

References

Bakke P 1957 *J. Fluid Mech.* **2** 467
Fothergill I R 1974 *PhD Thesis* University of Aston in Birmingham
Glauert M B 1956 *J. Fluid Mech.* **1** 525

Ionic thermometers

Miroslav Strnad
Power Research Institute, Prague, Czechoslovakia

Abstract. An original method of temperature measurement based on conductivity changes near the phase transition point of ionic compounds and suitable for the range from 200 to 700 °C according to the thermometric compound used, is given.

By choosing between two approaches it is possible to evaluate either a discrete value of temperature or continuous measurement in a range to about 50 °C below the phase transition point of thermometric compounds.

The extreme nonlinearity of conductivity of the chosen group of ionic crystals used as well as the technical applications developed in our laboratories have not previously been published.

The aim of the research is the application of this measuring method for temperature indication in nuclear reactors. Preliminary tests in radiation fields in an experimental reactor are yielding a real hope in this direction.

Our paper is concerned with the description of a new kind of temperature indicator, exploiting one of the physical properties of a certain group of ionic crystals which, as we have found, has demonstrated nonlinearity of electrical conductivity at the temperature of the phase change. The research work was originally directed towards the improvement of existing thermometry for the classical power plants.

We employed the resistivity method, and the output signal may be processed in the electronic evaluating equipment directly. In contrast to metal resistance thermometers and thermistors which have been known and used in technical practice until now and whose electrical conductivity is purely electronic the ionic crystals show the overwhelmingly ionic conductivity at the melting point. Therefore, the latter show a negative thermal resistance coefficient and their electrical resistance drops with increasing temperature.

At the melting point, or phase transition, of all three types of the conductors large changes in their electrical resistance occur. There are some which show exceptionally steep resistance changes within a relatively narrow temperature interval at the phase transition. These facts are well known from elementary physics and are sufficiently described, especially in the case of metals and semiconductors. Yet the technical literature lacks hitherto scientific as well as experimental information about the electrical conductivity of ionic crystals during the phase transition.

The research has been aimed at the determination of the change of the electrical resistance with temperature in a group of selected chemical compounds with an overwhelmingly ionic bond. This dependence has been studied in the neighbourhood of the phase transition temperature, particularly in the region of the melting point. The factors determining the suitability of substances for the purpose of temperature measurement were as follows:

(i) Relative magnitude of the electrical resistance change to the temperature change, ie the steepness of the resistance characteristics.
(ii) Reversibility of the physical phenomenon, ie with the temperature falling the substance must crystallize and return into the original insulating state again.
(iii) Reproducibility of the parameters sub 1.
(iv) Time stability of the parameters sub 1.

High chemical purity of the substances used is essential. Impurities and inclusions usually reduce the steepness of the resistance characteristics, causing instability and increase the conductivity in the solid state. The measuring electrical current passed through the substance, the ionic current, has to be alternating only without the DC component. The DC component would cause an electrolytic deterioration of the substance building a permanent conductive connection between the electrodes. The frequency used may be relatively low, 50 Hz and higher. The electrical current, being dependent on the geometrical dimensions of the actual sensing element, is usually within hundreds of μA.

From about 150 different chemical compounds the following substances with good thermometric parameters have been selected experimentally:

	Compound	Temperature interval of steep resistance change by six orders (10^7–10^1 Ω)	Melting point (°C)
1.	$CdBr_2$	566–567·5 °C	567
2.	$CdCl_2$	566–568	568
3.	CdI_2	385–388	388
4.	$K_2Cr_2O_7$	393–398	398
5.	RbI	641–643	642
6.	ZnI_2	444–446	446
7.	HgI_2	256–259	259
8.	KIO_3	462–464	560
9.	KI	682–684	723
10.	$TlNO_3$	203–206	206
11.	NH_4Br	533–535	subl. 541
12.	$NaC_2H_3O_2$	322–324	324

The actual behaviour of the resistance characteristics of two such compounds is illustrated in figures 1 and 2.

The research started in our laboratory bears a long-term character and will continue. At present some of the most significant results may be summarized:

(i) The influence of chemical purity on the form of the resistance characteristic is shown in figure 3. As an example the experiment with the compound $CdBr_2$ is shown. There is an obvious improvement in behaviour between the original material and after chemical purification. At this point it is necessary to add that every compound bears its specific chemical properties and thus requires its own individual purification process.

Ionic thermometers

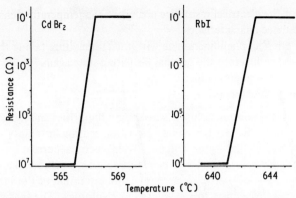

Figure 1. The resistance characteristic of $CdBr_2$.

Figure 2. The resistance characteristic of RbI.

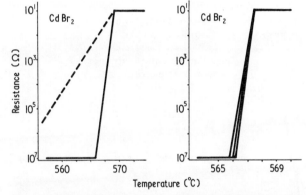

Figure 3. Effect of the chemical purity of $CdBr_2$ on the shape of its resistance curve. Broken line, original compound; full line, after purification.

Figure 4. Comparison of the resistance characteristics of three individual sensors filled with $CdBr_2$.

(ii) Studies of the reproducibility of the resistance characteristics of individual sensors containing the identical thermoactive compound were undertaken and are shown in figures 4 and 5. Examples show the parameters of three individual sensors containing the compounds $CdBr_2$ and KIO_3. The order of the temperature scattering along the temperature axis is in the range of tenths of a Kelvin as is shown in the diagrams. Maintaining a high standard in the technological process for production of the sensors is essential.

(iii) Long-term stability measurements of the thermometer parameters were made for limited number of sensors. The majority of the above-mentioned 12 compounds were subjected to long-term continual tests for 2000–3000 hours, ie with a continous temperature cycling in the interval of the phase transition temperatures and in the presence of nominal alternating electrical current passing through the substance. The results were satisfactory. As an example one might take the comparison of characteristics of two compounds – $CdBr_2$ and RbI – after temperature cycling for 8800 and 6500 hours (figures 6 and 7). The range of the differences along the temperature axis is tenths of a Kelvin.

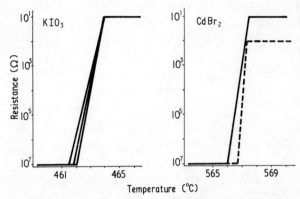

Figure 5. Comparison of the resistance characteristics of three individual sensors filled with KIO_3.

Figure 6. Comparison of the resistance characteristics of $CdBr_2$ after temperature cycling for 8800 hours (broken line). Full line, original curve.

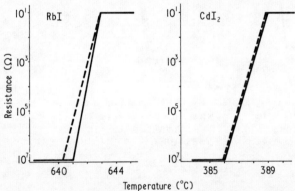

Figure 7. Comparison of the resistance characteristics of RbI after temperature cycling for 6500 hours (broken line). Full line, original curve.

Figure 8. Comparison of the resistance characteristics of CdI_2 before (full line) and after (broken line) irradiation by an integral dose of 5×10^{22} nm^{-2}.

(iv) Recently completed tests concerned the irradiation of the ionic thermometric compounds in the radiation zone of the nuclear reactor. Three compounds – CdI_2 (388 °C), KIO_3 (462 °C) and $K_2Cr_2O_7$ (398 °C) – were studied. The compound's behaviour was measured and recorded on an $X-Y$ recorder during the whole experiment. The integral irradiation dose was about the level of 5×10^{22} nm^{-2}. The results are shown in figures 8 and 9. The parameters of CdI_2 remained unchanged. In the case of $K_2Cr_2O_7$ a small shift of the resistance characteristic was recorded. The parameters of the third compound, KIO_3, remained stable. The conclusions of the first informative irradiation experiment have indicated the possibility of the exploitation of ionic compounds for the purpose of nuclear reactor thermometry.

For practical application of ionic compounds for temperature sensing, the ionic sensor illustrated in figure 10(a) was developed. Its construction is very simple. Essentially, it is a cylindrical flask made out of a special hard glass in which two sensing metal

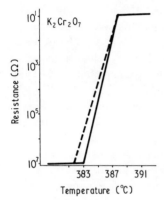

Figure 9. Comparison of the resistance characteristics of $K_2Cr_2O_7$ before (full line) and after (broken line) irradiation by an integral dose of 5×10^{22} nm^{-2}.

 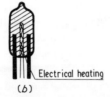

Figure 10. (a) Schematic design of two-electrodes (two-value) ionic sensor. (b) Schematic design of ionic sensor combined with electrical heating system (calorimetric ionic sensor).

electrodes are sealed. The inside of the flask is filled with a thermoactive ionic compound. After sealing this flask it is possible to measure the temperature by simply measuring the electrical resistance. A steep change of resistance indicates that the temperature interval under question has been reached. It is easy to apply this phenomenon to temperature control. The output signal of the ionic sensor is used to control the temperature within the defined temperature interval, ie at the phase transition temperature. The temperature discrimination ability is high, being given by the steepness of the resistance characteristics of the compound.

The design of technically applicable sensors must meet the following three requirements.

(i) Close reproducibility of the resistance characteristics which makes it possible to produce interchangeable sensors.
(ii) Sufficiently long life in the service conditions.

(iii) Long-term stability of the thermometric parameters, proved by precision measurements over several thousand hours.

The reproducibility of the resistance properties of the temperature sensors depends essentially on the homogenization of the compounds representing their thermoactive charge, on the filling technology and on the method of sealing the sensors.

The function life is determined by the chemical, mechanical and thermal resistance of the covering material used for sensors. Until now special hard glass has been used for this purpose.

The time stability of the important thermometric properties depends mainly upon the conservation of the original chemical purity of the thermoactive compound during long-term exposure of the sensors to the service conditions. The compound has to be inert towards the glass as well as towards the metal sensor electrodes. There is another condition, ie during the passage of electrical current through the compound no electrolytic effect must occur (ie, it must be pure alternating current).

The application could be broadened by construction of special calorimetric sensors in which an independent heater is placed in the centre of the sensor flask (figure 10b). The heating system is separated from the electrodes and from the salt charge galvanically. This design makes possible broader temperature sensing from above the original temperature interval down to about 50 K below. It enables us to use the sensors for measurement at a wider range of levels or for measurement along a decreasing temperature gradient. The sensor is automatically kept on the phase temperature by means of its own regulator — essentially it represents a miniature thermostat. The information about the temperature of the surroundings comes from measuring the electric power input to the sensor heating system. The steep change in conductivity of the thermometric compound, its physical stability and large amplification in the control loop makes a high sensitivity possible. The measurement precision depends on the stability with time of the resistance properties of the compound, on the stability of the thermal resistance between the sensor and its surroundings, and on the precision of measurement of the small electrical output, which is approximately 1 W.

The dimensions of the sensors (glass flasks) are:
    diameter   3 mm ± 5%,
    length   12–15 mm.

This applies for the simple two-electrodes (two-value) sensors as well as for the combined (calorimetric) sensors.

The new technical applications of this method for the temperature measurement are given in the following points:

(i) This method, being based on essentially simple physical phenomena of a group of chemical compounds, is absolute from the view point of the thermodynamic temperature scale. In the future it would be possible to exploit the properties of these compounds or similar new compounds for verification and calibration purposes. An electrical furnace controlled by ionic sensors was designed using this principle. The temperature in the furnace block was kept at the level of the phase transition temperature with accuracy below 0·1 K by means of this thermostat.

(ii) The application of the calorimetric sensors as fixed level control thermometers for protection against overheating of some technological structure. At present equipment installed in steam piping, where overheating is registered at a temperature of 545 °C, is under trial.

(iii) The use of the calorimetric sensors for adjustable level temperature control. At present an experiment in the control of the superheated steam temperature in one selected classical condensation power station is being prepared.

(iv) Another possibility represents the previously mentioned application of these sensors for the purposes of controlling and checking the temperatures in the inner parts of nuclear reactors for which this method appears to have important possibilities. Up to now the first steps only have been made. There are a great number of as yet unknown problems, to which the answers may be obtained by continuing research work in this field of thermometric technology applied to nuclear in-core instrumentation.

The discussed physical property of this group of ionic crystals, their nonlinearity of the electrical conductivity, represents a remarkable phenomenon not applied until now in technical practice. It seems that the results obtained promise some useful applications even if further research will be necessary.

Measuring of the temperature maxima and distribution on the surface of dynamically heated moving systems, using ^{85}Kr techniques

J Fodor, L Léb and G Muk

Research Institute of Automotive Industry, 1115 Budapest, Csóka Utca 9, Hungary

Abstract. The degassing properties of noble gas–metal systems are suitable for measuring the temperature maxima and in a ^{85}Kr–metal system it is easy to measure the degassing process, for example by following up the change in radioactivity. The method was used for measuring the temperature distribution of thermal loading in the combustion chamber of diesel engines and supercharged diesel engines.

1. Characteristic features of metal–gas systems

Incorporation of inert gas ions into solid matter was observed by Loeb (1939) in gas-discharge tubes filled with Ar, Xe and Kr. After describing this phenomenon Leck (1956) bombarded metallic targets with inert gas ions to analyse the inert gas absorption and retention. He stated that gas was incorporated into solid matter in the form of ions or neutral atoms, and by raising the bombarding voltage the quantity of gas bonded on the surface or inside the metal was also increased.

Since a relatively high temperature (450–550 °C) was needed to complete the desorption process within a reasonable time, he came to the conclusion that the binding energy of gas ions inside the metal considerably exceeded that of a merely physical adsorption (attractive forces of van der Waals).

The properties of incorporated gas atoms were analysed by Rimmer and Cotrell (1957). In their opinion, as a first stage, gas atoms are incorporated into the lattice points between the crystal planes of solid matter. Under the influence of internal tension, however, they tend to displace one of the metallic atoms from its place, actually replacing it, and in other words fitting into the crystal lattice by substitution.

Chleck and Maehl (1963) arrived at the conclusion that a krypton atom (having a diameter of 3–4 Å) was too large to be situated between the lattice points in metals, and for this reason was situated on the lattice points.

Almen and Bruce (1961) used an electromagnetic isotope separator to introduce inert gases into various materials. By means of this equipment they could use ions of given masses, charges, energies and penetration angles. Experiments were undertaken to define the saturation values of Kr incorporated into metals and concentrated by them. Saturation values were interpreted by a dynamic balance of the scattering of atoms diffusing inwards under the effect of further bombarding and of those already collected.

Davies et al (1960) investigated the depth of penetration. He found that krypton distribution was definitely uneven and at a given depth an asymmetrical concentration peak was to be observed.

2. Gas-releasing properties of metal—gas systems

If the gas-releasing properties of metal—gas systems are analysed in terms of diffusion laws at a temperature of 550 °C, gas release should be completed in a short time. However, test results do not confirm this: even at higher temperatures the activity of kryptonates does not decrease to zero. This means that the process need not be regarded as a simple diffusion determined by the diffusion coefficient

$$D = D_0 \exp(-E/RT)$$

independent of the concentration.

According to Chleck and Maehl (1963) kryptonates steadily preserve their krypton contents up to several months at room temperature if there are no chemical and physical effects. Because of the concentration of gas atoms on the surface, kryptonized surfaces, in turn, seem to be very susceptible to any surface phenomenon, such as, *inter alia,* temperature effects which are able to modify the metallic lattice.

3. Temperature measurements of machine parts

In the case of internal combustion engines, the aim of development is to obtain the maximum possible power output from the smallest possible volume; this entails ever increasing working temperature loads. It is, therefore, essential to be familiar with the thermal loads at higher power outputs particularly in the machine parts of the combustion chamber (pistons, nozzles, cylinder heads, valves). It proved to be rather complicated, or even impossible to measure with instruments the temperature and/or temperature distribution on moving parts. For this reason procedures were developed to measure temperature on the basis of properties irreversibly changing under thermal effect.

One of these methods is where kryptonates with ^{85}Kr isotope are applied for the purpose of measuring temperature maxima and temperature distribution.

4. Preparing kryptonates

Two methods are known for preparing a krypton—gas—metal system, namely diffusion and ion bombardment methods. The diffusion method is rather limited in its use, as the heat treatment for the diffusion process is not allowed to be applied to finished machine parts.

The ion bombardment method can be used on a wider range of materials. Either finished machine parts (nozzles, valves) are kryptonized, if needed, using 'screening' so that a given surface is kryptonized, or inserts are kryptonized for major machine parts, such as pistons and cylinder heads, for example. The principal scheme of such a kryptonizing device is shown in figure 1.

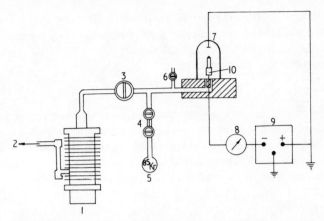

Figure 1. Scheme for preparing kryptonates. 1, Diffusion air pump; 2, forepump; 3, vacuum valve; 4, dose valves; 5, ^{85}Kr flask; 6, atmospheric valves; 7, reactor head; 8, ammeter; 9, supply voltage; 10, target.

From tracing techniques, it appears that the ^{85}Kr isotope used for kryptonizing purposes has favourable properties. Its half-life is fairly long (10·6 years), it has easy to measure beta-radiation (0·67 MeV) and it does not get incorporated into the human body.

On the basis of data in the literature, kryptonates can be expected to be suitable for measuring temperature maxima and heat distribution on machine parts in the combustion chamber of internal combustion engines.

5. Characteristics of kryptonates

In order to make temperature measurements it was necessary to know how long kryptonates could be applied since each series of experiments including mounting and preparing variants was time-consuming, taking up to about one month.

As illustrated in figure 2, kryptonates made of heat-resistant metals like Mo, Pt and alloy can be applied over a 1·5 month period because their krypton contents do not change considerably after the initial phase where the activity decrease was significant.

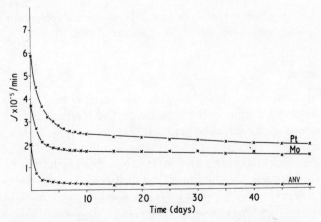

Figure 2. Krypton losses of kryptonates at room temperature.

Furthermore, property changes of kryptonates were studied under the effect of heat treatment in order to determine the temperature range where they can be applied. As to the degassing process of kryptonates made of heat resistant alloys under thermal effects, three phases of different rates can be observed: at a temperature range up to 200 °C, up to 700 °C and over 700 °C. The highest rate of degassing is observed in the range up to 200 °C, in the next range it becomes significantly lower, and finally over 700 °C there is hardly any change. From measurement aspects the second range (from 200 °C up to 700 °C) seems to be favourable.

In order to determine the experimental time required and tolerable for temperature measurements, kryptonates made of heat-resistent alloys were treated by heat three times at equal temperatures for 30 minutes every time, and the krypton release of the sample was measured. It was stated that during a test period of 1·5 hours the gas release of the samples was not significant. Consequently, thermal effects of a design and/or of a working parameter can be measured reliably by measuring the temperature of kryptonates during a period of 1·5 hours. The time needed to stabilize the thermal conditions resulting from the change was 10 minutes in the case of the diesel engine tested.

6. Measuring the temperature and its distribution

The temperature measurement procedure is shown on the flow diagram in figure 3. An ion bombarding method was applied to 'kryptonate' the original engine parts such as nozzles, needles, piston rings, valves, and/or inserts for pistons and cylinder heads (phase 1 in the flow diagram). To measure the surface temperature of nozzles and needles a point was kryptonated on the neck of the nozzle and the pin of the needle. After controlling the radioactivity of the kryptonated pieces these were put into the engine for the planned experiment.

The next phase (phase 2 in the flow diagram) is to have the engine run with the required technical parameters (eg maximum rev/min or maximum load) for 10–15

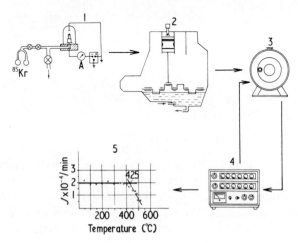

Figure 3. Flow diagram of temperature measurement by using kryptonates. 1, Preparation of kryptonate by ion bombardment; 2, experiment; 3, heat treatment; 4, measuring of radioactivity; 5, diagram for the evaluation of the temperature measurement.

minutes during which the thermal conditions to be measured are stabilized. This is the 'thermal exposure' which is equal to the dynamical heat balance in the tested combustion chamber. It is influenced by several factors such as the construction of the nozzle, the cooling effect of the fuel and the heat effect of the combustion inside the chamber. Under this thermal exposure the nozzle and the needle lose a significant part of their krypton content.

The tested pieces are dismantled from the engine nozzle and the needle and are put in a furnace in which the temperature is controlled by thermocouples. The reliability of the temperature-measuring system is critical to the evaluation of the kryptonate technique. This thermocouple controlling system entails an error of ±2%. The furnace is heated to the given temperature but this should not be as high as that of the 'thermal exposure'. Then a stepwise heat treatment and measurement of radioactivity is effected where the duration of treatment is 20 minutes, and the increase of temperature 5 °C in each step (phases are 3 and 4 in the flow diagram).

In the diagram of temperature versus radioactivity a steady activity level can be measured during the controlled repeated heating and counting process until the temperature of 'thermal exposure' is exceeded. Once this maximum is exceeded a significant loss of krypton begins.

The diagram of activity versus temperature plotted on the basis of these data permits us to determine the maximum temperature of the 'thermal exposure'. This is the breaking point in the curve figure 4.

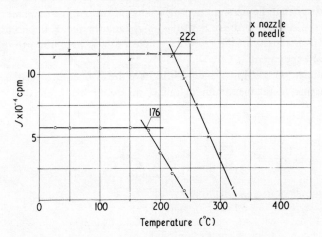

Figure 4. Temperature measurement of nozzle.

Temperature maxima were measured on nozzles of diesel engines of various types (temperatures of the nozzles, those of the needles), temperature distributions were measured on the piston heads in combustion chambers of different types, on valves and on cylinder heads. Similarly, temperature maxima on valve clamping springs of different air compressors were checked. Effects of supercharging of diesel engines on thermal loads of various engine parts were also analysed.

The distribution of the temperature of a piston head can be seen in figure 5. Photographs of machine parts on which the temperature was measured are shown in figure 6.

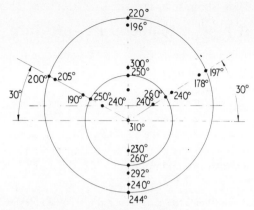

Figure 5. Heat distribution of a piston head.

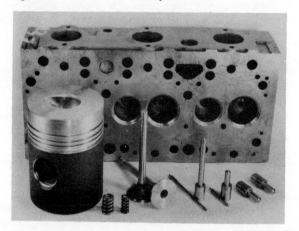

Figure 6. Machine parts on which temperature was measured.

References

Almen O and Bruce G 1961 *Nucl. Instrum. Meth.* **11** 257–78
Chleck D and Maehl R 1963 *Int. J. Appl. Radiat. Isotopes* **14** 593–8
Davies J A, McIntyre J D, Cushing R L and Lounsbury M 1960 *Can. J. Chem.* **38** 1535–46
Leck J H 1956 *Proc. Symp. on Chemisorption, Kiel* pp162–8
Loeb L B 1939 *Processes of Electrical Discharge in Gases* (New York:Wiley)
Rimmer D E and Cotrell A H 1957 *Phil. Mag.* **8** 1345

The gas-controlled heat pipe: a temperature–pressure transducer†

C A Busse, J P Labrande and C Bassani
Euratom CCR, 21020 Ispra, Italy

Abstract. The so-called gas-controlled heat pipe allows surfaces of very uniform temperature to be made and this temperature to be controlled by the pressure of a cold inert gas. In the ideal case the temperature–pressure relationship is given by the vapour pressure curve of the working fluid of the heat pipe. Deviations from this ideal case are discussed. Measurements along the axis of an argon-controlled water–copper heat pipe are presented, showing a temperature constancy of 10^{-3} °C over 30 cm. It is concluded that the gas-controlled heat pipe holds promise of realizing isothermic vessels with a temperature homogeneity and long-term reproducibility of some parts in 10^6, up to 2000 °C and large sizes. Such heat-pipe thermostats could serve as temperature standards, and they should be versatile tools for research where temperature criteria are of prime importance.

1. Introduction

The gas-controlled heat pipe is a rather new device for measuring and controlling temperatures. We shall explain its principle and discuss its performance limitations.

2. Description of the gas-controlled heat pipe

The heat pipe was developed in the 1960s as part of space programmes and has proven to be a unique self-contained apparatus for isothermal heat transfer from cryogenic to more than 2000 °C. Figure 1 shows the principle of the heat pipe (Gaugler 1944, Grover *et al* 1964). It is essentially a closed, evacuated chamber whose inside walls are lined with a capillary structure, or wick, that is saturated with a volatile working fluid. The operation of a heat pipe combines two familiar principles of physics: vapour heat transfer, and capillary action.

In the heated section of the heat pipe the working fluid vaporizes. The vapour flows to the cooled section where it is condensed, releasing the latent heat of vaporization which it has absorbed in the heated section. The liquid–vapour–liquid circulation loop

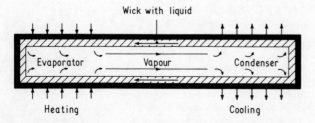

Figure 1. Heat pipe (schematic).

† Dedicated to Professor K H Höcker for his 60th birthday.

is closed by the return of the working fluid in the wick to the evaporator, under the simultaneous action of capillary forces and gravity.

Due to the reversibility of the evaporation and condensation process, the heat pipe is basically an isothermal heat transfer device.

The gas-controlled heat pipe is a variant of the normal heat pipe, which is obtained by introducing a non-condensing inert gas into the heat pipe. It was found experimentally that by the action of the two-phase circulation the gas is separated from the vapour and accumulated at the downstream end of the vapour flow, ie, in the cooled section of the heat pipe (figure 2). The result is a hot, nearly isothermal vapour zone as in a normal heat pipe, and a cold gas zone (Grover et al 1964).

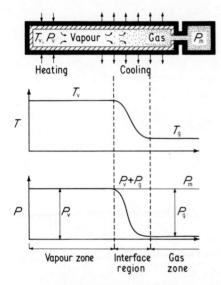

Figure 2. Gas-controlled heat pipe (schematic).

The interesting fact is that the pressure of the inert gas can be easily measured by connecting a manometer to the cold end of the heat pipe (Bohdansky and Schins 1965), and that this pressure is related to the temperature of the vapour zone. Thus, by measuring and controlling the pressure of the inert gas, the temperature of the isothermal vapour zone can be measured and controlled†.

This principle can be used to realize gas-controlled thermostats (figure 3). For this purpose the thermostatic chamber is inserted into the vapour. The ensemble may be conceived as a double-walled vessel, the space between the two walls acting as a heat pipe. In order to provide easy access to the thermostatic chamber, this assembly can be subdivided into two separate gas-controlled heat pipes, which enclose the thermostatic chamber. In order to obtain the same vapour temperature in the two heat pipes it is

---

† Similar systems for temperature control by a gas pressure were proposed some decades ago (eg, Gregory 1934, Bäckström 1943), the only difference being that these former devices did not have a wick. In practice, however, it is difficult to obtain high-temperature precision with wickless structures. This results from the fact that there the vapour is produced by boiling of a pool of liquid, which is a rather unstable, strongly fluctuating process compared with the smooth evaporation from the thin wick in a heat pipe.

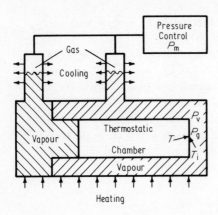

Figure 3. Heat pipe thermostat (schematic).

sufficient, as will be shown below, to connect them to the same gas pressure control system.

The following two basic questions will be discussed: how small can the deviations from the isothermal be made in the thermostatic chamber, and how accurately can its temperature be reproduced?

3. Performance limits of the ideal heat-pipe thermostat

Let us first consider an ideal case, which we define as follows (see figure 3):
The total pressure P is constant throughout the heat pipe,

$$P = P_m. \tag{1}$$

The vapour pressure P_v in the vapour zone is equal to the total pressure

$$P_v = P. \tag{2}$$

The saturation temperature T_s of the vapour is the saturation temperature T_{s0} over a plane surface of pure working fluid

$$T_s = T_{s0}(P_v). \tag{3}$$

The temperature T_i of the liquid–vapour interface on the outside of the thermostatic chamber is equal to the saturation temperature

$$T_i = T_s. \tag{4}$$

The temperature T at the inner side of the thermostatic chamber is equal to the temperature at the liquid–vapour interface

$$T = T_i. \tag{5}$$

From (1) to (5) it follows that

$$T = T_{s0}(P_m). \tag{6}$$

This means that in the ideal case the chamber temperature and the pressure at the pressure-measuring instrument are related by the vapour pressure curve of the utilized working fluid. Hence the chamber is perfectly isothermal, and the temperature reproducibility depends only on the pressure reproducibility (provided that the working fluid itself is stable).

From the Clausius–Clapeyron equation and the ideal gas law, one obtains

$$\frac{dT_{s0}}{T_{s0}} = \frac{RT_{s0}}{ML}\frac{dP_v}{P_v} \tag{7}$$

and from (6) and (7) we obtain

$$\frac{\Delta T}{T} = \frac{RT}{ML}\frac{\Delta P_m}{P_m}. \tag{8}$$

According to the Pictet–Trouton rule, $ML \approx 10RT_b$. The factor RT/ML can therefore be approximated as $0.1\,T/T_b$. In practice, heat pipes are generally operated in the pressure region between about 0.1 atm and 10 atm†, which corresponds to a temperature variation of only ±20% around the boiling point T_b. Therefore, in the first approximation, equation (8) may be written as

$$\frac{\Delta T}{T} \approx 0.1\frac{\Delta P_m}{P_m}. \tag{9}$$

The accuracy of pressure measurements varies with pressure. At 1 atm an accuracy of 2 parts in 10^5 can be reached (McIlraith 1973). Herewith it follows from (9) that with a gas-controlled heat pipe a temperature reproducibility of about 2 parts in 10^6 could be expected in the ideal case.

4. Performance limits of the real heat-pipe thermostat

In practice the ideal equation (6) has to be modified by a correction ΔT

$$T = T_{s0}(P_m) + \Delta T. \tag{10}$$

The correction is caused by the fact that the basic equations (1) to (5) are not strictly valid. We shall discuss the resulting consequences.

4.1. $P \neq P_m$

This inequality results from the dynamic pressure drop in the flowing vapour and from the hydrostatic pressure variation. The dynamic pressure drop leads to a positive ΔT. For example, for a cylindrical vapour channel (Cotter 1965, Busse 1967)

$$\frac{\Delta T}{T} \lesssim \left(\frac{RT}{ML}\right)^2 \frac{\dot{Q}}{P^2 d^4}\left\{\eta l\left[\frac{128}{\pi} + 0.242\left(\frac{\dot{Q}}{d\eta L}\right)^{3/4}\right] + \frac{2\dot{Q}}{L}\right\}. \tag{11}$$

This correction increases with the first to second power of the heat flux $\dot{Q}$. The minimum necessary value of $\dot{Q}$ depends on the heat losses of the ensemble, especially

† At lower pressures the heat transport capability becomes too small and vapour–gas separation difficult; at higher pressures container problems become more severe.

at its outer walls. As a rule, $\dot{Q}$ should be considerably larger than these losses in order to guarantee that the thermostatic chamber is always completely surrounded by vapour. Figure 4 shows a plot of the heat flux $\dot{Q}$ against the minimum diameter d, which is required to make the temperature correction smaller than 2×10^{-7} for a pipe length of 100 cm and a number of working fluids at 1 atm of pressure. The diameters, even in the kW region, are of the order of 5 to 10 cm. At a pressure of 10 atm they would be roughly a factor 3 smaller. The diagram demonstrates the extreme effective thermal conductivity of a vapour in two-phase equilibrium, which for the presented examples is nearly a factor 10^7 higher than the thermal conductivity of copper.

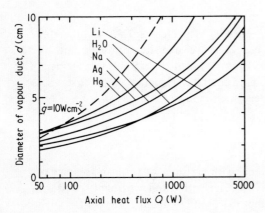

Figure 4. Dynamic temperature drop. Length $l = 100$ cm; pressure $P = 1$ atm, $\Delta T/T = 2 \times 10^{-7}$.

The hydrostatic pressure variation along a height h of vapour is

$$P = \rho g h. \tag{12}$$

This gives a temperature correction

$$\frac{\Delta T}{T} = \frac{gh}{L}. \tag{13}$$

Similarly, a column of gas on the vapour causes a correction

$$\frac{\Delta T}{T} = \frac{T}{T_g} \frac{M_g}{M} \frac{gh}{L}. \tag{14}$$

The tolerable height of the vapour column limits the diameter of the vapour zone (table 1). This effect, however, can principally be eliminated by a suitable wick structure on the thermostatic chamber (see below). The effect of the gas column can essentially be compensated by a correction of the pressure reading.

4.2. $P_v \neq P$

A partial pressure P_g of gas (including volatile impurities) in the vapour is caused by diffusion of the gas against the direction of the vapour flow, convective mixing, and gas solution in the liquid.

The gas-controlled heat pipe

Table 1. Height h of a column of vapour (at boiling point) or gas (at 20 °C), which produces a temperature correction of $|\Delta T/T| = 10^{-6}$.

Working fluid	h (cm)		
	Vapour	Argon	Helium
H_2O	23	8·1	81
Hg	3	7·0	70
Na	24	3·5	35
Li	216	6·8	68
Ag	39	12·7	127

As a result

$$P_v = P - P_g. \tag{15}$$

From the steady state condition, namely that the gas diffusion velocity and the average vapour velocity are equal, one obtains for the correction due to diffusion

$$\frac{\Delta T}{T} = -\frac{RT}{ML} \exp\left(-\frac{RT}{ML}\frac{\dot{q}z}{DP}\right). \tag{16}$$

Table 2 shows that with modest heat flux densities the diffusion correction can be made negligibly small. These minimum heat flux densities are compatible with the upper limits imposed by figure 4. The results of table 2 are nearly independent of pressure.

Table 2. Minimum axial heat flux density $\dot{q}$ required for reducing the temperature correction due to diffusion below $|\Delta T/T| = 2 \times 10^{-7}$ in a distance $z \geq 10$ cm from the gas plug, at $P = 1$ atm.

Working fluid	D (cm² s⁻¹)		$\dot{q}$ (W cm⁻²)	
	Ar	He	Ar	He
H_2O	0·363	1·34	0·6	2·3
Hg	0·475	2·52	0·7	3·7
Na	1·65	4·46	2·0	5·5
Li	5·23	10·7	9·3	19·7
Ag	4·67	17·5	7·5	27·6

Convective mixing can be induced by gravity effects. For example, in a horizontal vapour flow a gas, which is heavier than the vapour, will tend to slip below the vapour. In the same case, but with a vertical upward flow of vapour, the interface can be rather unstable and vapour may be carried far upward into the gas, eventually blocking the gas line after solidification. Therefore in a vertical upward flow of vapour the gas should be lighter than the vapour. For example, the combination Na–He is quite stable in this geometry (M Bader 1974 private communication). The most stable stratification vapour/hot gas/cold gas would be obtainable with a vertical downward vapour flow of vapour and a gas, which is heavier than the vapour. In general it seems that gravity-induced mixing effects can be minimized by suitable design of the gas-controlled heat pipe.

Gas can be carried by the liquid phase from the gas zone to the vapour zone and is released into the vapour by diffusion and evaporation, giving rise finally to a gas concentration x on the vapour side of the liquid–vapour interface. The resulting negative temperature correction is

$$\frac{\Delta T}{T} = -\frac{RT}{ML}x. \qquad (17)$$

Table 3. Temperature correction caused by a saturated solution of gas in the liquid phase at 1 atm gas pressure.

Working fluid	Saturation temperature (°C)	Argon Solubility (mole fraction)	$-(\Delta T/T)$	Helium Solubility (mole fraction)	$-(\Delta T/T)$
H$_2$O	0	4×10^{-5}	3×10^{-6}	7×10^{-6}	6×10^{-7}
	100	10^{-5}	10^{-6}	10^{-5}	10^{-6}
Na	99	10^{-14}	10^{-15}	10^{-12}	10^{-13}
	883	9×10^{-7}	10^{-7}	4×10^{-6}	4×10^{-7}
Li	181	–	–	5×10^{-10}	5×10^{-11}
	1340	–	–	6×10^{-9}	6×10^{-10}

Table 3 shows some values for the gas solution correction assuming that x is equal to the control gas solubility at 1 atm of gas pressure. The values are given for saturation both at the melting point and the boiling point of some working fluids of interest. The temperature corrections are generally very small. However, gas concentrations still higher than the solubility might be possible, due to gas trapping in the liquid near the vapour–gas interface under the action of the condensing vapour. Furthermore, some gas enrichment will occur on every condensing surface. These effects need further study.

4.3. $T_s \neq T_{s0}(P_v)$

This deviation is caused by the curvature of the liquid–vapour interface and by dissolved substances in the working fluid. For a static rise by a height h in an isothermal wick structure the resulting temperature correction is

$$\frac{\Delta T}{T} = -\frac{gh}{L}. \qquad (18)$$

The correction would cancel the hydrostatic correction (13) of the vapour column. In order to approach this static case it is, however, necessary to provide the entire surface of the thermostatic chamber with a wick structure of sufficient capillary rise and low liquid flow resistance.

A solute concentration y at the liquid side of the liquid–vapour interface leads to a temperature correction similar to (17), but with a positive sign

$$\frac{\Delta T}{T} = \frac{RT}{ML} y. \tag{19}$$

For condensing surfaces within the vapour zone this correction is essentially zero, due to the small concentration of non-volatile substances in the vapour, the steady regeneration of the surface and the small solubility of gas and volatile impurities in the liquid. In the evaporator, that part of the correction (19) which stems from dissolved gas or volatile impurities, cancels the correction (17). As to the non-volatile part of the solute in the evaporator, its surface concentration must be below 20 ppm (atomic) in order that $\Delta T/T \leq 2 \times 10^{-6}$. Because of the enrichment of non-volatile substances in the evaporator this will generally require that the solubility limits of the wall material and all non-volatile impurities in the working fluid are in total below 20 ppm, which in practice may be difficult to realize. The problem can probably be overcome by suitable design, eg, either by taking care that the surface of the thermostatic chamber is not an evaporating surface (eg, by avoiding that more vapour-superheat or heat radiation from the superheated surface of the evaporator reaches the thermostatic chamber than corresponds to the heat absorption of the chamber) or by steady purging of the chamber wall with a flow of fresh condensate liquid.

4.4. $T_i \neq T_s$

This inequality is caused by finite rates of condensation or evaporation. The resulting temperature correction (see, eg, Sukhatme and Rohsenow 1966) is

$$\frac{\Delta T}{T} = \pm \frac{2-\sigma}{\sigma} \left(\frac{\pi}{2}\right)^{1/2} \left(\frac{RT}{ML}\right)^{3/2} \frac{\dot{q}_n}{PL^{1/2}}. \tag{20}$$

The correction is proportional to the heat flux density $\dot{q}_n$ normal to the interface. It is positive for an evaporating surface, negative for a condensing one. This means that the

Table 4. Tolerable interface heat flux densities $\dot{q}_n$ for a temperature jump of $|\Delta T/T| = 2 \times 10^{-6}$ across the interface, at $P = 1$ atm.

Working fluid	T_b (°C)	σ	L (W s g^{-1})	ML/RT_b	$\dot{q}_n$ (W cm^{-2})
H$_2$O	100	0.04 1	2255	13.1	0.023 1.1
Hg	357	1	294	11.3	0.3
Na	883	1	3871	9.3	0.9
Li	1340	1	21157	10.9	2.7
Ag	2180	1	2352	12.4	1.1

vapour produced in the evaporator is superheated by ΔT and that the temperature of a condensing surface drops by ΔT below the saturation temperature of the vapour. The small amount of superheat can be eliminated by passing the vapour along a slightly cooled surface before it reaches the thermostatic chamber. The subcooling of the condensing surface of the chamber can be kept small by keeping the heat flux density into the chamber small, ie, by using sufficiently large chambers for a given heat loss from the chamber (eg, by a black body hole). Table 4 shows that the tolerable heat flux densities for a subcooling of $\Delta T/T = 2 \times 10^{-6}$ are of the order of $1\,\text{W cm}^{-2}$.

4.5. $T \neq T_i$

Heat conduction through the wick and the wall of the thermostatic chamber leads to a negative correction

$$\frac{\Delta T}{T} = -\frac{w\dot{q}_n}{\lambda T}. \tag{21}$$

Table 5 shows that the tolerable heat flux densities are of the order of $0.01\,\text{W cm}^{-2}$, ie, about 100 times smaller than those imposed by the condensation non-equilibrium according to table 4.

Table 5. Tolerable heat flux densities across the wall of the thermostatic chamber for $\Delta T/T = -2 \times 10^{-6}$, with a wall thickness $w = 2$ mm.

Material	T (°C)	λ (W cm^{-1} K^{-1})	$\dot{q}_n$ (W cm^{-2})
Cu	100	3·95	0·014
Inconel 600	1000	0·31	0·004
Mo	1500	0·86	0·015
W	2000	1·00	0·023
W–26 Re	2000	0·68	0·015

5. Experimental results

Figure 5 shows two examples of temperature distributions in the axis of an argon-controlled cylindrical heat pipe, built from copper and using water as the working fluid (Labrande 1975). The heat pipe is 50 cm long, with a vapour channel diameter of 1·2 cm. The temperatures are measured in a tube open at both ends and situated along the axis of the heat pipe (outer diameter 0·5 cm). The capillary structure consists of a screen wick on the outer tube and a thread on the inner tube. The temperature-measuring probe contains two platinum resistances of 1 cm length 3 cm apart (from centre to centre). The difference between the values of the two resistances is directly measured with a special bridge, eliminating systematic errors by combining each measurement with a measurement in an inverted position of the resistances (by introducing the probe from the other side of the heat pipe). The measuring system has a sensitivity of about 2×10^{-4} °C. The curves shown were obtained by adding the measured temperature differences, ie, the absolute position of the curves in the diagram

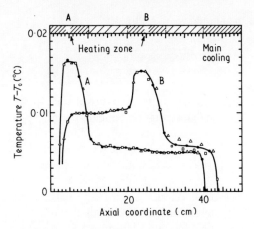

Figure 5. Axial temperature distribution in a gas-controlled heat pipe (Cu/H$_2$O/Ar). $T_0 \approx 100\,°C$.

is arbitrary. Curve A (total heat flux about 75 W) was measured with heating at one end and (main) cooling at the other end of the heat pipe. Between the evaporator and the (main) condensor it has a zone 30 cm long where the temperature is constant to about $10^{-3}\,°C$. In the evaporator there is a temperature increase by $10^{-2}\,°C$, caused probably primarily by heating of the central tube by the surrounding superheated evaporator surface.

In curve B heating is done in the middle of the heat pipe (heat flux towards the left about 20 W, towards the right about 120 W). There is a zone of constant temperature on each side of the heating zone. Their absolute temperature level differs by about $4 \times 10^{-3}\,°C$. This is attributed to a difference of the partial gas pressure and the respective temperature correction (17) in the two zones. It is interesting that the higher temperature in the left-hand zone indicates a lower gas content. This is plausible from the fact that during the measurement the left end of the heat pipe was hot, ie, that essentially no gas plug can have been present and therefore gas solution effects in the left-hand zone should have been much smaller than in the right-hand zone. Whether the temperature level of the left-hand zone corresponds to truly pure vapour has, however, still to be investigated. In any case, the arrangement of the thermostatic chamber in the vapour of the 'backside' of the heating zone could be an interesting possibility to reduce the gas solution effects.

6. Conclusions

The gas-controlled heat pipe holds promise of realizing isothermic vessels with a temperature homogeneity and reproducibility of some parts in 10^6. The dimensions of the vessels can nearly be varied at will. Long time operation up to temperatures of 2000 °C seems possible with present heat-pipe technology (see, eg, Quataert *et al* 1973).

Such heat-pipe thermostats could serve as temperature standards, and they should be versatile tools for research where temperature criteria are of prime importance.

Acknowledgments

The authors are indebted to M Bader for stimulating discussion. They gratefully acknowledge the technological support of C Cappelletti, F Geiger and J Loens in the realization of the experimental device and the assistance of A Pirovano in the measurements.

Appendix. Nomenclature

d, diameter of vapour duct
D, diffusion constant of gas in vapour
g, acceleration of gravity
h, height
l, length of vapour duct
L, specific heat of vaporization
M, molar mass of vapour
M_g, molar mass of gas
P, total pressure
P_g, gas pressure
P_m, total pressure at pressure-measuring instrument
P_v, vapour pressure
$\dot{q}$, axial heat flux density
$\dot{q}_n$, heat flux density normal to evaporating or condensing surface
$\dot{Q}$, total heat flux in heat pipe
R, gas constant
T, temperature at the inner side of the thermostatic chamber
T_b, boiling temperature of working fluid
T_g, gas temperature
T_i, temperature of liquid–vapour interface
T_s, saturation temperature of the vapour
$T_{s0}(P_v)$, saturation temperature of vapour of pressure P_v over a plane surface of the pure working fluid
w, thickness of wall of thermostatic chamber with wick
x, mole fraction of gas (and volatile impurities) in the vapour at the liquid–vapour interface
y, mole fraction of dissolved substances in the working fluid at the liquid–vapour interface
z, distance from gas plug
Δ, small variation of a quantity
η, viscosity of vapour
λ, effective thermal conductivity of wall of thermostatic chamber with wick
ρ, density of vapour
σ, condensation coefficient (the fraction of the molecules striking the surface which actually do condense)

References

Bäckström S M 1943 *US Patent No* 2581347
Bohdansky J and Schins H E J 1965 *J. Appl. Phys.* **36** 3683–4
Busse C A 1967 *IEEE Conf. Record of the Thermionic Conversion Specialist Conf., Palo Alto* (New York: IEEE) pp391–8
Cotter T P 1965 *Los Alamos Scientific Report* LA-3246-MS
Gaugler R S 1942 *US Patent No* 2350348
Gregory L S 1934 *US Patent No* 2049699
Grover G M, Cotter T P and Erickson G F 1964 *J. Appl. Phys.* **35** 1990–1
Labrande J P 1975 *PhD Thesis* ENSMA, UER Sciences, Poitiers
McIlraith A H 1973 *Conf. on Precision and Force Measurements, NPL, February 1973*
Quataert P, Busse C A and Geiger F 1973 *Int. Heat Pipe Conf., Stuttgart, 1973* (Düsseldorf: VDI, Fachgruppe Energietechnik)
Sukhatme S P and Rohsenow W M 1966 *ASME J. Heat Transfer* **88c** 19–28

Inst. Phys. Conf. Ser. No. 26 © 1975: Chapter 7

Intercalibration of temperature transducer with a heat pipe furnace

P Coville and A Laurencier
Société Anonyme d'Études et Réalisations Nucléaires (SODERN), 10, Rue de la Passerelle, Suresnes, France.

Abstract. We describe a heat pipe furnace in which channels are provided to accommodate a number of thermocouples or resistance thermometers for comparison. These channels are isothermal and the temperature is constant over several hours. Furthermore, the temperature reached by the system when heated from room temperature is always the same within a few tenths of a degree centigrade without reference to another temperature transducer. This property can be compared to the fixed point obtained by freezing of pure metals or compounds.

1. Introduction

When we have to calibrate temperature transducers three conditions at least must be fulfilled. Indeed in any case what we measure — the output of the transducer — is related to its temperature and we assume this is also the temperature of the furnace or of another transducer used as a standard.

Firstly, the furnace must be uniform in temperature over dimensions which are large in comparison with those of transducers under tests. Secondly, the temperature must be constant so that measurements can be performed in steady state conditions. Finally, the temperature of the furnace must be measured with a good accuracy by another transducer calibrated and checked from time to time for its stability. The fixed freezing point of a metal is another way to get a well defined temperature.

This paper deals with a simple device which meets these three constraints.

2. Heat pipe operating with a gas buffer

It is a well known property of the heat pipes that, when they are closed with some quantity of non-condensable gas, the vapour and this gas will separate in operation and their pressures come into equilibrium. The temperature of the heat pipe is determined by this pressure (Feldman and Whiting 1968, Dutcher and Burke 1970, Reay 1972). A schematic view of gas-controlled heat pipe is shown in figure 1. When heat is applied to the insulated part, the liquid is evaporated and the vapour pushes the residual gas out of the heated zone. As heat losses in this section are small, the temperature rises, limited only by transport of latent heat of evaporation. Out of the insulant, heat losses become higher and, power being limited, the length of the condensing zone is also limited, the more so if that cooling can be improved by a heat sink or forced convection. If the

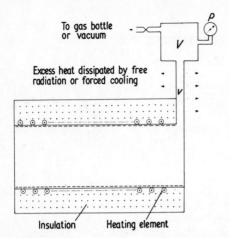

Figure 1. Heat pipe with a gas buffer.

heating power is increased this length tends to increase giving more cooling surface where the excess power is liberated. The gas therefore occupies a reduced volume and the pressure grows, but if the ratio of total volume V to displaced volume v is large enough, temperature variation can be very low.

3. Application to a 'fixed point' furnace

Figure 2 shows, in more detail, our application of the above principles.

A number of small tubes closed at one end (A) are sealed inside a larger tube (B) which is mounted like a heat pipe with capillary structure on the walls and a small quantity of liquid to wet this structure. The lower part of the main tube is insulated (C) and heated from the outside (D). A large gas holder (E) is connected at the top by a small tube (F) and maintained in melting ice. In order to minimize dead volumes the open ends of the small tubes are only a few centimetres above the insulated section while the connecting tube is put ten centimetres higher so that no vapour can reach and condense in the gas holder. A manometer (G) and a small valve (H) permit control of initial gas pressure.

As explained previously, the temperature of the system is mainly fixed by the total amount of gas and by the pressure reached when this gas is pushed out of the insulated

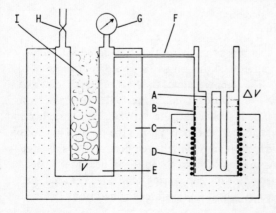

Figure 2. Fixed point furnace. A, well; B, heat pipe; C, insulation; D, heater; E, gas reserve; F, connecting tube; G, manometer; H, valve; I, ice.

section of the heat pipe. For this reason the main gas holder is held at ice temperature so that this pressure cannot change with ambient temperature. The influence of this parameter is restricted to the connecting tube and this dead volume was made as small as possible.

In this way we can calculate the stability of the system and compare it with experimental values. With a good approximation all vapour pressure/temperature relations can be expressed in the form:

$$\operatorname{Ln} p = \frac{-A}{T} + B \tag{1}$$

where p is the pressure in mm of mercury, A and B are the constants for a given fluid. So we can write

$$\frac{\Delta p}{p} = \frac{A \Delta T}{T^2}. \tag{2}$$

At the same time pressure and volume of the gas are related by pV = constant or

$$\frac{\Delta p}{p} = \frac{-\Delta V}{V}. \tag{3}$$

So the temperature variation of the heat pipe is given by the relation

$$\Delta T = \frac{-T^2}{A} \frac{\Delta V}{V}.$$

Table 1 gives values of A and ΔT for useful fluids at 1% change in volume.

In the furnace (figure 2) V = 1800 cm^3 and ΔV is 10 cm^3 for one centimetre displacement of vapour front giving $\Delta V/V$ = 5·6 × 10^{-3}. We will now explain why such a small variation does not need a very high degree of power control.

In our experiment the insulated section was 15 cm long and the power used to maintain 700 °C was 200 W. Sodium was used as working fluid. At this temperature radiation losses in the non-insulated section are about 4 W cm^{-2}. The radiating area is nearly

Table 1. ΔT for 1% change in volume of gas.

	Sodium†	Potassium†	Caesium†	Dowtherm	Silicon oil
Boiling temperature	890 °C	780	700	270	440
A	1·24 × 10^4	1·02 × 10^4	8·84 × 10^3	9·4 × 10^3	5·6 × 10^3
ΔT for $\Delta V/V$ = 10^{-2}	7·6 × 10^{-1} at 700 °C	9·2 × 10^{-1} at 700 °C	6·8 × 10^{-1} at 500 °C	4 × 10^{-1} at 200 °C	4·8 × 10^{-1} at 400 °C

† Derived from values published in *Handbook of Chemistry and Physics* 46th edn (page D 124). Other values taken from our own measurements.

14 cm²/cm length, therefore 1 cm displacement of vapour front means a difference of 56 watts in power dissipation and, from table 1, the temperature will change by

$$\Delta T = 76 \times 5 \cdot 6 \times 10^{-3} = 0 \cdot 4 \,°C.$$

Figure 3 gives the temperature–power relation for a sodium-filled system closed at 100 mm Hg and we can see a total variation of 1 °C when power input goes from 150 to 250 watts.

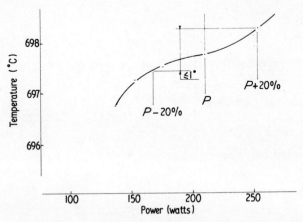

Figure 3. Temperature–power relation.

Our first experiments have been done with the heater put directly on the mains and on the records we can only see fluctuations of the output of thermocouple less than ±10 μV which comes from rapid changes in mains voltage (see figure 4). In further tests we have used a small voltage controller which is usually sold for TV and these fluctuations have been cut down by a factor of 5, the temperature being constant over several hours at ±0·1 °C. This time depends only on the quantity and the insulation of the ice bath. With less than 1 litre of ice we get 5 to 6 hours of stability without changing the bath.

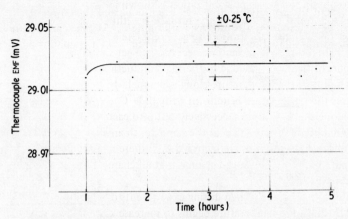

Figure 4. Stability test without power control.

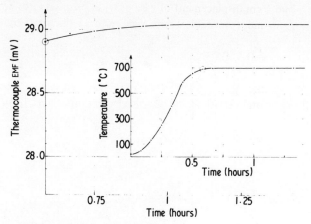

Figure 5. Temperature stabilization.

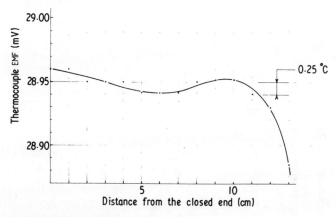

Figure 6. Thermal profile.

The time to obtain temperature stabilization is shown in figure 5. After half an hour, starting from room temperature, the furnace is within 5 °C of its final temperature. Nevertheless, it takes about 1 hour more to get complete stabilization in the range of 600 to 800°C. This comes from the time necessary to get stable exchange conditions in the upper part of the heat pipe due to thermal conduction along the main tube.

Figure 6 is a plot of the temperature along a channel. In the centre section over 5 to 6 centimetres the temperature is uniform with ±0·1 °C. Figure 7 gives the EMF measured with one thermocouple successively put into each of the eight channels. On each channel measurements are made at the same depth and after 5 minutes stabilization fluctuations over 10 minutes are recorded. All these measurements lay between 33·125 mV and 33·135 mV so we can assume that all channels are at the same temperature within ±0·125 °C.

Reproducibility has been checked during experiments on stability. When all the ice has melted in the bath the temperature begins to increase since the gas reserve is heated. But from day to day if we make a new ice bath we return to the original equilibrium

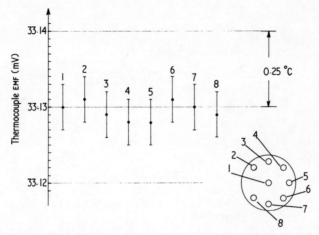

Figure 7. Record of temperature fluctuations in eight channels.

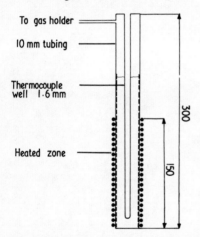

Figure 8. Schematic view of small fixed-point heat pipe.

temperature. For example during a three-day experiment we have obtained the following values with a chromel–alumel thermocouple: 29·025, 29·030 and 29·020 mV. At this level of uncertainty we are on the limit of our measurement capability.

4. Experiments with other filling materials

Figure 8 shows a small heat pipe we have used to test various liquids to work at different temperatures. Each one has a limited range of use if we limit the vapour pressure to 760 mm Hg and we have found that the heat pipe does not work with too low a pressure. Our experience is that this lower limit is $\simeq$50 mm of mercury.

Table 2 gives the useful temperature range for those liquids we have tested.

Uniformity and stability have been checked specially on a Dowtherm-filled heat pipe at a pressure of 300 mm Hg. On this heat pipe ten measurements have been performed after heating and half an hour stabilization, giving

$$E = 8 \cdot 91 \text{ mV} \pm 5 \text{ } \mu\text{V},$$

Table 2.

Perchlor-ethylene	Dowtherm	Oil DC 702	Oil Siss 763	Caesium	Potassium	Sodium
60–120 C	165–270	300–400	325–435	465–680	540–760	650–880

in other words, all values were within the limits of our measurement capability.

In conclusion we think this kind of furnace is able to give a good intercalibration facility since, from one point to another, the temperature difference is about one-tenth of a degree. Furthermore, a completely closed system can be calibrated for its equilibrium temperature and used thereafter as a temperature reference with the same uncertainty of ±0·1 °C.

References

Feldman K T and Whiting G H 1968 *Mech. Eng.* (November)
Dutcher C H and Burke M R 1970 *Electronics* Feb 16
Reay D A 1972 *Electron. Eng.* (August) 35–7

Heat pipes for the realization of isothermal conditions at temperature reference sources

G Neuer and O Brost

Institut für Kernenergetik der Universität Stuttgart, 7000 Stuttgart 80, Pfaffenwaldring 31, W Germany

Abstract. The calibration of temperature sensors and radiation detectors requires reference sources in the form of isothermal cavities. The realization of isothermal conditions employing conventional techniques is complicated and expensive, especially in the temperature range where liquid baths can no longer be used. Applying the principle of the heat pipe for this purpose, the heating problem is reduced since the temperature equalization is accomplished in a closed vapour space surrounding the cavity wall.

For laboratory standards, e.g. used for secondary calibration purposes, rather simple designs of heat pipes can be employed to realize temperature differences below 1 K at the cavity wall. The design principles are described for the temperature range 750–1350 K. Heat pipes using sodium as a working fluid were built and constructional guidelines are given. Results of experiments carried out to determine temperature profiles and lifetime at temperatures up to the gold point are presented.

1. Introduction

The employment of isothermal cavities is desired for calibrating temperature sensors and radiation detectors. In the temperature range up to 500 K, low-temperature gradients along the cavity may be realized relatively easily through the use of liquid baths or circulating systems. At higher temperatures, the problem is far more complicated. Heaters consisting of several separately regulated zones or directly heated cavities, long enough to have a central isothermal section, are generally used at these temperatures. A graphite cavity black body radiator for use at temperatures above 1200 K developed at IKE (figure 1) has its geometric profile so shaped that the heat sinks at the cooled supports and at the current leads are compensated as near to the end of the rod as possible (Groll and Neuer 1972). Such a cavity radiator having a hole diameter up to 15 mm has been used successfully for calibration of total radiation detectors in connection with total emittance measurements (Neuer 1971). However, the investigation of a suitable rod profile for such radiators is rather complicated necessitating the use of a computer program, though the black body itself is compact, simple to operate and control, and has a small time constant.

Another possible way to realize isothermal cavities is by using the heat-pipe principle. Based on the present technological understanding of the heat pipe, heat-pipe temperature reference sources in the range 750–1350 K are attainable. For purposes of secondary calibration, rather simple designs can be employed. The present paper describes the principle and operation of such heat pipes.

Heat pipes for temperature reference sources 447

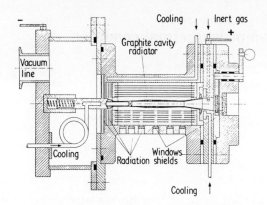

Figure 1. Schematic cross section of a black body device using an electrically heated graphite rod.

2. Design principles

A heat pipe essentially consists of an evacuated hermetically sealed container, the inner walls of which are lined with a capillary structure, such as screen wire mesh, felt metal or sintered powder, or it may have fine grooves or threads cut on it (Cotter 1965, Grover *et al* 1964). The capillary structure is saturated with a liquid, the choice of the liquid depending on the temperature range of interest. Some typical heat-pipe fluids and the temperature range in which they may be used are shown in table 1. When heat is applied to any part of the container, the liquid in the capillary evaporates, the vapour flows under a slight pressure gradient towards the colder parts of the system and condenses there, rejecting its latent heat of evaporation to the cooler wall. The condensate returns to the heated section owing to the surface tension forces available in the capillary medium as a consequence of the varying curvature of the meniscii in the evaporation and condensation sections of the heat pipe. Owing to the high evaporation and condensation coefficients, the heat pipe transports heat from the heated to the cooled

Table 1. Working fluids and compatible wall materials for heat-pipe black bodies.

Working fluid	Wall material	Temperature range (K)†
Liquid nitrogen, LN_2	Stainless steel	70–110
Methane, CH_4	Copper, aluminium	100–150
Carbon tetrafluoride, CF_4	Copper, aluminium	100–200
Freons, $CCLF_3$, ...	Copper, aluminium	120–300
Ammonia, NH_3	Stainless steel, nickel, aluminium	230–330
Acetone, C_3H_6O	Copper, stainless steel	230–420
Methanol, CH_4O	Copper, nickel	240–420
Water, H_2O	Copper, nickel, titanium	300–550
Organic fluids	Stainless steel, super alloys, carbon steel	400–600
Mercury + additives, Hg + ...	Stainless steel	450–800
Potassium, K	Stainless steel, nickel	700–1000
Sodium, Na	Stainless steel, nickel, Inconel	800–1350

† Recommended temperature range limited by fluid properties and strength of wall materials.

section of the system in the presence of very small temperature gradients. Moreover, as the heat pipe operates without any auxiliary part (eg a circulation pump), it can render long trouble-free service.

The temperature gradient in a heat-pipe system is essentially a function of the heat to be transported, the radial thermal resistance between the source and the sink and the heat-pipe vapour space, and the axial thermal resistance between the evaporator and the condenser sections. The radial resistance depends on the material and thickness of the wall, and the external heat transfer coefficient between the source/sink environment and the container outer surface. The axial thermal resistance is influenced by the axial resistance of the heat-pipe wall and the pressure drop for vapour flow from the hot to the cold end, the latter affecting the saturation temperature of the operating fluid in the two sections.

The temperature gradients in a heat-pipe system used as a black body may hence be minimized by incorporating the following measures:

> good thermal insulation around the system
> uniform heating
> short distances for flow of liquid and vapour of the working fluid
> homogeneous thickness of the wall surrounding the cavity.

All these factors were incorporated in the heat-pipe development for use during the present study. Figure 2 shows a sodium heat pipe constructed from 321 stainless steel. The heat pipe has an outer diameter of 60 mm, a uniform wall thickness of 2 mm and a length of 190 mm. The cavity depth is 160 mm, the cylindrical part of the cavity has an inner diameter of 40 mm and the cavity orifice is 14 mm in diameter. The temperature range of operation of the heat pipe is controlled by the strength behaviour of the wall material, and is hence restricted to 1100 K. A similar sodium-filled heat-pipe black body fabricated using Inconel-600, a nickel-based alloy, and having a wall thickness of 3 mm could be used up to a temperature of 1300 K. For short times, the operat-

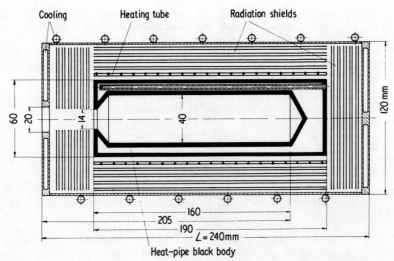

Figure 2. Cross section of a heat-pipe black body for use in vacuum.

ing temperature could be raised to 1350 K. Both heat pipes have short distances for vapour flow between the outer and the inner wall and hence negligible vapour pressure losses. Four arteries parallel to the axis provided short paths for the returning liquid and good azimuthal distribution of the liquid was achieved by sectioning the flow regimes into four parts. Uniform heating was made possible using a specially constructed furnace and radiation shields provided a good insulation and hence fewer losses.

3. Construction examples and measurement results

3.1. Annular heat pipes

Tubular furnaces having a simple but uniformly wound heating coil can be used for temperature calibration if, for example, an annular heat pipe is inserted into such a furnace as shown in figure 3. An isothermal zone may be so produced which can be used for suitable inserts, for example diaphragms and discs as black body or adjusting plates for thermocouples. The length of the heat pipe in figure 3 was one-half the furnace

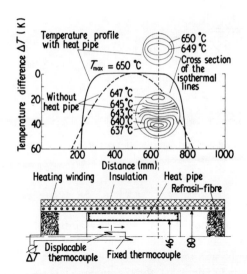

Figure 3. Annular heat pipe for use in a tubular furnace, also showing measured temperature profiles with and without heat pipe.

length. Consequently the temperature profile inside the heat pipe could be examined at different cross sections of the heat pipe. Measurements showed the temperature profile along the heat-pipe axis to be nearly the same if the heat pipe was positioned at the end or in the centre of the furnace as shown in figure 3. This figure also shows the temperature difference between a fixed and a displaceable thermocouple as a function of heat-pipe length. The broken line represents the temperature profile, measured along the axis of the tubular furnace without a heat pipe, and the full line is that measured inside the heat pipe. Since the temperature of the displaceable thermocouple is influenced by convection and heat exchange with the colder environment, several thermocouples were fixed at the inner wall of the heat pipe. Thereby it could be established that the maximum temperature difference at the heat-pipe wall was about 1 K. In addition, figure 3 shows the isotherms at a cross section about 150 mm from the centre, again without and with the heat pipe. The maximum radial temperature difference was 10 K without and only 1 K with the heat pipe.

3.2. Heat-pipe black body

A complete heat-pipe black body device, constructed for use in vacuum is shown in figure 2. Some constructional details have been discussed earlier in §2. The inconel-600 heat pipe was used in a tubular furnace in a manner similar to the annular heat pipe (figure 4). From the bottom of the cavity up to a distance of about 100 mm above it,

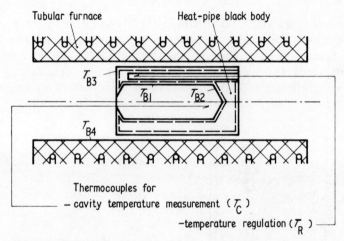

Figure 4. Heat-pipe black body, heated in a tubular furnace and temperature measurement positions for thermocouple and pyrometer measurement.

no discernible temperature difference could be established. Thermocouple measurements near the cavity opening are uncertain on account of heat exchange with the colder environments. At cavity radiators only the wall temperature is of interest. Using a calibrated spectral pyrometer for measuring the wall temperatures and a Pt/Rh–Pt thermocouple for cavity temperature measurement (figure 4), the thermocouple temperature T_C, the brightness temperatures, T_{B1}, T_{B2} measured at different positions inside the cavity, T_{B3} at the front face and T_{B4} at the cylindrical wall of the tubular furnace 1 cm from front face were compared. The results for a cavity temperature just above the gold point were:

Cavity temperature $\qquad\qquad\qquad\qquad\qquad\qquad\qquad T_C$ = 1355 K (±0·5 K)
Wall temperature (brightness temperatures, λ = 0·65 μm) $\quad T_{B1} = T_{B2}$ = 1354 K (±2 K)
$\qquad\qquad\qquad\qquad\qquad\qquad\qquad\qquad\qquad\qquad\quad T_{B3}$ = 1340 K (±2 K)
$\qquad\qquad\qquad\qquad\qquad\qquad\qquad\qquad\qquad\qquad\quad T_{B4}$ = 1205 K (±10 K)

T_{B1} and T_{B2} are equal and the measured brightness temperature is the same as measured by means of the thermocouple. If we assume that the real temperature T_{B3} at the front face is nearly the same as T_{B2}, then the resulting spectral emittance is 0·84. This could be realistic considering the scale formation on the heat-pipe surface at high temperature and radiation exchange with the surrounding furnace wall.

For the lifetime of a heat pipe, the compatibility between the working fluid and the structural materials and the stability of the wall material in air at high temperature are decisive. An additional problem is that the vapour pressure of sodium increases with increasing temperature whereas the strength of Inconel decreases rapidly. Strength calculation of the heat pipes built have been carried out but experience in operation at temperatures above 1300 K was not available. Therefore, the heat-pipe black body has been working for several months with interruptions and at different temperatures. Up to the end of February 1975, the total operating time was more than 1000 h. The number of hours of operation at the different temperatures is shown in figure 5. The tests, especially at temperatures above the gold point, are continuing.

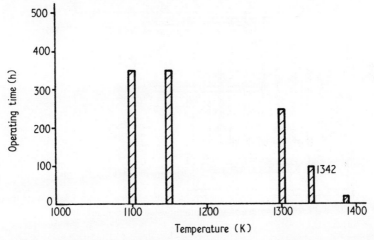

Figure 5. Test duration at different temperatures (Inconel-600 heat-pipe black body, total operating time = 1020 h, life test continuing).

3.3. Heat-pipe insert for calibration of thermocouples

A device for calibration of thermocouples is at present in a conceptual stage at IKE. A heat-pipe system similar to that shown in figure 6 appears to be a simple way for calibrating thermocouples. Several holes can be used to enclose the thermocouples under well controlled isothermal conditions. By using a calibrated reference thermocouple, it would then be simple to determine the deviations of the thermocouples under test.

4. Conclusions

Heat pipes are seen to be well suited to serve as cavity radiators for calibration of pyrometers as well as thermocouples. Results obtained during the present study indicate that these devices could operate as simple units, for example as calibration devices with good precision for laboratory standards.

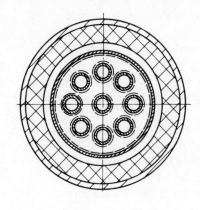

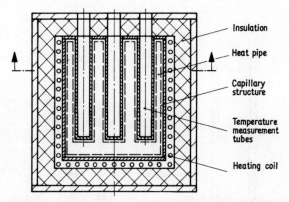

Figure 6. Calibration unit for thermocouples using a heat pipe.

References

Cotter T P 1965 *Los Alamos Scientific Laboratory report* LA3246MS
Groll M and Neuer G 1972 *TMCSI* **4** part 1 449–56
Grover G M, Cotter T P and Erickson G F 1964 *J. Appl. Phys.* **35** 1930
Neuer G 1971 *Wärme- und Stoffübertragung* **4** no 3 133–41

Author Index

Actis A, *398*
Agerskans J, *375*
Allnutt L A, *341*
Ancsin J, *57*
Anderson R L, *38*
Andrews J W, *256*
Ashcroft I R, *321*

Bassani C, *428*
Berry K H, *32*
Berry R J, *99*
de Bie J R, *306*
Bigge W R, *389*
Blahová M, *65*
Brenez J, *203*
Brost O, *446*
Bugden W G, *181*
Burley N A, *162, 172*
Burns G W, *144, 162*
Busse C A, *428*

Cagna G, *117*
Cezairliyan A, *287*
Coates P B, *238, 256*
Compton J P, *91*
Coslovi L, *287*
Coville P, *439*
Crovini L, *107, 398*

Dorda G, *131*
Durieux M, *17*

Egan T M, *211*
Eisele I, *131*

Falla A, *203*
Fenton A W, *195*
Fodor J, *422*
Fothergill I R, *409*
Furukawa G T, *389*

Heimann W, *219*
Hurst W S, *144*

Jansák L, *65*
Johnston J S, *80*
Jones T P, *172, 211*
Jung H J, *278*

Kaufmann H J, *244*
Klomp W G, *306*
Kordoš P, *65*
Kunz H, *244, 273*

Labrande J P, *428*
Lapworth K C, *341*
Laurencier A, *439*
Léb L, *422*
Lechner W, *297*

Marcarino P, *107*
Mathieu F, *203*
van der Meer L W, *315*
Meier R, *203*
Mester U, *219*
Muk G, *422*

Neubert W, *38*
Neuer G, *446*
Norris P A, *321*

Pavese F, *70, 117*
Powell R L, *162*
Preston R C, *348*

Quinn T J, *1*

Reilly M L, *389*
Richter J, *329*
Ricketson B W A, *135*
Riddle J L, *389*
Righini F, *287*
Rosso A, *287*
Ruffino G, *264*
Rusby R L, *44, 125*

Schob O, *297*
Schooley J F, *49*

Selman G L, *181*
Shawyer R E, *368*
Strnad M, *415*
Sutton G R, *188*

Thomson A, *195*

Tomlinson J A, *181*

de Vries J, *315*

Ward S D, *91*
Wende B, *358*